AF180183

40 Years of Particle Physics

Group photograph of the conference participants.

40 Years of Particle Physics

Proceedings of the International Conference to Celebrate the 40th Anniversary of the Discoveries of the π- and V-particles, held at the University of Bristol, 22–24 July 1987

Edited by B Foster and P H Fowler
H H Wills Physics Laboratory, University of Bristol

Adam Hilger

Adam Hilger
Bristol and Philadelphia

Copyright © 1988 by IOP Publishing Ltd and individual contributors. All rights reserved. Multiple copying of the contents or parts thereof without permission is in breach of copyright but permission is hereby given to copy titles and abstracts of papers and names of authors. Permission is usually given upon written application to IOP Publishing Ltd to copy illustrations and short extracts from the text of individual contributions, provided that the source (and, where appropriate, the copyright) is acknowledged. Authorisation to photocopy items for internal use, or the internal and personal use of specific clients in the USA, is granted by IOP Publishing Ltd for libraries and other users registered with the Copyright Clearance Center (CCC) Transactional Reporting Service, provided that the base fee of $2.50 per copy per article is paid direct to CCC, 27 Congress Street, Salem, MA 01970, USA.
$2.50+.00

British Library Cataloguing in Publication Data

International Conference to Celebrate the
40th Anniversary of the Discoveries of the
Pi- and V- Particles *(1987: Bristol, England)*
40 years of particle physics.
1. Elementary particles. Theories, to 1987
I. Title II. Foster, B. III. Fowler, P.H.
539.7′21′01

ISBN 0-85274-354-8

Library of Congress Cataloging-in-Publication Data are available

Organising Committee
P H Fowler (Chairman), B Foster (Secretary), R C E Devenish, J Malos, N McCubbin, P Murphy, D H Perkins, V J Smith, J J Thresher, R Williamson

Published under The Institute of Physics imprint by IOP Publishing Ltd
Techno House, Redcliffe Way, Bristol BS1 6NX, England
242 Cherry Street, Philadelphia, PA 19106, USA

Printed in Great Britain by J W Arrowsmith Ltd, Bristol

Contents

List of Participants

Name	Institution
G Alexander	Tel-Aviv University, Israel
E Amaldi	Rome University, Italy
R Armenteros	CERN, Switzerland
A Astbury	Victoria University, Canada
J Ballam	SLAC, Stanford University, USA
D M Binnie	Imperial College, London, UK
N Blackett	Llanfrothen, UK
D J Broadhurst	Open University, UK
Sir C Butler	Loughborough University, UK
R G Chambers	Bristol University, UK
V Clapham	Bristol, UK
D C Colley	Birmingham University, UK
M Conversi	Rome University, Italy
M Crozon	College de France, France
R H Dalitz	Oxford University, UK
R R Daniel	Tata Inst., Bombay, India
J Davies	Bristol University, UK
D H Davis	University College, London, UK
M De Maria	Rome University, Italy
R C E Devenish	Oxford University, UK
N Doble	CERN, Switzerland
J G M Duthie	Alabama University, USA
A G Ekspong	Stockholm University, Sweden
A Engler	Carnegie-Mellon University, USA
D Evans	Durham University, UK
D F Falla	Aberystwyth, University of Wales, UK
G Fidecaro	CERN, Switzerland
B Foster	Bristol University, UK

P H Fowler	Bristol University, UK
Sir C Frank	Bristol University, UK
M Friedlander	Washington University, USA
H Froehlich	Liverpool University, UK
I Gambaro	Genoa University, UK
Y Goldschmidt-Clermont	CERN, Switzerland
P Harper	Bath University, UK
A J Herz	CERN, Switzerland
F F Heymann	University College, London, UK
R R Hillier	Bristol University, UK
D Hodgkin	Bristol University, UK
P E Hodgson	Oxford University, UK
A L Hodson	Leeds University, UK
J E Hooper	Neils Bohr Institute, Denmark
M Ianniello	Rome University, Italy
W Jentschke	Hamburg University, FRG
C Jones	CERN, Switzerland
B Jongejans	NIKHEF, Amsterdam, Netherlands
G Kalmus	Rutherford Appleton Laboratory, UK
P I P Kalmus	Queen Mary College, London, UK
N Kemmer	Edinburgh University, UK
D T King	Tennessee University, USA
Sir J Kingman	Bristol University, UK
J Laberrigue-Frolow	LPHNE, Paris, France
A Lahiri	Calcutta University, India
C M G Lattes	Las Campinas University, Brazil
W O Lock	CERN, Switzerland
J Malos	Bristol University, UK
R E Marshak	Virginia Polytechnic Inst., USA
N A McCubbin	Rutherford Appleton Laboratory, UK
M G K Menon	New Delhi, India

Sir A Merrison	Bristol University, UK
W Mitchell	Science and Engineering Research Council, UK
A Montwill	Institute of Advanced Studies, Dublin, Eire
H Muirhead	Bristol, UK
J H Mulvey	Oxford University, UK
P Murphy	Manchester University, UK
R M Payne	Exmouth, UK
Sir R Peierls	Oxford University, UK
D H Perkins	Oxford University, UK
C Peyrou	CERN, Switzerland
K Pinkau	MPI Garching, FRG
J Pniewski	Warsaw University, Poland
Mrs I Powell	Bristol, UK
H Rechenberg	MPI, Munich, FRG
D M Ritson	SLAC, Stanford University, USA
G D Rochester	Durham University, UK
W G V Rosser	Exeter University, UK
A Russo	Palermo University, Italy
J Sacton	Brussels University, Belgium
R D Sard	Illinois University, USA
D H Saxon	Rutherford Appleton Laboratory, UK
V J Smith	Bristol University, UK
S O Sorensen	Oslo University, Norway
G H Stafford	Oxford University, UK
C Sutton	Didcot, UK
R E Taylor	SLAC, Stanford University, USA
V L Telegdi	ETH, Zurich, Switzerland
J J Thresher	CERN, Switzerland
K Tindall	Bristol University, UK
P Tolhun	Ankara University, Turkey
G Tomasini	Genoa University, Italy

S N Tovey	Melbourne University, Australia
G Vanderhaeghe	CERN, Switzerland
L Voyvodic	Department of Energy, Washington, USA
C Waller	Essex, UK
J W Webb	Nottingham University, UK
R Williamson	Manchester University Institute of Science and Technology, UK
J Zakrzewski	Warsaw University, Poland
G T Zorn	Maryland University, USA

Preface

The fortieth anniversary of the discoveries of the π meson, found by Powell and co-workers in Bristol, and the V particles, discovered by Rochester and Butler in Manchester occurred in 1987. In many ways these discoveries marked the beginning of particle physics as a subject and have been of immense importance in its subsequent development. In addition they led to a great upsurge of interest in cosmic ray studies which has continued to this day.

Many of the original researchers involved with these discoveries are still alive and that, together with the growing interest in the study of the history of particle physics, led to the idea of holding a conference to review the circumstances and significance of this work. This was eventually held in July, 1987. The first day was devoted principally to the discovery of the π meson, including an outline of work done in the ten years previous to the discovery, both theoretical and experimental, an account of the discovery itself and talks bearing on the foundation, history and present-day uses of the emulsion technique. The following round table gave a chance to discuss some of the issues raised during the day and led into the conference banquet. In many ways the discovery of the π meson led to the development of international collaboration and the foundation of CERN. It was therefore appropriate that, following the banquet, a talk on the foundation and early days of CERN, the European Centre for particle physics, was delivered by one of the fathers of the organisation. The second day was devoted to a discussion of the discovery of the V particles at Manchester by the original workers, followed by talks on the subsequent development and uses of the cloud chamber. The conference was rounded off by a talk summarising the theoretical importance of these discoveries and by an overview of their general significance.

This volume contains written versions of each of the main talks as provided by the speaker. In addition there is a verbatim record of the invariably lively discussion session which followed each of the contributions, together with a transcript of the round table discussion. The contents of these transcripts have not been individually checked with the contributors, and any inaccuracies are the responsibility of the editors.

The organisation of a conference such as this is an immense task, and the editors are glad to be able to thank those persons who contributed so much to its success. First the editors wish to thank their colleagues on the Organising Committee, R C E Devenish, J Malos, N McCubbin, P Murphy, D H Perkins, V J Smith, J J Thresher and R Williamson, who unsparingly gave of their valuable time and effort in attending committee meetings and in organisation. The editors thank the staff of Bristol University who were vital to the success of the meeting; the staff of the Senate Finance Office dealt with the accounts with a minimum of red-tape; K Tindall, the departmental superintendent and his staff including D Flower and N Swatton ensured that the audiovisual aids were problem free and G Keene took many memorable photographs of delegates during the meeting; the particle physics group research students, particularly H F Fawcett, A J Martin and F F Wilson, were very material in ensuring the smooth running of the conference sessions; finally the Physics Department secretarial staff were invaluable, particularly J Blomfield and D Foster, who manned the Conference desk and did immense amounts of typing work. The other contributors to the administrative services were kindly provided by the Rutherford Appleton Laboratory's Scientific Administration; the editors wish to thank M Sherwin and, in particular, J Lawrence and J Mills, whose contributions to all aspects of the conference administration are difficult to overestimate.

It is impossible to run an international conference such as this without substantial financial support for travel and administration. We are grateful to The Bristol and West Building Society and Rolls-Royce plc for their contributions; to The Institute of Physics and its publishing house, Adam Hilger, and the Lord Mayor and City of Bristol for their substantial help; finally we express our deepest gratitude to the Science and Engineering Research Council for their extremely generous support of this conference. Without their support, this meeting would never have been held.

Finally we recall the words of the Vice-Chancellor of the University of Bristol, Sir John Kingman, in his welcoming address. He exhorted the delegates to think about the future of particle physics as well as the past. At present funding for the subject is under severe pressure and withdrawal from CERN is a serious possibility. Almost all delegates signed a letter to the Prime Minister, the Rt. Hon. Margaret Thatcher, MP, urging her to ensure that the UK remains a full participant in CERN and maintains its vigorous tradition in particle physics research. This seems a fitting way to end a conference substantially devoted to honouring Cecil Powell, a great internationalist whose support for CERN and international cooperation in science never wavered. These proceedings are dedicated to his memory.

B Foster
P H Fowler

From the discovery of the mesotron to that of its leptonic nature

Marcello Conversi
Department of Physics, University of Rome "La Sapienza", Rome, Italy.

ABSTRACT: About ten years of the early history of particle physics are reviewed in order to tell, with some personal recollections, what one might call "the tangled tale" of the muon, first observed as an unstable particle of intermediate mass in the hard component of cosmic rays, for long erroneously identified with the Yukawa carrier of the strong nuclear forces, subsequently recognized as a second charged lepton ~ 200 times heavier than the electron, subject to weak processes of spontaneous decay and nuclear capture similar to β decay, then found as the daughter particle of the pion discovered at Bristol and finally identified as a fermion.

1. INTRODUCTION

The present talk covers a comparatively short period in the history of cosmic rays: the decade ranging from the discovery of the mesotron (defined as a particle of mass intermediate between the electron and proton masses) and the discovery of its "non-hadronic" nature, which led to the concept of universal weak interaction and opened the field of lepton physics. This period represents a natural introduction to the present Conference which celebrates the 1947 discovery of the charged pion at Bristol. But I wish to apologize for inevitable repetition of statements already made in previous Conferences of historical character, for incomplete quotations of the important work carried out in those years, and also for reporting in greater detail the work in which I have been personally involved.

It may be convenient briefly to recall first the main steps which preceded the discovery of the mesotron. Compelling evidence for the existence of charged penetrating cosmic-ray particles (the hard component of cosmic radiation at sea level) was obtained nearly sixty years ago in a pioneering experiment by Bothe and Kolhörster (1929), at a time when cosmic rays were believed to consist of high energy photons and Compton electrons. The result of this experiment, which combined Bothe's "time-coincidence method" with the wire-counter technique just developed by Geiger and Müller (1928), was promptly confirmed and extended by Rossi (1934) and his collaborators using GM counters and the electronic coincidences just invented by Rossi (1930). Several years elapsed before one understood the nature of this hard component, which turned out to constitute nearly 80% of cosmic rays at sea level.

After the discoveries of the positron (Anderson 1932) and of electron-positron pairs (Blackett and Occhialini 1933), and after the observation of the "Rossi transition curve" with a "triangular" arrangement of GM counters (Rossi 1932; 1933), Bethe and Heitler (1934) developed a theory of quantum electrodynamics which enabled them to calculate both the energy loss of an electron by bremsstrahlung and the cross-section of high energy γ rays for pair production. Without introducing the hypothesis of some new particle the theory of course could not explain the penetrating power of the hard component of cosmic rays which, as established much later, is made up of "weakly interacting" muons. The times were not mature to conclude that the penetrating cosmic rays are not electrons and most physicists, rather than think in terms of new unknown particles, thought for some time that the Bethe and Heitler theory broke down for high energy electrons. Notice that just because electrons suffer strong

© 1988 IOP Publishing Ltd

radiation losses, they were observed with comparatively small momentum (say less than ~300 MeV/c) in the cosmic radiation present at sea level. On the other hand most of the muons reaching sea level have momenta well in excess of 300 MeV/c.

By 1935 it was made clear, in particular by Auger (1935), that there are two components in cosmic radiation: the easily absorbable "soft" component made of electrons and high-energy photons which produce showers, and the penetrating "hard" component made of "something else" whose nature no one understood. Actually a long road of hard theoretical work was trodden to arrive at a full understanding of the phenomenon of the showers of particles first observed by Skobeltzin. The phenomenon was finally explained with the cascade theory of Bhabha and Heitler (1937), and of Carlson and Oppenheimer (1937), based on the quantum electrodynamical theory of Bethe and Heitler (1934). By then the nature of the soft component of cosmic radiation was conclusively clarified.

But what about the nature of the penetrating component?

2. DISCOVERY OF THE MESOTRON

The discovery of the mesotron, unlike that of the positron, was not sudden and unexpected but was rather the culmination of a vast amount of careful experimental investigation.

Neddermeyer and Anderson (1937) were the first to report conclusive evidence that the penetrating charged cosmic-ray particles have a mass intermediate between the electron and the proton masses. This conclusion was reached after a long series of measurements made near sea level and on the summit of Pike's Peak (about 4.500 m a.s.l.) in the years 1934-36. Use was made of a counter-controlled cloud chamber placed in a strong magnetic field and containing a 1 cm thick platinum plate across the middle so that the particle tracks could be observed both before and after passing through it. By measuring the curvature of each track on both sides of the plate Anderson and Neddermeyer could derive the incident momentum, p, and the momentum loss, -dp/dx, of the corresponding particle.

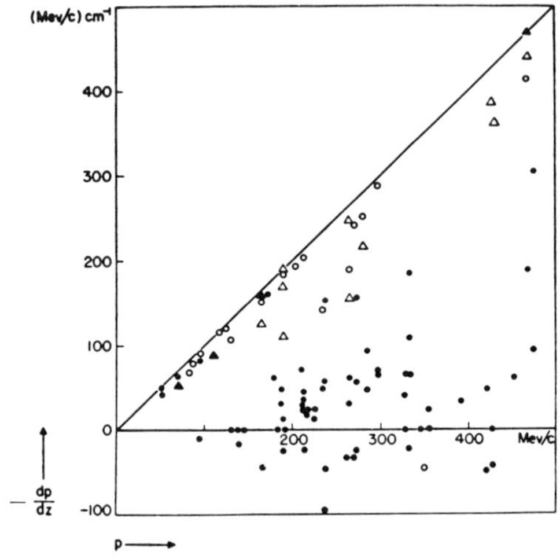

Fig. 1. Momentum loss per cm of platinum (-dp/dz) vs incident momentum (p) of cosmic-ray particles. •: single particles which traverse the platinum plate without producing interactions. ▲: single particles which produce showers in the plate. o: particles occurring in groups which traverse the plate without producing interactions. Δ: particles occurring in groups which produce showers in the plate. (From Neddermeyer and Anderson 1937).

In a plot of -dp/dx vs p (Fig. 1) the points were marked differently for "single particles" (i.e. particles occurringly singly and traversing the plate without producing showers) and for particles belonging to a shower or producing showers ("shower particles"). The momentum loss, which for the former particles appeared in the plot to be essentially constant within the large dispersion, was for the latter particles nearly proportional to the incident momentum, as expected for electrons well above the critical energy. Moreover single particles never produced showers in traversing the plate whereas particles belonging to a shower often did so. Finally, by comparing the results of the measurements performed at sea level and on the summit of Pike's Peak, the increase in the number of particles with altitude was found to be about a factor 3 larger for shower particles than for single particles. These striking differences were correctly interpreted on the assumption that the shower-particles are electrons and positrons, and that the single penetrating particles (the hard component of cosmic rays) have a mass much larger than the electron mass. On the other hand the single-particles could not be protons for various reasons. The most serious objection against such an interpretation lay in the fact that the energy distribution of the observed secondary "knock-on" electrons was much richer in high energy electrons than expected for primary protons (particles of similar masses tend to share their energy equally in knock-on collisions). Another difficulty was that, since the single particles had both positive and negative charges, the aforementioned interpretation would imply the existence of negative high energy protons in number nearly equal to that of the positive ones. Moreover, identifying the single penetrating particles as protons would require that there be a considerable number of particles slow enough to ionize more than the minimum, as any proton of momentum below about 1 GeV/c would be heavily ionizing. Many particles of momentum less than 1 GeV/c were instead observed with ionization close to the minimum. Finally, in many cases the ionization-curvature relations of the cloud chamber tracks clearly indicated that the single particles could not be as massive as protons. (These cases could be interpreted in terms of electrons, but only if electrons ceased to undergo appreciable radiation losses above an energy of some 100 MeV, as if the Bethe-Heitler theory broke down at high energies). The final conclusion reached through this investigation was that the hard component of cosmic rays contained particles heavier than an electron but lighter than a proton, of mass somewhere in the range from 100 to 300 electron masses. The name mesotron was proposed by Anderson and Neddermeyer in a note to Nature, after consultation with Millikan.

Soon after and independently of the work discussed above (their article was received on August 28, 1937, by The Physical Review) Nishina, Takeuchi and Ichimiya (1937) reported further evidence for the existence of particles of intermediate mass using again the technique of the magnetic counter-controlled cloud chamber with a metal plate inserted across the middle. In their own statement their «results could most naturally be explained if one assumes the existence of new particles of mass heavier than that of an electron and lighter than that of a proton». The results were in fact consistent with the assumption that the new particles had a mass between 1/7 and 1/10 that of the proton. These authors concluded their paper by stating that «although the exact determination of the composition of the penetrating component of cosmic-ray particles has not yet been possible, its large part no doubt consists of the above new particles, through the existence of which various difficulties in connection with cosmic ray phenomena ... find a natural explanation».

Although very strong, the evidence for the existence of the mesotron provided by these experiments was of a somewhat negative nature in that the observations could not be apparently explained except by the introduction of such a new particle. More direct evidence came from magnetic cloud chamber work by studying the tracks of particles of momentum low enough that they would be heavily ionizing if they were protons.

In a letter of October 6, 1937, to The Physical Review, Street and Stevenson (1937) reported the photograph of the track of a particle of intermediate mass shown in Fig. 2. By means of a veto-counter system the chamber, placed in a magnetic field (H = 3500 gauss), was triggered mostly by particles near the end of their range. Its expansion was delayed by about 1 s to allow the ions to diffuse for a more precise determination of the primary ionization by droplet counting. In the specific case the ionization of the particle was about 6 times that of a minimum ionizing particle and the product $H\rho$ (ρ being the measured radius of curvature) was 96 Kgauss · cm. Assuming that the particle came from the top of the chamber, it had to be a negative particle of mass ~ 130 electron masses with an error of ± 25% (evidently underestimated).

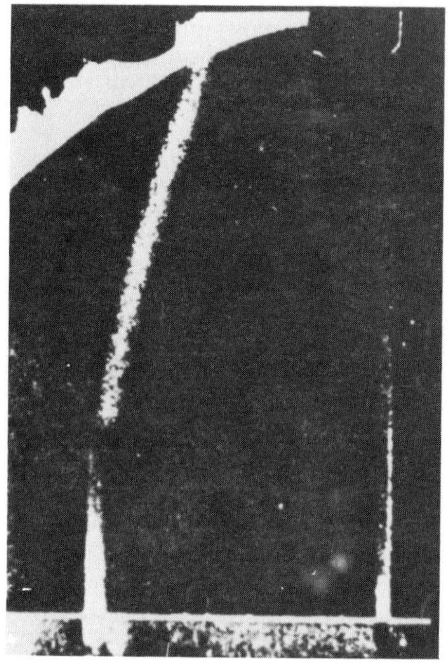

Fig. 2. Cloud-chamber picture of a mesotron. If coming from the top the particle has a negative electric charge. Its mass is derived from curvature and ionization measurements. (From Street and Stevenson 1937).

The apparatus employed in subsequent work (Street 1939) was the forerunner of the large experimental set-ups used later in cosmic-ray research. It consisted of two cloud chambers one on top of the other. The top chamber was in a magnetic field to provide momentum measurements and to determine the sign of the particles; the bottom one contained lead plates. The chambers were triggered by a system of GM counters. A lead plate at the top selected penetrating particles. In several instances particles were observed which had a range larger than that which a proton would have for the measured momentum, and which did not show the heavy ionization expected for protons. They did not produce showers in the plates of the bottom chamber and had also too big a range for electrons even at energies where the Bethe-Heitler theory had been shown to be valid.

For the sake of historical thoroughness the photograph of an event published as early as 1933 (Kunze 1933) with the remark that the blacker track ionizes more than an electron and less than a proton is reproduced in Fig. 3. This event remained unoticed for many years.

In 1938 Neddermeyer and Anderson (1938) published a striking picture of their cloud chamber in the middle of which a GM counter had been installed (Fig. 4).

The picture shows the track of a positive non-heavily ionizing particle which, after loosing energy in traversing the counter walls, comes out as a heavily ionizing particle and stops in the gas. Curvature and range give for the particle a mass of about 240 electron masses. Unfortunately the chamber was not sensitive enough to permit observation of the decay electron from the stop.

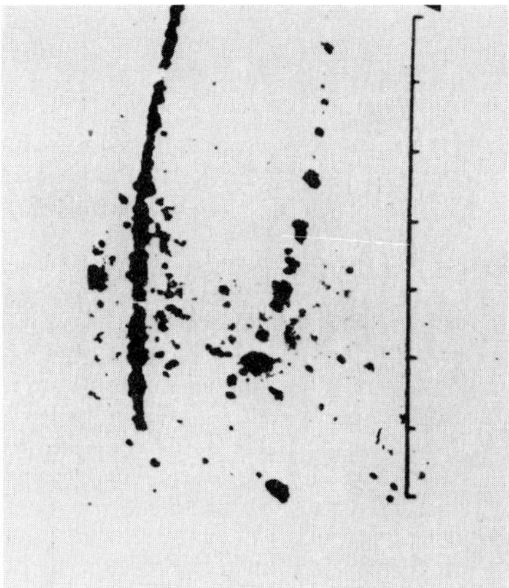

Fig. 3. Picture published as early as 1933 by Kunze (1933) with the remark that the blacker track corresponds to a particle which ionizes more than an electron and less than a proton.

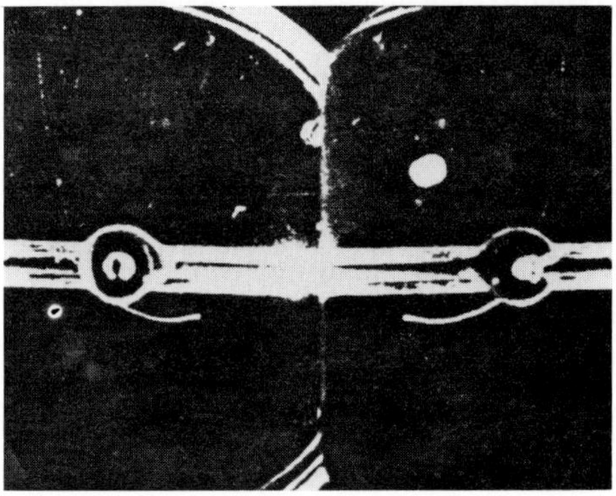

Fig. 4. Cloud-chamber picture of a positive mesotron which traverses a GM counter inside the chamber and stops underneath in the gas. Its mass (~240 m_e) is derived from curvature and range measurements. The chamber sensitivity was insufficient to observe the decay positron. (From Neddermeyer and Anderson 1938).

At this time the existence of the mesotron was established beyond any doubt and the main object of the subsequent experimentation was to check whether its properties fit the predictions of the "meson theory" started by Yukawa (1935).

3. THE YUKAWA AGENT OF NUCLEAR FORCES, THE MESON

In fact, as is well known, fiftytwo years ago Hideki Yukawa (1935) - then a young and obscure theoretical physicist at the University of Osaka - published a fundamental paper, totally neglected for more than two years, proposing that the short-range nuclear forces involve the exchange of "heavy quanta" between nucleons. In his paper Yukawa pointed out that the heavy quanta would not be observed directly in nuclear transformations because the energy required for their creation (their rest energy being about 100 MeV) was not available. The experimental demonstration that these quanta do indeed exist came out twelve years later in cosmic ray studies through the Bristol discovery of the charged pion to which this Conference is dedicated. But for a long time the Yukawa heavy quantum - often referred to as meson - was erroneously identified with the cosmic ray mesotron, as proposed by Yukawa himself (1937) as soon as the experimental evidence for the existence of the mesotron came out.

The essential part of the Yukawa meson theory - which eventually became an archetype for the theory of elementary particles - may appear today as a straightforward application of quantum field theory to the case of nuclear forces. But it could not have appeared so at the time of Yukawa's work. Three years later Wick (1938), using Heisenberg's uncertainty principle and the concept of virtual particle, provided a clarification of the meaning of the range-mass relationship (range of the nuclear forces equal to the Compton wavelength of the exchanged meson). Using only this relationship and the data then available on the range of the nuclear forces the mass of the meson was estimated by Yukawa to be about 200 electron masses.

In order to explain the phenomenon of the nuclear β decay Yukawa assumed that the meson was unstable and decayed into an electron plus a neutrino. In a later paper (Yukawa et al. 1938) the mean lifetime for a free meson at rest was estimated to be in the neighborhood of a microsecond. So mass and lifetime of the meson as derived from these early theoretical estimates turned out to have values close to those eventually observed for the cosmic ray mesotron, thus strongly reinforcing the assumption of the identity of the two particles.

A few years later Tomonaga and Araki (1940), starting from a previous work of Yukawa and Okayama (1939), investigated theoretically the effect of the nuclear Coulomb field on the behaviour of positive and negative mesons slowed down in matter. They pointed out that because of this field «the competition between nuclear capture and spontaneous decay must be different for mesons of different sign». They first demonstrated that «the capture in almost all cases does not take place before mesons come to rest». Then they calculated the capture probability per unit time for mesons at rest in various materials (lead, aluminium, air) showing that in all instances such a probability is negligibly small for the positive mesons (which are repelled by positively charged nuclei and compelled to wander around in matter until they decay), whereas for negative mesons it is always much larger than the probability of spontaneous decay (Table 1).

TABLE 1 - Capture probability per unit time of negative mesons in various materials according to Tomonaga and Araki (1940). The probability of spontaneous decay is always much smaller ($\sim 4.5 \cdot 10^5\ s^{-1}$).

Material	Lead	Aluminium	Air
Capture probability	$2.5 \cdot 10^{12}\ s^{-1}$	$1.2 \cdot 10^{11}\ s^{-1}$	$3 \cdot 10^7\ s^{-1}$

Even though no "Z^4 effect" is predicted [an effect pointed out later by Wheeler (1947)] the capture probability is seen to be quite different in the various materials, but always much greater than the probability of decay.

After this theoretical work a direct experimental investigation on the behaviour of positive and negative mesotron at rest in matter became a challenge for the experimentalist willing to check the assumed identity of mesotrons and mesons. Indeed this assumed identity appears in retrospect somewhat unjustified for it clashed with the experimental fact that the cosmic ray mesotrons were more penetrating than one would expect for strongly interacting particles such as the one invented by Yukawa to explain the strong nuclear forces. But it is a fact that for about a decade physicists - with possibly some exceptions - were convinced of this identity.

4. MASS AND LIFETIME MEASUREMENTS OF THE MESOTRON

These days, accurate determinations of the mass, μ, and lifetime, τ, of the mesotron (the lepton-μ) appear as essential steps to arrive at a precise knowledge of a fundamental constant of Nature, the Fermi coupling constant related to μ and τ by the well known (approximate) formula ($c = h = 1$) $G^2 = 192\ \pi^3/\tau\mu^5$. [With kinematic and radiative corrections, and using the precise today-values of τ and μ, $G = (1.16634 \pm 0.00003)\ 10^{-5}\ GeV^{-2}$]. This of course was not known at the time of the discovery of the mesotron and the efforts directed to obtain experimental determinations of μ and τ were motivated largely by the desire to check whether the properties of the cosmic ray mesotron were in line with those predicted for the Yukawa meson on theoretical grounds.

For some time after the discovery of the mesotron it was not clear whether or not the experimental data concerning mass measurements could be interpreted in terms of a unique mass μ of the order of 200 electron masses. Bethe (1946) made a keen analysis of the data available at the time pointing out the importance of the contribution of scattering in the cloud chamber gas to the apparent curvature of the particles. His conclusion was that when scattering was properly taken into account «all existing data are compatible with a unique mass of the meson, of about 200 electron masses». Meanwhile the question was conclusively settled especially through the elaborate measurements of Fretter (1946) who by the momentum-range method already adopted by Street (1939), obtained $\mu = 202 \pm 5$ in electron mass units [see also Brode (1949)].

The story of the determination of the mesotron mean lifetime τ is more complex. Early studies of the intensity of the hard component of cosmic radiation (Ehmert 1937; Auger et al. 1937) had shown an "anomalous absorption" of this component: the mesotrons were apparently more strongly absorbed by a given long thickness of air than by an "equivalent" short thickness of condensed matter representing the same quantity of matter traversed.

Kulenkampff (1938) and Euler and Heisenberg (1938) were the first to point out that such an anomalous absorption could be interpreted in terms of a relativistic effect by the spontaneous decay of the mesotrons produced in the upper layers of the atmosphere by primary cosmic rays. In fact, a mesotron travelling with velocity βc and momentum $p = \beta c \mu \gamma$ (μ being the mesotron rest mass and γ the Lorentz dilatation factor) has a decay mean free path $\lambda = \beta c \gamma \tau = p\ (\tau/\mu)$ if τ is its meanlife at rest. Hence the intensity of mesotrons of momentum p travelling over a length h is attenuated by the factor $\exp(-h/\lambda) = \exp(-h/p)(\mu/\tau)$. This is an important attenuation factor for a large fraction of the momentum spectrum of cosmic ray mesotrons if h represents the large distance (typically of the order of a km) between two measuring stations: e.g. one at sea level and the other at mountain altitude. On the other hand the attenuation factor can be ignored for the small thickness of condensed matter (typically some 10 cm) to be installed in the upper station to compensate for the absorption in the column of air between the two stations.

The considerations just outlined provide a method to derive the ratio τ/μ from attenuation measurements if a momentum spectrum of the cosmic ray mesotrons is assumed. Many authors carried out measurements of this type reporting results not always in agreement within the quoted errors. As an example the results of the accurate measurements of Rossi et al. (1940) at various altitudes are reported in Fig. 5. A graphite absorber was used in this experiment as a dense compensating material. The relevant fact was that more than twice as many mesotrons disappeared in the long expanse of thin air between any two stations, as disappeared in the short distance of the compensating dense carbon absorber. This could not be accounted for in terms of the so-called "density effect" (Fermi 1940; Halpern and Hall 1940) but only by the disappearance of mesotrons through decay in flight.

These experiments on the anomalous absorption of the mesotrons provided an indirect means to obtain the value of the meanlife τ for, as already stated, the value of the mass μ was roughly known ($\mu = \sim 200$ m). Curiously enough, the wrong identification of the mesotron with the meson, and the wrong assignments for the meanlife and mass of the latter, conspired to provide a conclusion essentially correct, namely that the cosmic ray mesotron (the muon) is unstable with a mean lifetime of about 2 μs, as reported in the above mentioned experiment of Rossi et al. (1940).

Up to that time there was no direct evidence for the disintegration of the mesotron, for the few cloud chamber pictures of mesotrons stopping in the chamber failed to show any convincing evidence of emitted electrons (Neddermeyer and Anderson 1938, Nishina et al. 1939, Maier-Leibnitz 1939). However in 1940 a cloud chamber picture by Williams and Roberts (1940) (see also Williams and Evans 1940) showed a track identifiable as that of a mesotron coming to rest in the gas of the chamber, and the track of what appeared to be a positive electron coming out of the end of the mesotron track.

Fig. 5. Illustrating the "anomalous absorption" of cosmic-ray mesotrons. The solid line represents the logarithm of the mesotron intensity vs atmospheric depth (x). The dashed lines represent the result of absorption measurements in carbon. (from Rossi et al. 1940).

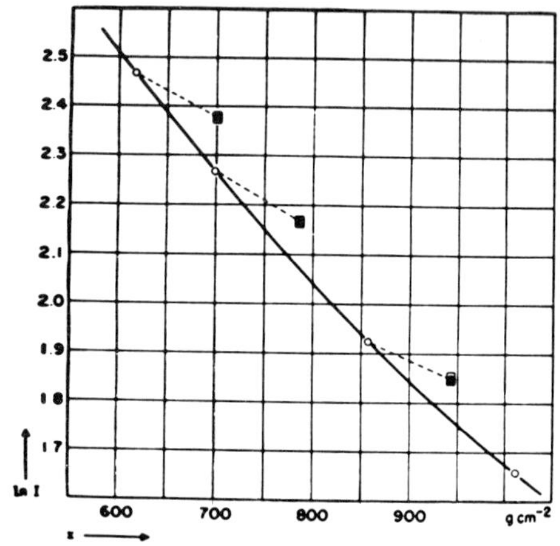

By 1940 conclusive evidence that the mesotron is an unstable particle decaying with a lifetime τ of a few microseconds had been therefore achieved; but a direct measurement of τ was still missing. Such a measurement had been attempted in fact but without success. For example Montgomery et al. (1939) studied the "delayed coincidences" between two layers of GM counters separated by a thin mesotron-stopping lead plate; but they failed in observing a signal of mesotron decay because of the large fraction of spuriously delayed counter pulses (~ 6% of all non-stopping mesotrons).

The first successfull measurement of τ was made by Rasetti (1941), on the assumption of an exponential decay and deriving the value $\tau = (1.5 \pm 0.3)$ μs from the logarithmic decrease relative to two points of the assumed decay curve. The geometry of the apparatus was very carefully designed but at the price of some relevant loss in counting rate. A well collimated mesotron beam (selected by coincidences between the pulses of GM counters and suitable absorbers) traversed an aluminium absorber in which a small fraction of the incoming mesotrons stopped. A few veto counters placed at the bottom of the aluminium absorber signaled the mesotron stops. The decay electrons were detected by counters placed at the sides of the aluminium absorber out of the solid angle of the beam. Hence, unless scattered in the absorbers, the incoming mesotrons never traversed the electron counters and the danger of spuriously delayed pulses from the latter was potentially eliminated. Rasetti avoided the task of developing fast delayed coincidences by measuring simultaneously the rates of three prompt coincidences with resolving times of 0.95, 1.95 and 15 μs. Since virtually all mesotrons decay before 15 μs, the differences D_1, D_2, between the rate measured with 15 μs resolving time on one side and, on the other side, the rates measured with 0.95 μs and 1.95 μs, give in practice the numbers of mesotrons still present after these latter times. The meanlife τ is then derived from the measured ratio D_1/D_2 $[1/\tau = \ln (D_1/D_2)\ \mu s^{-1}]$.

The spurious delays of the GM counter pulses, which prevented success in the earlier attempts to measure τ (Montgomery et al. 1939), were probably also the cause for obtaining smaller values for τ than the real one (nearly 2.2 μs) in Rasetti's experiment and in the nearly contemporary experiment of Auger et al. (1941; 1942) who reported (Auger et al. 1942) the value $\tau = 1.3$ μs $\pm$ 50%. In a subsequent publication of the work by the French group (Chaminade et al. 1944), which like the experiment by Oreste Piccioni and myself referred to below was carried out during the difficult years of the war, the value $\tau = (2.2 \pm 0.2)$ μs was reported.

An accurate determination of τ had been made in the meantime in the USA by Rossi and Nereson (1942) who had measured the delay between the mesotron stop and the electron emission on single mesotron decay events by the clever and elegant technique of the "time-to amplitude converter". The exponential decay curves obtained by these authors for mesotrons stopped in various materials are shown in Fig. 6. The value of τ reported in the final paper (Nereson and Rossi 1943) is $\tau = (2.15 \pm 0.07)\ \mu s$.

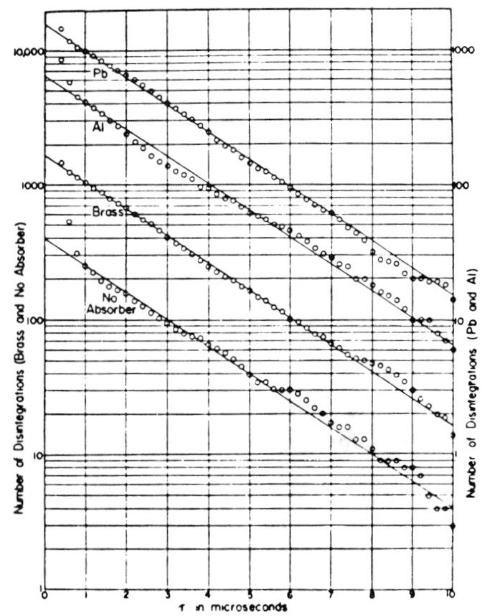

Fig. 6. Integral decay curves of cosmic-ray mesotrons stopped in various materials. Each point represents the number of observed decay events for which the decay time is greater than t. (From Nereson and Rossi 1943).

5. SOME RECOLLECTIONS OF WORK MADE IN ROME IN THE EARLY 1940's

I will refer now in more detail to the work carried out in Rome during the 2nd World War in which I became personally involved. In fact, as I stated already elsewere (Fermilab Symposium, 1980), my career as a physicist started in those times. Because of a strong amblyopia to my left eye I was rejected for military service and so avoided participation in the war. This allowed me to do the work described below.

By mid-July of 1940 I graduated in Physics at the University of Rome and early in September I was back in the Physics Institute. I was lucky because Gilberto Bernardini (then a charming young professor of physics at the University of Bologna spending a sizeable fraction of his time in Rome) invited me to join him in a new attempt to observe the effect of magnetized iron bars on the high-rigidity cosmic rays. Similar attempts had been made in the past (Skobeltzyn 1929, Rossi 1931, Mott-Smith 1932) but unsuccessfully, mostly because of the nearly equal numbers and energy distributions of positive and negative cosmic-ray particles at sea level. These attempts were made at a time when cosmic rays were supposed to be all negative electrons. Hence the common expectation was to find a large effect. The negative results of the attempts raised then the question of whether the field acting on the charged particles traversing a magnetized iron bar (the "interior field" $\mathbf{H}_i$) was close to the magnetizing field, **H**, rather than to the magnetic induction, **B**. The question was solved on theoretical ground in favour of **B** by von Weizsäcker (1933); but an experimental confirmation of this theoretical conclusion did not come soon. In fact, in a specific experiment "on the interior magnetic field in iron" Alvarez (1934) reached the conclusion that «the data obtained so far are in definite disagreement with the classical theory of magnetism, which postulates an interior field = **B**. The experimental results indicate that the deflecting field in iron is less than 1/3 of **B**. This is consistent with the findings of Mott-Smith (1932) and Rossi (1931) ...». Two years later

Danforth and Swann (1936) made another investigation on the "effective magnetic vector within the iron", $\mathbf{H}_i$, without reaching a definite conclusion. To the best of my knowledge the first experimental evidence that $\mathbf{H}_i$ had to be close to **B** came out of the work that Bernardini and I carried out in Rome in 1940. Incidentally this work led to my first scientific publication (Bernardini and Conversi 1940).

In this work we used "magnetic lenses" of a type first developed by Bruno Rossi (1931) «on a principle suggested by prof. Puccianti» (Luigi Puccianti, who had among his students Fermi, Rasetti, Bernardini, Pontecorvo, Piccioni, and other well known scholars of the celebrated "Scuola Normale Superiore di Pisa", held in those days the chair of Experimental Physics which I "inherited" 20 years later). These magnetic lenses are basically made up of two vertical iron bars with opposite horizontal magnetization, in a closed magnetic circuit (Fig. 7). The system behaves as a converging magnetic lens for charged particles of a given charge sign that traverse the bars incoming from the top, and as a divergent lens for particles of the opposite sign.

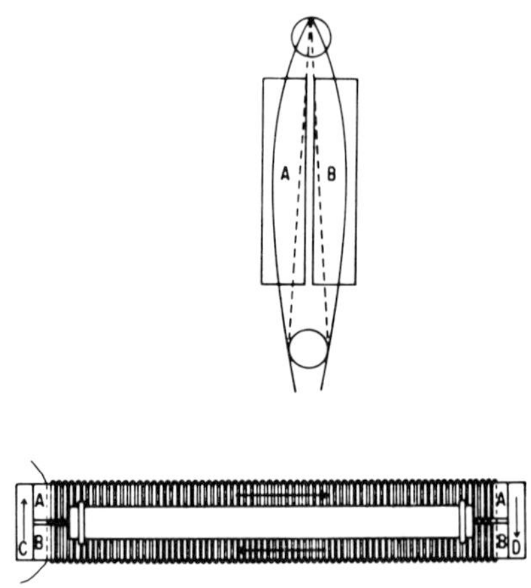

Fig. 7. Illustrating the principle of operation of the "magnetic lenses" capable of concentrating the particles of a given charge sign (solid line trajectories at top), while acting as a defocussing device for the particles of the opposite sign (dotted line trajectories). (From Rossi 1931).

The novelty of our work with respect to Rossi (1931) was to use two lenses, one on top of the other, so that the top lens would alter the natural charge (and energy) distribution of the cosmic-ray particles and the bottom lens could be utilized to analyze the altered distribution.With such a "doublet" we recorded the numbers of threefold coincidences among three thin GM counters placed just on top, in the middle of, and below the doublet. By counting these threefold coincidences with the field in the bottom lens alternately "parallel" and "antiparallel" with respect to that in the top lens, a large difference in the two counting rates was recorded: a 35% effect which could only be accounted for in terms of an interior magnetic field $\mathbf{H}_i$ close to **B**, as found after that Giancarlo Wick had solved the non-trivial task of developing the theory of the instrument. Moreover this large effect clearly indicated the possibility of further applications of the method to cosmic ray studies; studies which were made in fact in the following years (Bernardini et al. 1941; Conversi and Scrocco 1943; Bernardini et al. 1945).

While this work was in progress I was lucky again when late in 1941 or perhaps early in 1942 Piccioni - then a brilliant young physicist clearly destined to become a top experimentalist - proposed that we work together. He had been impressed by Rasetti's work, and thought we could perform a new experiment capable of providing a more conclusive demonstration of the muon decay and a precise determination of its mean lifetime τ through the detection of several points of the muon exponential decay curve. As I stated elsewhere ("Winspread Conference") «the task of developing an adequate technique was already formidable, and it was made eventually even more difficult, of course, by the special conditions of the war. But we felt from the beginning that the possibility of investigating directly the fate of mesons coming to rest in matter would have rewarded any effort by opening to us a new fascinating line of research.» And in fact we were rewarded at the end of a long road of hard experimental work by the discovery of the non-hadronic nature of the mesotron.

Piccioni (1983) vividly recalled at the 1980 Fermilab Symposium the struggle that led to this discovery; later he discussed in the Paris Colloquium (Piccioni 1982), and especially in the Wingspread Conference (Piccioni 1984), all the main steps related to the accomplishment of our four consecutive experiments:
1) Direct measurement of the mesotron lifetime (Conversi and Piccioni 1944a)
2) Test of Tomonaga-Araki theory without magnetic lenses (Conversi and Piccioni 1944b)
3) Test of Tomonaga-Araki theory with magnetic lenses (Conversi, Pancini, Piccioni 1945)
4) Test with magnetic lenses and graphite absorber (Conversi, Pancini, Piccioni 1947).
I will therefore only briefly describe below part of this experimental work. But I wish to make clear immediately that the first experiment, in which Piccioni had a leading role, represented by far the greatest effort in the whole series providing the basis for all subsequent work. He also made essential and original contributions to the development of the tools we needed to reach our goal, such as (to quote just an example) the fast "series coincidences" (Piccioni 1943) based on the use of secondary emission vacuum tubes.

6. RECOLLECTIONS OF WORK MADE AT THE "VIRGILIO"

The first experiment was prepared at the Physics Institute. However, on July 19, 1943, when the electronic system was nearly completed, Rome was bombed for the first time by the American air force. Nearly 80 bombs fell within the perimeter of the University campus, and one of them fell just outside the window of our laboratory, a few minutes after I had left it, fortunately having moved the electronic system far away from the window. Afraid that other bombing would follow making the continuation of our work impossible, we decided to move as soon as possible to some place near the Vatican City, presumably more protected from the air-raids. So towards the end of July 1943, with the help of a few friends including professor Edoardo Amaldi, we transported all our equipment to a semi-underground class room of the "Liceo Ginnasio Statale Virgilio" (in the old Via Giulia), whose deputy headmaster, prof. L.Fagiolo, was an acquaintance of mine. The class-room was soon transformed into a laboratory with the essential equipment. We completed the installation of the counters and electronics and started to test the apparatus. But we were soon interrupted by the Italian armistice (September 8, 1943) and the consequent occupation of Rome by the German troups. Work was resumed about one month later under very difficult and sometimes dramatic conditions. Late in 1943 we finally started to run our first experiment.

Fig. 8 shows the experimental arrangement of the GM counters and absorbers used for this and the subsequent experiment. The first and second sets of six counters from the top detected an incoming muon. If the muon stopped in the absorber and a decay electron was emitted later, a delayed coincidence was recorded by the third set of six counters from the top. The anticoincidence counters A vetoed a count if the muon traversed one of them. A 1 cm thick lead plate between the electron-decay counters and counters A prevented the decay electrons from vetoing themselves. Due to the large solid angles involved the rate of recorded decay electrons (typically 5/hr with a 5 cm thick iron absorber) was more than 20 times larger than in Rasetti's experiment. Early in 1944 Piccioni and I had collected enough data to reach the conclusion that the cosmic-ray mesotrons stopped in the iron absorber did indeed exhibit an exponential decay. We thought we had been the first to record the exponential decay curve of a free particle. Due to the total lack of communications under war time conditions, we did not know that on the other side of the Atlantic Ocean Rossi and Nereson (1942) had already reached this goal. The experiment I have briefly recalled led finally to the result $\tau = 2.3\ \mu s$ (± 6.5%) for the mean lifetime of the mesotron at rest in iron.

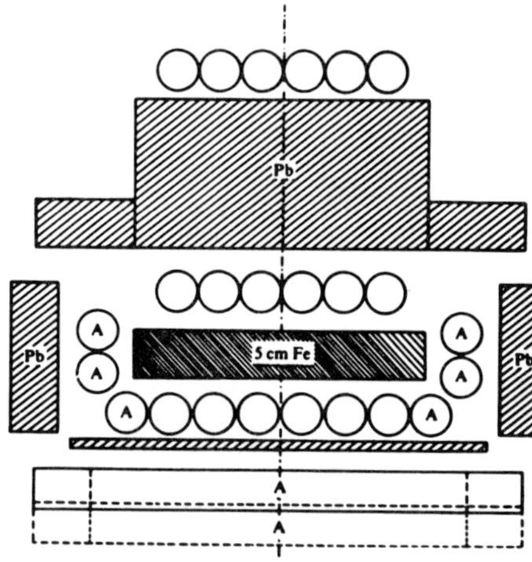

Fig. 8. Apparatus used at the "Virgilio" in the "first experiment" designed to obtain the exponential decay curve of cosmic-ray mesotrons stopped in the iron absorber. (From Conversi and Piccioni 1944). The same apparatus, but with a 0.6 cm thick Fe absorber, was used in the second experiment designed to obtain the ratio $\eta = \mu$-decays/μ-stops.

The second experiment aimed at testing the predictions of Tomonaga and Araki (1940) on the behaviour of positive and negative mesotrons at rest in matter. As I stated before, given the identity meson ≡ mesotron, these predictions were that of all mesotrons stopped in any material only the positive ones should decay with a meanlife τ of ~2.2 μs, the negative ones undergoing atomic and nuclear capture in a time much shorter than τ. Since there is a 20% excess of positive mesotrons at sea level, the fraction η of stopped mesotrons expected to decay should be accordingly about 0.55. Rasetti (1941) had made a rough experimental determination of η in aluminium finding $\eta = 0.42 \pm 0.15$. Maze and Chaminade (1942) reported an "approximative agreement" with $\eta \approx 1$, but without explaining how this result was reached. In both these experiments the measured values of the meanlife τ (already reported above) were small compared to 2.2 μs. This clearly indicated the presence of a background of spurious delays which rendered unrealiable the estimate of η, derived by a delicate comparison between the number of stopped mesotrons and the corresponding "absolute" number of decays. In fact to get this last number one needs indeed to know the "absolute delay" under the conditions of the experiment. Furthermore, in order to obtain the value of η one needs to know which fraction of the decay electrons is able to emerge from the absorber. Neither of these items was well known in the two experiments mentioned above. Hence not much importance could be attached to the possible apparent discrepancy between their results. Piccioni and I circumvented the experimental difficulties I have just outlined by making a very carefull determination of the absolute delay and by using an iron plate absorber thin enough (0.6 cm) to make sure that a very large fraction of the decay electrons from the stopped mesotrons would emerge from it. Actually, a comparison of the rates measured with this 0.6 cm iron plate and with the 5 cm thick iron absorber already used in the previous experiment, showed that the decay electrons in iron had a range considerably larger than 0.6 cm, which we estimated in fact to be about 2.5 cm. The result of our second experiment (Conversi and Piccioni 1944b) was $\eta = 0.49 \pm 0.07$, with an upper limit of 0.56 ± 0.08, well consistent with the theoretical predictions.

7. TEST OF TOMONAGA-ARAKI PREDICTIONS WITH MAGNETIC LENSES

After this result was achieved it appeared clearly desirable to make a direct and conclusive test of the theoretical predictions by concentrating positive and negative mesotrons separately on the absorber. The obvious next step was then to consider whether an efficient separation between positive and negative stopping mesotrons could be achieved by the magnetic lenses of the Rossi type successfully utilized in Rome since 1940. Piccioni and I discussed at length this possibility in terms of rejection power of the lenses against the particles of the unwanted sign and of the expected counting rate, which would have been considerably reduced, of course, with respect to that of the two previous experiments. Due to the difficulty to take correctly into account the effect of the multiple scattering at various depths in the iron, we could not provide a reliable calculation of the rejection power. For a particle trajectory of length L the cumulative effect of the multiple Coulomb scattering increases as $\sqrt{L}$, whereas the deflecting effect of the magnetic field increases faster with L. Hence the vertical thickness of the lenses had to be as big as compatible with the requirement of still having a detecting solid angle large enough to keep the counting rate of the delayed coincidences at a reasonable value, say about 0.5/hr. These considerations led us to adopt a set-up (see Fig. 9) in which two lenses of 20 cm vertical thickness were installed side by side (with veto counters in between) in order to double the counting rate. The choice of the parameters turned out to be satisfactory from the standpoint of both counting rate and discrimination against the stopping mesotrons of the unwanted charge sign. Piccioni and I projected this third experiment at the Liceo Virgilio where we had "seen" the mesotron decay ... But in the meantime Rome had been liberated by the Allied troups (June 5, 1944) and we went back to our Institute at the University, on the other side of the city.

Long before, our late friend Ettore Pancini - a man of rare intellectual perspicacity - had expressed his interest in the line of research undertaken by Piccioni and myself. However, he was first in the war and later, after the Italian armistice, in the CNL (National Committee for the Liberation of Italy)*, with no chance to work as a physicist during this long period of hard life. Early in 1943, during a convalescence leave from the army spent in Rome (he had been wounded in the war) he showed a keen interest in the work of Tomonaga and Araki mentioning the possibility of investigating experimentally the behaviour of positive and negative mesotrons with the help of magnetic lenses. Shortly after the liberation of Northern Italy (April 25, 1945) Pancini was back in Rome, where he settled definitively in September. Piccioni and I were very glad to have him with us in the performance of our third experiment though this was near completion.

In fact the result of this third experiment (Conversi, Pancini and Piccioni 1945), carried out using the same electronic system as in the two previous experiments and a 3 cm thick iron absorber, was published in a letter to the Physical Review received on October 15, 1945. In the conclusion of this paper we stated that «the prediction of Tomonaga and Araki seems to be confirmed experimentally»: we had recorded in fact a hourly rate of 0.33 ± 0.04 delayed coincidences for positive particles concentrated in the iron absorber against 0.07 ± 0.02 for the negative ones.

Then we decided to replace the iron with a low-Z absorber which turned out to be made of graphite rods, as shown in Fig. 9. One essential reason for this choice, which Piccioni recalls as the "unique" reason (Piccioni 1984), was the hope to observe high energy γ rays possibly emitted after the nuclear capture of the negative mesotrons. [A few years later Piccioni (1948) carried-out an elegant experiment to search for this effect reaching the proof that high energy γ-rays are not emitted by nuclei after μ^--capture]. In my recollections I wanted to extend the test to small Z also for the sake of experimental thoroughness. Even though the conclusions of Tomonaga and Araki (1940) were that meson capture should prevail largely over decay for negative mesons at rest in any material, their work clearly indicated a decrease of the capture probability with decreasing Z (Table 1). The choice of carbon was good anyway. Graphite rods were the only possible low-Z material quickly obtainable on the market in those days and graphite has a good stopping power for muons and a small absorption cross section for γ-rays.

* A leader in the "Partisan Movement", Pancini fought in the Italian Resistance under the name of "Achille".

Fig. 9. Model of the apparatus used in the "Rome experiment", stored at the Smithsonian Institution in Washington DC. Counters, magnetic lenses and absorbers are clearly visible. Also clearly visible are the graphite cylinders where the effect (non-capture of negative mesotrons) was unexpectedly observed.

Table 2 shows the results of this fourth experiment. The decay rates reported in the last column were obtained as the difference between threefold and fourfold coincidences, with a small correction for counter inefficiency. The measurements with 5 cm iron - essentially a replica of the previous experiment - show the high rejection efficiency of the magnetic lenses against the particles of the unwanted charge sign.

TABLE 2 - Results of the fourth "Rome experiment".

Stopped Mesotrons	Absorber	Decay rate (events/10hr)
Positive	5 cm Fe	6.7 ± 0.65
Negative	5 cm Fe	0.3 ± 0.25
Negative	None	-0.1
Positive	4 cm C	3.6 ± 0.45
Negative	4 cm C + 5 cm Fe	2.7 ± 0.35
Negative	6.2 cm Fe	0

The 5 cm Fe-absorber was placed on top of the 4 cm C-absorber (5th row of Table 2) as a precaution to guard against the scattering of very low energy mesotrons which might destroy the magnetic selection. Of course the decays come only from C since negative mesotrons are captured in Fe (2nd row). Data were taken with a 6.2 cm Fe-absorber (last row) because this thickness of iron has the same stopping power as 4 cm C + 5 cm Fe. The last two rows of the Table provide the demonstration of the non-capture in graphite. In fact they show that negative mesotrons stopped in the 4 cm C (+ 5 cm Fe on top of it) do decay at a rate of 0.27 ± 0.035 events/hr, whereas no decays are observed with an equivalent iron absorber.

This result was entirely unexpected and we thought at first that something was wrong with our set-up. But all tests we did and repeated gave the confirmation that the apparatus worked properly. There was no instrumental explanation of the simple and clear fact that delayed coincidences were counted when we concentrated negative mesotrons on the graphite absorber, whereas they disappeared when the graphite was replaced with iron. This appeared astonishing to us because, to the best of our understanding, the predictions of Tomonaga and Araki clearly indicated that negative mesons slowed-down in any material should undergo nuclear capture in a time much shorter than 2.2 μs.

Of course this result could have been obtained indirectly also without the magnetic lenses, as shown by Sigurgeirsson and Yamakawa (1947) who promptly gave the first confirmation of our fourth experiment. But I do believe that the magnetic lenses did not only add "aesthetics" to the latter : they provided the means of a direct, straightforward, convincing proof of a new and entirely unexpected effect, showing that negative mesotrons stopped in carbon did undergo spontaneous decay whereas they did not when stopped in iron.

8. INTERPRETATION AND MAIN CONSEQUENCES OF THE "ROME EXPERIMENT"

The startling result of our fourth experiment, which from now on I will call the "Rome experiment" for short, was made known to Enrico Fermi via Amaldi towards the end of 1946, soon after we had reached our final results. Fermi immediately realized the great relevance of our finding and in December 1946 he gave a famous talk at the "Thursday Physics Seminar" at the University of Chicago, where, as a joke, he started by writing our names on the blackboard commenting he would not dare to pronounce them ...

The analysis which followed (Fermi, Teller and Weisskopf 1947) led to the startling conclusion that there was a factor 10^{12} discrepancy between the meson's production cross section and the mesotron's absorption cross-section. This conclusion was reached after showing that the time required by a fast mesotron to be brought to rest and to be captured in an atomic K orbit is much shorter than the ~ 2 μs time required for its spontaneous decay. Doubts were raised (Fröhlich 1947, Fröhlich et al. 1948) on the validity of the model of the solid adopted to calculate the slowing down of mesotrons with an energy below 2 keV (that is with smaller velocity than that of the electrons of the solid). However, through a complex and detailed calculation of the over-all time required by a fast negative mesotron to slow down and be captured in a K-orbit, Fermi and Teller (1947) ruled out the remote possibility that this time be very much longer in graphite than in iron.

Clearly the cosmic ray mesotron was an entirely different object from the Yukawa meson. What was it? And where was the Yukawa agent of nuclear forces that physicists for ten years had thought to have found in cosmic rays?

The answer to the latter question came very soon here at Bristol with the discovery that we are now celebrating. Here, incidentally, the mesotron was rebaptized as the muon, μ. But before this great discovery was made known, important steps were accomplished in trying to understand the meaning of the Rome experiment and new chapters of particle physics were opened by the latter.

(i) At the Shelter Island Conference in early June of 1947 Marshak and Bethe (1947) put forward the bold "two-meson hypothesis". They explained the 10^{12} discrepancy factor by having the strongly-interacting Yukawa meson produced in the upper atmosphere decay into the lighter weakly-interacting mesotron observed at sea level. Indeed this hypothesis, quite close to the experimental facts emerging in those days at Bristol, had already been made in Japan by Sakata and Inoue (1946), even before the conclusion of the Rome experiment, in an important work prompted by the attempt to explain the small cross-section of mesotrons in interaction with matter. In this work the secondary weakly-interacting particle had been correctly assumed, by sheer intuition, to have spin 1/2.

(ii) Immediately after the conclusion of the Rome experiment Wheeler made another important theoretical cointribution deriving the "Z^4 law" for μ^- capture by the following simple argument. A μ^- slowed down and stopped in a material of atomic number Z is eventually captured into a Bohr orbit of radius $R = \hbar/Z\mu e^2 = \sim 250/Z$ fermi, μ being the μ^- mass and e the elementary electric charge. In light elements R>> nuclear radius = ~ $1.75\ {}^3\sqrt{Z}$ fermi. Hence the wave function ψ at the nucleus is ~ ψ (0), which is proportional to $Z^{3/2}$. Then the capture probability Λ_C per unit time, which is proportional to $|\psi|^2$ times the nuclear volume, turns out to be proportional to Z^4 and can be written as $\Lambda_c \equiv 1/\tau_c = (1/\tau)\ (Z/Z_o)^4$ where τ_c is the μ^- mean lifetime for capture, τ the free muon meanlife and Z_o a normalization constant found empirically to be close to 11.

Notice that since the decay and capture processes are in competition, the apparent mean lifetime, τ^-, of disappearence of a μ^- at rest in nuclear matter is given by $1/\tau^- = 1/\tau + 1/\tau_c = (1/\tau)\ [1+ (Z/Z_o)^4]$. The fraction, f, of spontaneous decays, $f = (1/\tau)/(1/\tau^-) = \tau^-/\tau$, is therefore $[1+(Z/Z_o)^4]^{-1}$, in essential agreement with the results of the Rome experiment in iron (Z = 26; f = .92) and carbon (Z = 6; f = .03) (Table 2).

Ticho (1948) using an apparatus similar to that of the Rome experiment, and Valley and Rossi (1948) using a magnetic cloud chamber triggered by delayed coincidences among GM counters [see also Valley (1947)], carried out important experimental investigations on the problems of μ^- capture just outlined above providing good quantitative fits. As an example the decay curves obtained by Ticho (1948), for μ^+ and μ^- stopped in two materials of Z close to the critical value $Z_o = \sim 11$ for which $\tau^- = \tau/2$, are shown in Fig. 10.

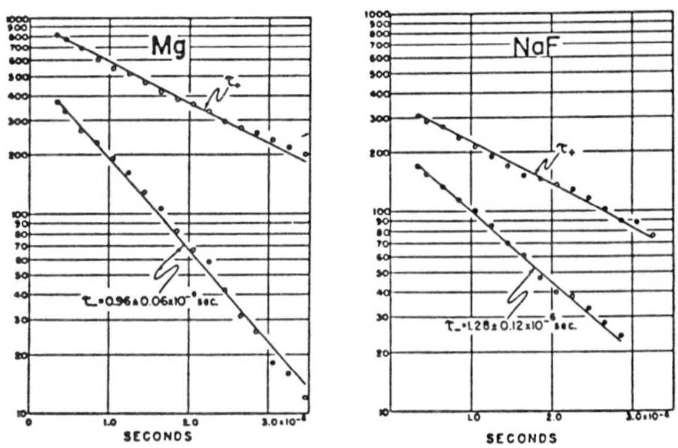

Fig. 10. Decay curves of positive and negative mesotrons stopped in NaF (Z = 10.1) and Mg (Z=12) absorbers (from Ticho 1948).

(iii) In my opinion the most far-reaching consequence of the Rome experiment was in the impact it had on the development of the concept of "Universal Fermi Interaction" [so called after Yang and Tiomno (1950)], intimately related to the lepton concept.

In a letter to Physical Review received on June 21, 1947, before the results of the Bristol group were known, Pontecorvo (1947) pointed out that on account of the result of the Rome experiment the mesotron cannot be the strongly interacting Yukawa particle so that the "two-steps" Yukawa theory of β decay is untenable. Therefore β decay ought to be described as a four fermion interaction by the original Fermi theory. He pointed out, furthermore, that the rate of mesotron capture by nuclei is of the same order of magnitude as that of atomic electron K capture if allowance is made for phase space and for the size of the Bohr orbits. The suggestion is then made that the mesotron has spin 1/2 and is captured with the emission of a neutrino ($\mu^- p \rightarrow n\nu$). I believe this is the first time that a generalization of the Fermi interaction to a particle other than the electron is proposed: it is the first step towards the concept of the universality of the weak interaction developed shortly after, as illustrated by a review talk at the "Wingspread Conference" (Tiomno 1984), with the contribution of various theorists:

Klein (1948), Lee, Rosenbluth and Yang (1949), Puppi (1948, 1949), and Tiomno and Wheeler (1949) (see also Tiomno, Wheeler and Rau, 1949).

(iv) Finally, another relevant consequence of the Rome experiment worth mentioning is the opening of the field of "mesic atoms", a new chapter of particle physics.

9. AFTER THE BRISTOL DISCOVERY...

After the Bristol discovery achieved with the splendid technique of the nuclear emulsion (Lattes, Muirhead, Occhialini and Powell 1947; Lattes, Occhialini and Powell 1947), the following conclusions - which will be discussed at length in the subsequent talks at this Conference - were reached: (i) There are two particles of intermediate mass: π and μ; (ii) The π, slightly heavier than the μ, decays into the latter with the emission of a unique neutral particle (presumably a neutrino: $\pi \rightarrow \mu\nu$) since the μ emitted in the decay has, within the straggling errors, always the same range; (iii) The best interpretation of the Rome experiment is that the π's are strongly interacting particles produced by primary cosmic rays in the upper atmosphere, which quickly decay into μ's. The μ's have weak interactions with nuclei, penetrate the atmosphere, and constitute the bulk of the penetrating component of cosmic rays near sea level (the mesotron component).

In my recollection of those days, which confirms that of Charles Peyrou (1982), a feeling of great euphoria dominated the world of physicists. Every difficulty seemed to be solved. The Yukawa meson was there: the π, from then on called <u>pion</u>. The problem of the nuclear forces would have soon become amenable to a quantitative treatment. At least for some time there seemed to be no great worry for some obscure points: The pion could not be the mediator of β decay; and what was the role of the μ, the <u>muon</u>?

In reality the problem of the nuclear forces could not be handled quantitatively. Nature was much more complex than was believed in those days. As we well know today the pion is a complex object made of a quark and an antiquark, and it is only the lightest of many similar but heavier structures. So it is still somewhat questionable whether it is appropriate to give it the priviledge of being identified with the Yukawa particle.

On the other hand, what about the muon? What was its role? Later, after years of hard and on some occasions extremely refined experimental work which can be exemplified by the extraordinarily accurate measurements of the "g-2" factors of the electron and the muon, it became clear that the muon is exactly like the electron, except for the mass: it is a "heavy electron" which "costs" ~ 200 times more energy to construct than an ordinary electron. A consideration that led Isidor Rabi to raise the question: «Who ordered it?», and Gell-Mann and Rosenbaum (1957) to write «The muon was the unwelcome baby on the doorstep, signifying the end of days of innocence».

From this situation originated by the Rome experiment slowly arose the concept of a "leptonic world". Indeed the word <u>lepton</u> was introduced by Rosenfeld (1948) «following a suggestion of Professor C. Møller, ... as pendant to "nucleon"... (from λεπτός, small, thin, delicate) to denote a particle of small mass irrespective of its charge». The lepton concept grew slowly and the definition today of lepton has nothing to do of course with the particle mass, but only with the fact that leptons - as known for the electron since the Fermi theory of β decay and as proven by the Rome experiment for the muon - are immune from strong interaction. Thus the muon was the second charged lepton to be discovered and nearly thirty years had to elapse before a third charged lepton - the τ, about 1.9 times heavier than the proton - appeared on the horizon through the discovery of Perl et al. (1975) at Stanford.

To complete this early story of the muon mention has to be made of important work, concerning its decay mode, carried-out soon after the Bristol discovery of the pion and the $\pi \rightarrow \mu$ decay.

The hypothesis of the $\mu \rightarrow e\gamma$ decay mode was seriously considered in those days. I believe Hincks and Pontecorvo (1948 a) were the first to search for such a process and to report a negative result, shortly followed by Sard and Althaus (1948) and by Piccioni (1948). Today we know this process is forbidden by the conservation of lepton numbers, and an upper limit close to 10^{-10} has been established for the probability of its occurrence relative to the ordinary decay mode ($\mu \rightarrow e\ \nu_e\ \nu_\mu, ...$).

The hypothesis that muon decay is a three-body process found some support in the work of Anderson et al. (1947), as it provided a possible interpretation of the observed decay of a mesotron giving an electron of 25 MeV only. A decay electron of energy as low as 15 MeV was also reported by Fowler et al. (1948). However, convincing evidence for a three-body decay came first out of the work of Steinberger (1948, 1949) who investigated the range spectrum of the electrons emitted from stopped cosmic-ray muons by counter techniques [see also Hincks and Pontecorvo (1948b)]. In this work the experimental data did not fit with a unique energy of the secondary electrons, whereas they did appear to be quite consistent with a distribution of energies as calculated from a three-particle decay process. Further evidence for a continuous energy spectrum of the electrons from muon decay was obtained shortly after by Leighton et al. (1949) using a counter-controlled magnetic cloud chamber. The statistical errors of the experimental points of the energy spectrum were too large to derive from this distribution a reliable value of the "Michel parameter" (Michel 1949). Worth quoting as an additional early proof of the three-body decay process is the work of Davies et al (1949) who measured the small-angle scattering of decay-electron tracks in nuclear emulsions.

The conclusion derivable from an analysis of the above mentioned work was that most probably the muon decay could be described by the three-body process $\mu \rightarrow e + \nu + \nu'$ where ν and ν' are two neutral and nearly massless particles, possibly neutrinos with spin 1/2. Then the muon must be a fermion, most probably of spin 1/2.

At this point I shall close what I like to call the "tangled tale of the muon", first observed as the costituent of the hard component of cosmic rays, then as an unstable particle of intermediate mass, for long erroneously identified with the Yukawa agent of nuclear forces, subsequently recognized as a sort of "heavy electron" (the first heavy lepton in today's words) subject to the weak processes of spontaneous decay and nuclear capture quite similar to β decay (universal weak interaction); then found, here at Bristol, as the daughter particle of the strongly interacting π meson, and finally identified as a fermion subject to the weak three-body decay $\mu \rightarrow e\ \nu\nu'$.

REFERENCES

Alvarez L 1934 Phys. Rev. 45 225
Anderson C D 1932 Science 76 238
Anderson C D, Adams R V Lloyd P E and Rau R R 1947 Phys. Rev.72 724
Auger P 1935 Nature 135 820
Auger P, Ehrenfest P, Freon A and Fournier A F 1937 Compt. Rend. 204 257
Auger P, Maze R and Chaminade R 1941 Compt. Rend. 213 381;
Auger P, Maze R and Chaminade R 1942 Phys. Rev 62 307
Bernardini G and Conversi M 1940 La Ric. Scient. 11 858
Bernardini G, Conversi M, Pancini E and Wick G C 1941 Ric. Scient. 12 1227
Bernardini G, Conversi M, Pancini E, Scrocco E and Wick G C 1945 Phys. Rev. 68 109
Bethe H A 1946 Phys. Rev. 70 821
Bethe H A and Heitler W 1934 Proc. Roy. Soc.A146 83
Bhabha H J and Heitler W 1937 Proc. Roy. Soc. A157 432
Blackett P M S and Occhialini G P S 1933 Proc. Roy. Soc. A139 699
Bothe W and Kolhörster W 1929 Zeits. f. Physik 56 751
Brode R B 1949 Phys. Rev. 75 904
Carlson J F and Oppenheimer J R 1937 Phys. Rev. 51 220
Chaminade R, Fréon A and Maze R 1944 Compt. Rend. 218 402

Conversi M, Pancini E and Piccioni O 1945 Phys. Rev. 68 232
Conversi M, Pancini E and Piccioni O 1947 Phys. Rev. 71 209 and 557
Conversi M and Piccioni O 1944a Nuovo Cimento 2 40
Conversi M and Piccioni O 1944b Nuovo Cimento 2 71
Conversi M and Scrocco E 1943 Nuovo Cimento 1 372
Danforth W E and Swann W F G 1936 Phys. Rev. 49 582
Davies J.H., Lock W O and Muirhead H 1949 Phil. Mag. 40 1250
Ehmert 1937 Zeit. f. Physik 106 751
Euler H and Heisenberg 1938 Ergeb. d. Exakten Naturwiss. 17 1
Fermi E 1940 Phys. Rev. 57 485
Fermi E and Teller E 1947 Phys. Rev. 72 399
Fermi E, Teller E and Weisskopf V C 1947 Phys. Rev. 71 314
Fowler E C, Cool R L and Street J C 1948 Phys. Rev. 74 101
Fretter W B 1946 Phys. Rev. 70 625
Fröhlich H 1947 Nature 160 255
Fröhlich H, Huby R, Kolodziesjski R, and Rosenberg R L 1948 Nature 162 450
Gell-Mann M and Rosenbaum E P 1957 Scientific American (July 1957) p. 72
Geiger H and Müller W 1928 Naturwiss. 16 617
Halpern O and Hall H 1940 Phys. Rev. 57 459
Hincks E P and Pontecorvo B 1948a Phys. Rev. 73 257
Hincks E P and Pontecorvo B 1948b Phys. Rev. 74 697
Klein O 1948 Nature 161 897 (June 5, 1948, issue)
Kulenkampff 1938 Verh. Deutsch Physik Ges. 19 92
Kunze P 1933 Zeits. f. Physik 83 1
Lattes C M G, Muirhead H, Occhialini G P S and Powell C F 1947 Nature 159 694
Lattes C M G, Occhialini G P S and Powell C F 1947 Nature 160 453 and 486
Lee T D, Rosenbluth M and Yang C N 1949 Phys. Rev. 75 905
Leighton R B, Anderson C D and Seriff A 1949 Phys. Rev. 75 1432
Maier-Leibnitz1939 Zeit f. Physik 112 569
Marshak R E and Bethe H A 1947 Phys. Rev. 72 506
Maze R and Chaminade R 1942 Compt. Rend. 214 266
Michel L 1949 Nature 163 959
Montgomery C G, Ramsey W E, Cowie D B and Montgomery 1939 Phys. Rev. 56 635
Mott-Smith L M 1932 Phys. Rev.39 403
Neddermeyer S H and Anderson C D 1937 Phys. Rev. 51 884
Neddermeyer S H and Anderson C D 1938 Phys. Rev. 54 88
Nereson N C and Rossi B 1943 Phys. Rev. 64 199
Nishina Y, Takeuchi M and Ichimiya T 1937 Phys. Rev. 52 1198
Nishina Y, Takeuchi M and Ichimiya T 1939 Phys. Rev. 55 585
Perl M L et al. 1975 Phys. Rev. Lett. 35 1489
Peyrou Ch 1982 Journ. de Physique 43 C8-7
Piccioni O 1943 Nuovo Cimento 1 56
Piccioni O 1948 Phys. Rev. 74 1754
Piccioni O 1982 Journ. de Physique 43 C8-207
Piccioni O 1983 in The Birth of Particle Physics (edited by L.M.Brown and L. Hoddeson) Cambridge University Press, p.222
Piccioni O 1984 Proceedings of "Wingspread Conference" p. 486
Pontecorvo B 1947 Phys. Rev. 72 246
Puppi G 1948 Nuovo Cimento 5 587
Puppi G 1949 Nuovo Cimento 6 194
Rasetti F 1941a Phys. Rev. 59 613
Rasetti F 1941b Phys. Rev. 60 198
Rosenfeld L 1948 Nuclear Forces, North Holland, p.XVII
Rossi B 1930 Nature 125 636
Rossi B 1931 Nature 128 300
Rossi B 1932 Phys. Zeits. 33 304
Rossi B 1933 Zeits. f. Physik 82 151

Rossi B 1934 in Proc. Int. Conf. on Physics, London, Vol. 1, ed. by J.h.Awberry (London: Phys. Soc. 1935) p. 233
Rossi B, Hilberry N and Hoag J B 1940 Phys. Rev. 57 461
Rossi B and Nereson N 1942 Phys. Rev. 62 417
Sakata S and Inoue T 1946 Prog. Theor. Phys. 1 143
Sard R D and Althaus E G 1948 Phys. Rev 74 1364
Sigurgeirsson T and Yamakawa A 1947 Phys. Rev. 71 319
Skobeltzyn D 1929 Zeit. f. Physik 54 686
Steinberger J 1948 Phys. Rev. 74 500
Steinberger J 1949 Phys. Rev. 75 1136
Street J C 1939 Journ. Franklin Society 227 765
Street J C and Stevenson E C 1937 Phys. Rev. 52 1003
Ticho H K 1948 Phys. Rev. 74 1337
Tiomno J 1984 Proceedings of "Wingspread Conference" p. 82
Tiomno J and Wheeler J A 1949 Rev. Mod. Phys. 21 153
Tiomno J Wheeler J A and Rau R R 1949 Rev. Mod. Phys. 21 144
Tomonaga S and Araki G 1940 Phys. Rev. 58 90
Valley G E 1947 Phys. Rev. 72 772
Valley G E and Rossi B 1948 Phys. Rev. 73 177
Weizsäker von C F 1933 Am. Physik 17 869
Wheeler J A 1947 Phys. Rev. 71 320
Wick G C 1938 Nature 142 994
Williams E J and Evans G R 1940 Nature 145 818
Williams E J and Roberts G E 1940 Nature 145 102
Yang C N and Tiommo J 1950 Phys. Rev. 79 495
Yukawa H 1935 Proc. Phys. Math. Soc. Japan 17 48
Yukawa H 1937 Proc. Phys. Math. Soc. Japan 19 712
Yukawa H and Okayama T 1939 Scient. Papers Inst. Phys. and Chem. Research 36 385
Yukawa h, Sakata S and Taketani M 1938 Proc. Phys. Mat. Soc. Japan 20 319

Note

Reference has been made in the text to the following Meetings dedicated to the history of particle physics:

(i) "Fermilab Symposium" (May 1980), which originated the book The Birth of Particle Physics edited by L.M. Brown and L. Hoddeson, Cambridge University Press, 1983.

(ii) "Paris Colloquium" (July 1982). The proceedings of this "International Colloquium on the history of Particle Physics" appeared in "Journal de Physique", Colloque C-8, Supplement n° 12, Décembre 1982.

(iii) "Wingspread Conference" (May-June 1984). The proceedings of this Conference appeared in the book 50 Years of Weak Interactions, edited by D.B. Cline and G. M. Riedasch, University of Wisconsin, Madison, Wisconsin, 1984.

The book Fifty Years of Weak Interaction Physics, edited by A. Bertin, R.A. Ricci and A. Vitale on the occasion of the fiftieth anniversary of Fermi's theory on nuclear beta-decay (Italian Physical Society, Bologna, April 1984) should also be mentioned as a reference of historical interest.

Discussion

Marshak

I have a comment. I would like to say a few words about what was going on right after the famous Italian experiment. Professor Conversi pointed out that Amaldi carried the news to Fermi who quickly understood the dramatic consequences of their work. What happened was that Fermi, Teller and Weisskopf published their note on the factor 10^{12} in February 1947, and immediately afterwards Oppenheimer, who was organising the first Shelter Island Conference at the beginning of June, wrote to the future participants, about 25 of us. There isn't time to quote from the letter, but he said that this was such a major thing that we may have to give up microscopic reversibility etc. At the time of the Shelter Island Conference, June 1-3 1947, this was one of the two major topics of discussion, the other being the Lamb Shift which led to QED. This was when Weisskopf, who had worked with Fermi and Teller, came up with the "Meson-pregnant nucleus" as an intermediate step to break the chain and avoid giving up microscopic reversibility. This led to what I thought was a simple hypothesis, the two-meson theory. The idea was to have the heavy meson strongly interacting and the light one weakly interacting. Let me just get straight the drama of what happened. The point was that the decays discovered by Lattes, Muirhead, Occhialini and Powell were published in the May 24th, 1947 issue of "Nature". But in those days we had no Airmail and this issue didn't get to the U.S.A. until mid-June. The way I found out about it was that I had to attend a political meeting in mid-June (I was President of the Federation of Atomic Scientists) and Phil Morrison met me on the train and said: "Hey, did you see those beautiful photographs from the Bristol Group?" This was about 10 days after the conference and I said: "By God, they must be the $\pi - \mu$ decays". I asked Bethe to join in writing the paper because, among many things, he knew the cosmic ray data very well. In our paper we used the fact that there was a break in the absorption curve of the penetrating component at 250 m W.E. underground and interpreted that as a decay which led to a value of about 2.10^{-8} secs. for the lifetime, very close to the accepted value. Another estimate using the equivalent of the Goldberger-Treiman Relationship gave a value of about 10^{-8} secs. That paper was sent to Powell, and I have here a letter from Powell dated August, 1947, which there isn't time to read. I visited Bristol in November 1947 and we compared our views of things. In his review paper of 1950 Powell noted that we had complete convergence on what was going on in the $\pi - \mu$ decays. For about a year, however, there was some doubt cast on the $\pi - \mu$ interpretation, i.e. that the π was strongly interacting and the μ was a heavy lepton. There were two experiments involved in this. Firstly, in the same November 1947, I discussed this with Rochester and Butler when I visited them in Manchester. They had their beautiful photographs of the two V particles which looked good, but they also said that in the same experiment, they had seen direct production of μs in the plate. That was the reason why, when I got back, I didn't advertise the V^0s very much because I was sure that they were wrong about the other observation, because there were no directly produced muons according to our theory, and of course, one believes one's own theory. I speculated that perhaps the reason for the confusion was that at the time the Bristol Group had an m_π to m_μ ratio of 1.7, the original fading problem. The other experiment was the Berkeley confusion propagated at the second Shelter Island Conference. I fought it vigourously, but for a while people thought that Berkeley must be right. Lattes had come to Berkeley

© 1988 IOP Publishing Ltd

to show my U.S. colleagues how to develop emulsion. They found lots of pions, but they also thought, and Alvarez agreed, in the Spring of 1948, that they were seeing muons directly from the production point. Again, the problem was the π - μ decay and the magnetic field bending the muon back to the source and causing confusion. It took about a year before everything was put straight.

One final thing I'd like to put straight, because there's been a lot of confusion and even ill-feeling over the years about the Sakata and Inoue and the Tanikawa work. During the war years the Japanese had a Meson Club and were working on many meson problems. I don't know how they managed it, but they did. There were these Japanese papers by Sakata and Inoue and Tanikawa where they introduced the two-meson theory to understand the low cross-section in the J.G. Wilson work and for the scattering of cosmic ray mesons in the plates which was about 1% of that expected. Of course, with only a 10^2 discrepancy they predicted a lifetime of 10^{-21} secs. That's understandable. If they had had the Italian data they would have got the right answer. I really want to make clear once and for all that that work was published in Japanese and circulated in their community. It was published in English in "Progress of Theoretical Physics" in 1946. But there were many disruptions in the war, and this was one of them. If we had had these papers at the Shelter Island Conference we would have acknowledged them immediately, but they were not delivered to the U.S. until the end of 1947. We had the excitement, because of your experiment, of fighting for a whole day at Shelter Island and finally coming out with the two-meson theory as the only way not to have to give up the basic principle of microscopic reversibility, which would have made the Chairman's father very unhappy! Thank you.

Ekspong

Your experiment in Rome was a turning point in the history of Physics and the beginning of modern particle physics. There is one thing which Professor Marshak just touched on which is still a bit puzzling. When Anderson and Neddemeyer showed that QED could not be the reason for penetrating particles in cosmic rays, they called them "heavy electrons". I had an interview with Anderson a few years ago when he confirmed this with a wink. History had to go a long way before we came back to this point of view. We knew in 1934 that these particles could penetrate more than 1 m. of lead. A student of today would think that we knew that these particles were not strongly interacting. This is the method used in all present-day equipment like UA1 to identify μ's by their penetration. So in 1934 we knew that in flight they were weakly interacting. The Rome experiment showed that they were not strongly interacting at rest. My question is: was theory in such bad shape in those days that we could only make calculations at rest to show that these particles were not strongly interacting? It was theorists who identified the penetrating particles with the Yukawa particle, Oppenheimer and Yukawa among them, and the rest of the physics community followed in line based on the mass and lifetime. But the cross-section for absorption was so tiny that it's amazing that this didn't seem a problem.

Conversi

Firstly, concerning the name. "Mesotron" was agreed on by Anderson with Millikan, as I read in a paper from Anderson. They sent a letter to "Nature" proposing the name mesotron, so the "heavy electron" was earlier, but the final name was given by them. Secondly, you are entirely right to be surprised that for ten years, people accepted this particle as a Yukawa particle despite its penetration. But a pyschological justification for this can be found in the fact that the mass and the lifetime of the particle as measured, was just in line with that expected for the Yukawa particle. Further, in cosmic rays a factor of 10 discrepancy was accepted, though not a factor 10^{12}!

Marshak

But let's get the order straight. The 10^{12} didn't come in until your experiment. What you're talking about is the factor of 100 in the cross-section. The theory was in rather poor shape; during the war Heisenberg and others thought this discrepancy could be cured by the pseudoscalar theory if a size was put in for the nucleus. So the factor of 100 wasn't so critical. The Japanese Meson Club and Tomonaga were trying to understand this factor of 100 via this mechanism, rather than the two-meson theory. That's why the factor 10^{12}, which didn't come until 1947, was the turning point.

Waiting for the pion

N. Kemmer, Professor (Emeritus)

35 Salisbury Road, Edinburgh

ABSTRACT: The author reports, very largely from personal experience, on the earliest developments in our understanding of the nature of nucleons and their interactions. Starting with Chadwick's discovery of the neutron and the first steps into nuclear theory taken by Heisenberg and Majorana, the emphasis of the talk is on the search for the agent that transmits nuclear forces. Yukawa's hypothesis, the discovery of the "mesotron" by Anderson and the subsequent burst of, principally theoretical, work up to the outbreak of war is described. While at that juncture the main goal of the work remained unachieved, some of the concepts developed by then - notably isospin - proved to be of lasting importance.

1. INTRODUCTION

We are here celebrating a great event in the history of physics. Some of the talks we shall be hearing to-day and tomorrow will tell us about the days of the great discovery, others will describe the new chapters of physics that began with it. The part that I have been given the honour to play is a rather different one. I can only talk of things that happened well *before* the date we are remembering. In fact, due to the intrusion of world events in a very different sphere my story almost terminates about eight years earlier. The discovery of the pion is an epilogue to it.

I can see only a few people with us here to-day who lived through the developments of physics of which I shall speak - though enough of them to put me right, should my memory lead me astray!

*There is nothing in this talk that has not been fully documented in numerous textbooks and monographs. The particular approach presented here duplicates with only little change of emphasis what I presented to the International Colloquium on the History of Particle Physics in Paris in 1982, Journ. de Physique 43 Suppl. C8 No. 12, 1982, p. 359, which includes an extensive bibliography.

A still closer parallel to this talk was published in German in Physikalische Blätter No. 39 (1938) p. 170 (speech of acceptance on receiving the Max Planck Medal). I do not think a bibliographical list here would serve any useful purpose, but a set of references is given on page 32.

© 1988 IOP Publishing Ltd

2. NUCLEAR INVESTIGATIONS BEGIN

Inevitably, my story must start with another most memorable event that took place fifteen years earlier - Chadwick's discovery of the neutron. That marked the start of Nuclear Physics as we see it to-day. It was followed very quickly by the first steps towards a theory of nuclear structure. These were taken by Heisenberg who, in a series of papers, gave a first version of what still forms the contents of what one finds as introductory material in textbooks on Nuclear Physics to-day.

The form in which Heisenberg began to state his idea was as an application of the new Quantum Mechanics that had grown from the seed he himself had sown only seven years earlier. He saw that the key fact on which a description of nuclear structure must be based had to be a strong, short-ranged force between protons and neutrons and one that would only involve nucleons (to use modern terminology) that were near neighbours within a nucleus.

Heisenberg recognised that such a force could be produced by a quantum mechanical mechanism that he himself had proved to be effective to describe the structure of the ionised hydrogen molecule H_2^+. There he had shown that the single electron had to be shared or exchanged between the two protons and thus would produce an attractive force between the protons only over a distance of the order of the dimensions of a hydrogen atom in its ground state and would not lead to high density crowding in heavy nuclei.

3. EXISTENCE OF AGENT SURMISED

For the case of the nuclear force Heisenberg was searching to define a similar mechanism, but was unable to be precise about the nature of the agent to be made responsible for the "exchange force". It would clearly have to be a charged particle but the electron, the only conceivable candidate at the time, could not be consistently assigned to the rôle. Thus, in Heisenberg's papers references to this hypothetical agent are quite vague and guarded. In any case, a modification of Heisenberg's expression for the exchange force was soon introduced by Majorana, which accounted better for the details of observed nuclear structures (in particular for the special stability of the α-particle), pushed Heisenberg's exchange picture of the binding mechanism completely into the background. Majorana's theory did involve a formal exchange operator, but this was one in which the *position* coordinates of the two interacting particles were exchanged in the expression for the interaction. Such a notion made the description an entirely phenomenological one - the search for an "agent" responsible for nuclear binding was, for the time being, abandoned.

Majorana also explicitly dropped another feature of Heisenberg's earliest papers which was at the time generally felt to be something of an aberration in his statement of his problem. My own personal memory tells me that Pauli himself looked upon it in this way, although it was Pauli's own work from which Heisenberg had taken the formalism he used. This was his use of 2×2 matrices to label the proton and the neutron "states" of the nuclear particles and to express the exchange character of the interaction between them. These matrices, which Heisenberg called ρ (and we call τ) are identical in structure with Pauli's spin matrices but have nothing whatever to do with real 3-dimensional space or with rotation in it. For several years that formalism dropped out of nuclear theory altogether.

4. DESCRIPTIVE DETAILS ACCUMULATE

Following the start of phenomenological nuclear theory given by Heisenberg and Majorana much experimental and theoretical effort was spent on trying to determine details of the form of the proton-neutron interaction potential. Bethe soon proved that much effort was being wasted, when he showed that no study of low energy p-n scattering data could produce anything more than values of two quantities for each of the spin states of the p-n system - a "scattering length" and an effective range. These soon became known with great accuracy. More difficult was the problem of determining whether there also existed nuclear forces between pairs of protons or pairs of neutrons (in the former case - in addition to their Coulomb repulsion). Ever since Heisenberg's first work, the possible existence of such forces had been under discussion, but the prevailing feeling was that they ought to be small compared to a dominant Majorana type p-n force. What hadn't come to the forefront of one's ideas was the fact that, unlike the p-n system, a system of two like nucleons is subject to Pauli's exclusion principle, so that in particular there was nothing with which the binding in the (totally symmetric) ground state of the deuteron can be compared.

In 1936 very precise results on proton-proton scattering revealed that there was strong evidence for "charge-independence" of the nuclear force: - when the Coulomb interaction between protons was subtracted (and only states of the same symmetry were compared), the nuclear force was universal - whether a nuclear was charged or not was irrelevant. This immediately brought Heisenberg's ρ-matrices back into use. (They were, unhappily introduced under the name "isotopic spin", fortunately now changed to isospin). To the study of nuclear structure this presented a very powerful new tool, used particularly effectively by Wigner and by Hund. However, that flourishing new subject will no longer enter my story.

5. FALSE TRAIL, MORE DETAILS ARE ASSEMBLED

Meanwhile the study of the puzzling phenomenon of β-decay had entered a new phase, not long after the discovery of the neutron - at the expense of having to accept Pauli's neutrino as a new, mysterious and almost in principle undetectable constituent of matter. That, the discovery of the positron and the emergence of Dirac's relativistic theory of the electron in its final form, offered Fermi the theoretical wherewithal to formulate a theory of β-decay which in its essentials remains the way we describe the phenomenon to-day.

Fermi postulated that nucleons are capable of emitting, and in principle absorbing, pairs of what we now call leptons. So there seemed to be a chance that the transmitting agent for nuclear forces that Heisenberg had sought was now available. Practically simultaneously (but independently) two Russian physicists, Tamm and Ivanenko, brought this idea forward. Sadly their hopes of triumph were very short lived. Soon it became clear that it was highly unlikely that the numbers involved in describing an effect so weak as β-decay could be related to those that account for the strongest forces we know. Also, it became obvious almost at once that the application of the mathematical rules we had at the time to formulate a theory of nuclear forces based on the Fermi-Tamm-Ivanenko field could not produce any consistent results anyhow. Even so, the Tamm-Ivanenko idea plays a small part in the story I am telling.

I think I can claim for myself the credit for linking the ideas of charge independence and isospin with the attempts to establish the nature of the field of nuclear forces. The otherwise useless Tamm-Ivanenko idea gave me the chance to do this. Their attempt to derive a proton-neutron force had begun with taking as an interaction term in their Hamiltonian just the type of expression that Fermi had introduced in his description of β-decay. It required no great effort to answer the question as to what modification, or rather addition, to that interaction Hamiltonian would be needed to produce interactions between pairs of protons and pairs of neutrons. One simply had to postulate that nucleons could also emit electron-positron pairs and/or neutrino pairs. Then the rules of isospin manipulation virtually forced on one the possible details of the shape of the total interaction Hamiltonian.

I published that result, with full stress that this was not a realistic theory. Incidentally what I wrote down then as a general form of that interaction Hamiltonian did not, as far as I know, reappear in print again until, many years later, "neutral currents" were introduced into particle theory. For me, however, that short paper proved to be a springboard on to more solid ground.

There is one more piece of entirely theoretical work that I must relate before I come to the central event of my story. Pauli and Weisskopf investigated what one gets when one quantises the Klein-Gordon equation (which we then used to call the relativistic Schrödinger equation) according to the rules that were first developed for the new subject of Quantum Electrodynamics. The purpose of their work was to produce a counterpart involving massive charged *bosons* of the formalism that was just being developed for the electron-positron field. The beautiful result was that their method gave an almost exact copy of the fermion theory - except for the absence of spin - with the same features of pair creation and annihilation as in the other case. However this was all done just as a formal exercise, since nobody had any reason to believe in the existence of massive charged bosons in the real world. I was on the spot when Pauli and Weisskopf produced this work and was greatly fascinated by it. By coincidence I soon stumbled on a formalism - also in a purely academic study - of a set of very closely related equations which described bosons of spin 1. I realised in due course that these equations had been used previously in various contexts, but I found them written up in precisely the form that became interesting to me by Proca (although that author's erroneous intention had been to use them to describe the electron). At some stage - I do not recall precisely when - I proved to myself that the canonical method of quantising that Pauli and Weisskopf had applied to their spinless bosons could be taken over very easily to produce a theory of spin 1 charged bosons - but whenever that was, I did not then publish that work.

6. TIP FROM EAST IGNORED UP TO SIGHTING IN WEST

Let me now come to the central chapter of my story, which is set in Japan, in surroundings entirely separated geographically and intellectually from the scene of everything of which I have been speaking so far. Hideki Yukawa, virtually in isolation from all other workers on the subject, had been pondering on the same problem that Heisenberg had tried to tackle without success some three years earlier. He knew that the world of physics had had to accept recently that the electron and the proton were not the only *fermions* in existence, but the only elementary boson that one knew was the photon and its status was a very different one from that of the fermions - perhaps it was misleading to class it as a true particle at all. Yet Yukawa realised that if there was any truth in Heisenberg's picture of a particle transmitting nuclear forces, that particle would have to obey Bose statistics and carry electric charge. Yukawa had no knowledge of the recent formal work in which Pauli and Weisskopf had introduced a charged, massive boson. So when he ventured, hesitantly to publish his speculative idea that nuclear interaction should be accounted for by the existence of a charged boson, he sought to formulate his theory in as familiar a form as possible - which was in analogy with Maxwell's equations. However, the outstanding novel feature of Yukawa's work was his recognition that his nuclear force agent had to be massive and, in particular, that its mass could be predicted within fairly narrow limits from the known range of the neutron-proton force.

At least this one central prediction in Yukawa's paper should have been noticed in the scientific world at large when it first appeared in print in 1935. Though the paper was published in a Japanese journal, it was written in English and not at least obscure in its content. In those days people like myself had no difficulty in going to a good scientific library and perusing *all* scientific new papers appearing in any week, including those in the less well known foreign journals. I believe Oppenheimer remembered seeing Yukawa's paper at the time, but I did not then meet anyone who was aware of this new idea having been published.

When however, two years later, the news broke that Carl Anderson had detected in cosmic radiation a new charged particle with mass in a range where nobody had expected to see a particle - around 200 electron masses, somebody quickly noticed that this seemed to confirm Yukawa's prediction startingly well.

7. GREAT CHASE LEADS TOWARDS YUKON

Yukawa himself couldn't have required better encouragement to get back to work on the theory he had published so tentatively two years earlier - now with a growing team of collaborators, but now also with others in the West competing in the development of his ideas.

Of course everybody began reading Yukawa's single, short early paper. I myself got caught up in the tide of new developments - the chase for the "Yukon", which some sceptical people like Pauli expected to be a passing craze (somebody called it "yukosis"). To-day you can hardly expect a fully balanced account of this from me - but I'll try.

The first thing that needs saying is that I don't think I was alone in misreading Yukawa's first paper. It seemed to me quite clear that the field variable he had defined was supposed to obey precisely the same Klein-Gordon equation that Pauli and Weisskopf had studied - i.e. a scalar field describing a boson of *zero* spin.

It was no great step to confirm Yukawa's prediction that the spherically symmetric solution of the equation defining the nuclear potential was $C \cdot e^{-\lambda r}/r$ as he had stated. However he seemed to me to be giving the constant C the wrong sign. I only realised many years later that I had skipped some of the details of what Yukawa was saying in his paper. He had started with the notion that he must use an analogue of Maxwell's equations, but was only going to deal with the equivalent of the scalar-potential in his first paper. That explained what I thought to be his wrong sign.

Neither Yukawa's procedure in the approximation he had adopted nor a theory based on the Pauli-Weisskopf (scalar) field was capable of predicting a nuclear force that had the spin-dependence required to account for the structure of the deuteron, the α-particle etc. When in 1937 Yukawa resumed work in collaboration with Sakata and Taketani he used the unabridged form of his Maxwell-like equations - which were precisely the same with which I had familiarised myself and which, when eventually my work was published, I called the Proca equations. Thus I had the luck of being able to come out in print with the full apparatus of the spin 1 theory virtually simultaneously with Yukawa and his collaborators. My work appeared in the main as part of a paper written in collaboration with Fröhlich and Heitler who had been attacking the same problem in Bristol. They had begun their study using the scalar equation, intending to introduce spin dependence by means of a spin-orbit type of interaction. When I met Heitler by chance in London I pointed out that their proposed interaction would involve parity violation. I also introduced them to the Proca equations. This started our collaboration.

We did not know at the time that Homi Bhabha had also been working on the same problem in Cambridge. He and I published practically identical preliminary results in the same issue of 'Nature' and his full paper appeared in the Proceedings of the Royal Society very shortly after our collaborative paper.

The paper by Yukawa, Sakata and Taketani came out almost simultaneously with these two from Britain. In effect the world received the account of spin 1 theory of nuclear interaction in three very similar versions almost all together.

8. EFFORTS DISAPPOINT. HOSTILITIES STOP ACTION

Thus there seemed to be a unanimous view that Anderson's particle had to be a charged boson of spin 1, describable by the Maxwell-like Proca equations. Unfortunately it became clear very soon that the truth was very different. Hardly anything that theorists could derive from that theory fitted observations and measurements on Anderson's particle.

There were three reasons why this had to be so - two of which are very well known. The first was that we didn't really have a workable theory - not even to the extent that quantum electrodynamics in the state then existing could be so regarded. Neither theory could be handled other than by perturbation theory and only up to first order - with infinite coefficients appearing beyond that. But quantum electrodynamics was at least plausibly credible because the expansion there was in powers of a *small* parameter. In the "Proca" theory the corresponding parameter was huge. The second reason - the one that has brought me here to talk to you to-day - was that relating Anderson's particle to Yukawa's was a monumental mistake! It was perhaps just as well that a divergence of nomenclature arose which in the main divided experimentalists from theorists. Following Anderson the former named their particle "mesotron", while we theorists in Britain disliked that construction and launched the name "meson". In fact that name was agreed on by Homi Bhabha, Maurice Pryce and myself and happily caught on in due course. (If the naming had occurred to-day I am pretty sure we would be talking of "yukons"!)

I just said that there were *three* reasons why all attempts to reconcile spin 1 vector meson theory with experimental were doomed to failure. The third reason is not as well known and is one for which history can point an accusing finger at me.

Please forgive me for detaining you briefly with this, my own story. In all the three papers that launched spin 1 meson theory virtually the same remark is made in passing - namely that what was being presented was meant to account for the proton-neutron interaction. At that time the charge independence idea was well established, so that all the authors were well aware that a full theory would have to postulate an uncharged counterpart to the new particle to get p-p and n-n forces right. None of the papers went further than making just one brief reference to this fact. However, in the case of Fröhlich, Heitler and myself it could have been different. I was expecting to contribute a chapter to the paper in which a full isospin invariant form of the theory would have been developed. I was well ahead with writing that up - my previous study of this aspect of the abortive Tamm-Ivanenko theory found me entirely prepared to do this. This did not happen, because Heitler felt that the paper was getting too long. He suggested that I should publish that work as a separate paper. That turn of events proved to be far from damaging to my reputation! On the other hand, if the isospin consideration had been included in the joint paper, this *might* have had the result that our theory would have looked significantly different from Yukawa's and Bhabha's - and would have been more prophetic.

This is why: In the same issue of Proc. Roy. Soc. in which the triple paper appeared there was also another one by me alone that was part of the background to the joint paper. In it I presented the details of the quantisation procedure for both the Proca and Klein-Gordon fields *and* also of the alternative spin 1 *pseudo*vector and spin 0 *pseudo*scalar fields, which are essentially different from the then more familiar fields of opposite parity in the form of the interaction terms by which the mesons couple to the nucleons. I examined what form these alternative fields would give to the proton-neutron interaction and concluded that the pseudovector case would lead to an entirely wrong shape of interaction, while in the pseudoscalar case there would be no static (velocity independent) interaction at all. However, in *that* paper I did not introduce any study of charge independence or isospin.

Many years later Pauli wrote to me to point out that had I done so, the equations for the nucleon-nucleon potential would have differed from the ones we adopted by a factor of the form $\underline{\tau}_1 \cdot \underline{\tau}_2$, which for the deuteron ground state would have the value -3! Thus in a charge independent theory our matching of theory to experimental data would have been entirely different - and by more careful study we should have concluded that our new particle was pseudoscalar!

On the other hand, I think it is fair to say that Yukawa deserves to be remembered as the man who, long before anybody else, initiated the study of charged vector fields. As we all know, these fields are to-day particularly important in a different context.

I could continue the sad story of how, right through the whole pre-war period the evidence grew ever stronger that our work was getting us no nearer either to success in understanding nuclear forces nor to any theoretical justification for the existence of the Anderson particle. I think it would be out of place here to describe in detail how all the attempts to save the situation seemed to come to nothing - I certainly do not feel competent to do so.

9. ACTIVITIES RESUMED: π REVEALED

After the war I viewed the rapid new developments with much greater detachment - primarily through interaction with my students. I had to begin teaching my subject with the story that ended with the pre-war chase leading to nothing, even though it had started with a most promising and beautiful idea. Fortunately I did not have to continue that tale for long. Soon I could report a breakthrough in quantum field theory which would allow an approach to the old problems with much improved understanding and much better tools. I had the great satisfaction of seeing quite a few of the students I was teaching contribute significantly to the new advances. Then, suddenly, the *physical* picture that had bewildered us for so long changed also - and that is what we are celebrating to-day.

You will appreciate that the opening of the new vistas afforded me even more grounds for personal satisfaction than for most. Some small seeds, sown so long ago, had germinated and were growing.

A few weeks ago a young worker visiting the Department of which I had been in charge gave a talk on the place of mesons in the scheme of physics as he saw it now. I asked him to let me have a copy of the introductory transparency that he had shown us. Here it is: I had to confess that what I knew, or had known of mesons related entirely to the one entry, bottom left, that I have marked with a large arrow!

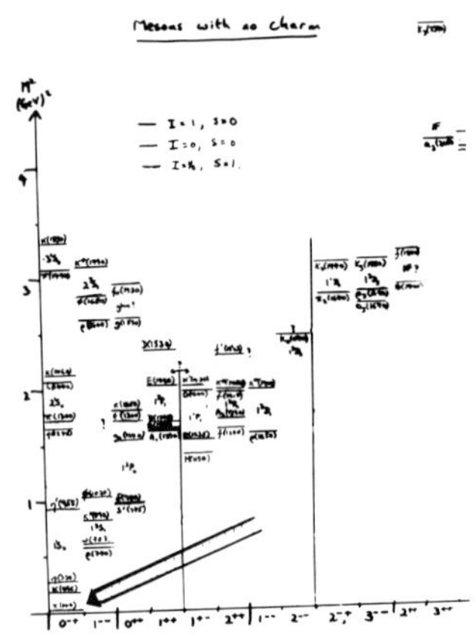

REFERENCES

CHADWICK PRS A136 (1932)705; HEISENBERG ZfP 77 (1932) 1; MAJORANA ZfP 82 (1933) 137; FERMI ZfP 85 (1934) 161; TAMM Nature 133 (1934) 981; IVANENKO Nature 133 (1934) 981; TUVE et al. PR 50 (1936) 806; BREIT et al. ibid. 825; CASSEN & CONDON ibid. 846; WIGNER PR 51 (1937) 106; KEMMER PR 52 (1937) 906; PAULI & WEISSKOPF Helv.PA 7 (1934) 709; PROCA J.Ph.Rad. 7 (1936) 347; YUKAWA Pr.PhMS Jap. 17 (1935) 48; NEDERMEYER & ANDERSON PR 51 (1937) 1005; FROHLICH, HEITLER & KEMMER PRS A166 (1938) 154; BHABHA PRS A166 (1938) 501 26; YUKAWA et al. Pr.PhMS Jap. (1938) 319; KEMMER Pr. Camb.PS 34 (1938) 354.

Discussion

Fröhlich

All of these difficulties, and there were some others, made me finally fed-up with all this and I decided that field theory without difficulties could be found in solid state physics. This was the birth of field theory in solid state physics and the first thing we did was what is now called the polaron. That is now of course an enormous subject, and it arose, however, originally from the difficulties, being fed-up with the difficulties. In solid state physics we didn't have these difficulties.

Kemmer

There is no answer to that.

Marshak

I just want to cover the couple of years right after which confirmed Kemmer's theory and the excitement connected with the experiments of Fermi, Anderson and others after the war on π-nucleon scattering when isospin invariance became popular and really started modern particle physics, which goes back to your work. It was the second Rochester Conference at the end of 1951 when Fermi came and reported on the 9:2:1 ratio of the π^- on proton giving π^0 and π^- giving π^- etc. The 9:2:1 ratio was an immediate consequence of isospin invariance if you assumed that the isospin 3/2 channel dominated over the isospin 1/2. There were then resonance indications that that was a good assumption. That was really a turning point and was one of the basic exciting fields of that second Rochester Conference in 1951.

Kemmer

I won't answer that, but I've remembered something worth mentioning in that context. I already mentioned the postcard I got from Pauli which asked why I didn't look at the equality of proton and neutron forces, and if I can reproduce the text said "so that the proton and neutron are just one and the same thing". "Das hat eine eigene vernuft in sich" which I can translate as "That has a certain rational to it". Looking at it many year afterwards I think those are the first words on non-spatial-temporal symmetries.

© 1988 IOP Publishing Ltd

The π discovery

P H Fowler

H H Wills Physics Laboratory, University of Bristol, Tyndall Avenue, Bristol BS8 1TL

1. INTRODUCTION AND EARLY HISTORY

We have heard in the previous papers the experimental situation in cosmic rays and our understanding $\sim$ 1945 of the properties of μ-mesons and the contradictions that were involved. In this talk I must start with technical aspects concerning the detectors. The pion was discovered among cosmic ray tracks produced in special photographic emulsion which after processing was studied under high powered optical microscopes.

The standard instrument against which any track detector would properly be evaluated was the Wilson Cloud Chamber. This had been developed over a considerable number of years, and with counter control developed by Blackett and Occhialini, was a beautiful, elegant and powerful weapon. The pictures were so vivid that one was forced to believe in the reality of the particles!

Attempts to get photographic film to record tracks had also gone on for a long time. Figure 1 shows some very early results of Kinoshita and Ikeuti (1915), where α-particle tracks were obtained from the radon produced active deposit on a needle. The halo-like appearance is clear, but individual tracks are also in evidence. These results remained a curiosity, and progress was slow until special emulsions started to be produced in the 1930's.

By 1939 Ilford half-tone gave very reasonable proton tracks which allowed both neutron-proton scattering to be performed here by Powell and collaborators, and the study of angular distributions of reactions using Liverpool Deuterons such as:-

$$D_1^2 + O_8^{16} \rightarrow O_8^{17} + {}_1H^1$$

$$\rightarrow {}^{*}O_8^{17} + {}_1H^1$$

where both ground state and excited states of the product nucleus are involved. The emulsion was fully competitive with counters at that time in such experiments. I remember doing track measurements on these reactions for Powell here on the 4th Floor and at home in the long vacation in 1942.

After the war there was an immediate dramatic improvement in quality of the emulsions. Figure 2 illustrates this. Not shown here is the discrimination between, say, α-particles and protons. α-particle tracks are

© 1988 IOP Publishing Ltd

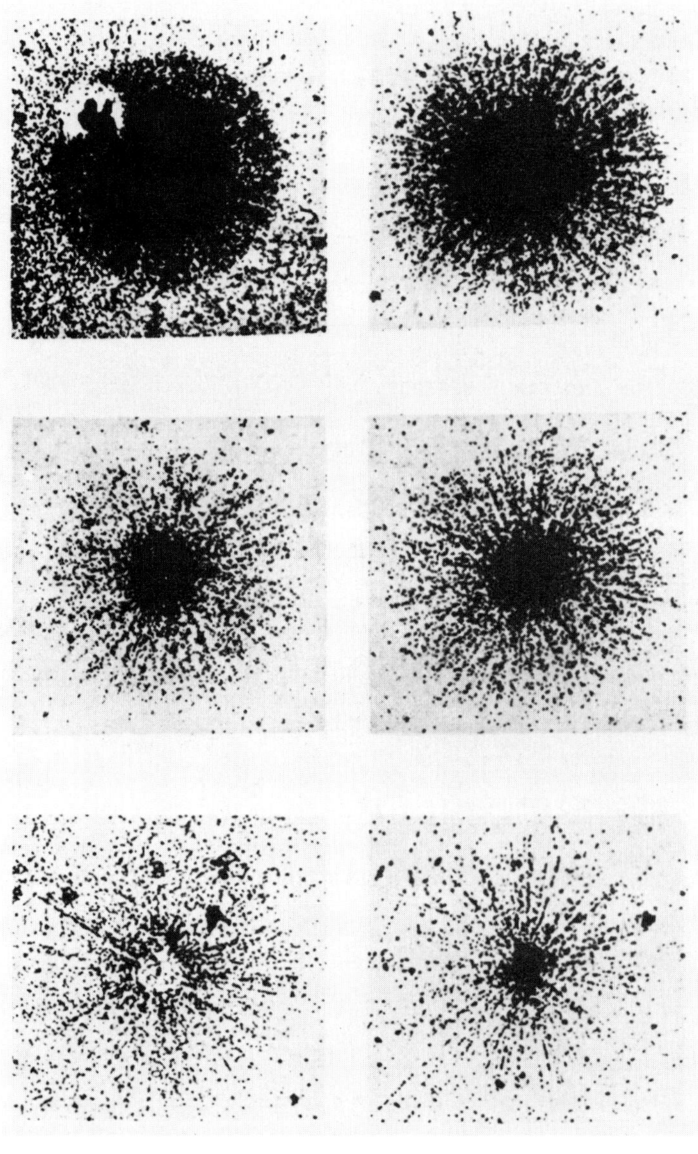

Wratten 'Ordinary' Plate Kinoshita and Ikeuti (1915)

Fig. 1. Halos due to radioactive infection of an emulsion

A series of exposures, of gradually increasing intensity, designed to show the development of 'halos' similar to those observed in minerals. The diameter of the solid central core depends upon the length of exposure and the number of α-particles emitted. The outer ring is due to constancy in range of the α-particles (from RaC') the mean diameter being ∿ 108 μ. The radioactive material was introduced by touching the plate with a needle point contaminated by exposure to radon.

saturated in the most sensitive emulsion, but are quite distinct in both the earlier less sensitive types.

Later speakers will, I know, spell out how these changes came about with special contracts to Industry. It was a joint effort - certainly quite a lot of the progress stemmed from work in University Laboratories - more especially in the processing techniques.

So, to sum up at this stage, immediately after the war the quality of emulsions both their sensitivity and the stability of the image improved dramatically. Highlights are given in the Table below.

TABLE 1

	Type	Typical Sensitivity
1939	Ilford B1 - half tone	1
1946	Ilford C2	∿ 4
1949	Kodak NT4 Ilford G5	(∿ 16 (Minimum ionisation (for Z = ± 1 visible

The most sensitive emulsion gave good tracks of relativistic particles, so charged particles of <u>all</u> velocities could be seen, if their range was sufficient. In sensitivity the emulsion was now the equal of the Cloud Chamber. Certainly, the emulsion techniques with the advent of the C2, and still more so, with the G5 and NT4 had met the severe criticisms made of the technique by Taylor in 1935:

"that it is impossible to deduce with any accuracy the energy of individual protons from a measurement of their range in the emulsion".

2. DISCOVERY OF THE π

Armed with the new emulsions, some of which were boron loaded which reduced fading of the latent image, exposures were made to the cosmic radiation at high altitudes, the Pic du Midi at 2800 m being one of a number of venues chosen. It was clear that the approximately 20 fold increase in flux of all energetic radiation was well worth the trouble. Nobody minded being sent to the mountains! Funnily enough I never went.

Late in 1946 after processing and then scanning the emulsions, it was clear there was a wealth of new material in them. Meson tracks were clearly recognisable in the Bristol data the breakdown of the 65 mesons was as follows:

TABLE 2

Meson tracks recorded from Lattes, Muirhead, Occhialini and Powell May 1947

40	'ρ' mesons, nothing at end of range
15	'σ' mesons, $\geq$ 2 prongs at end
8	'σ' mesons, with 1 prong at end
2	'Double' mesons
65	Total

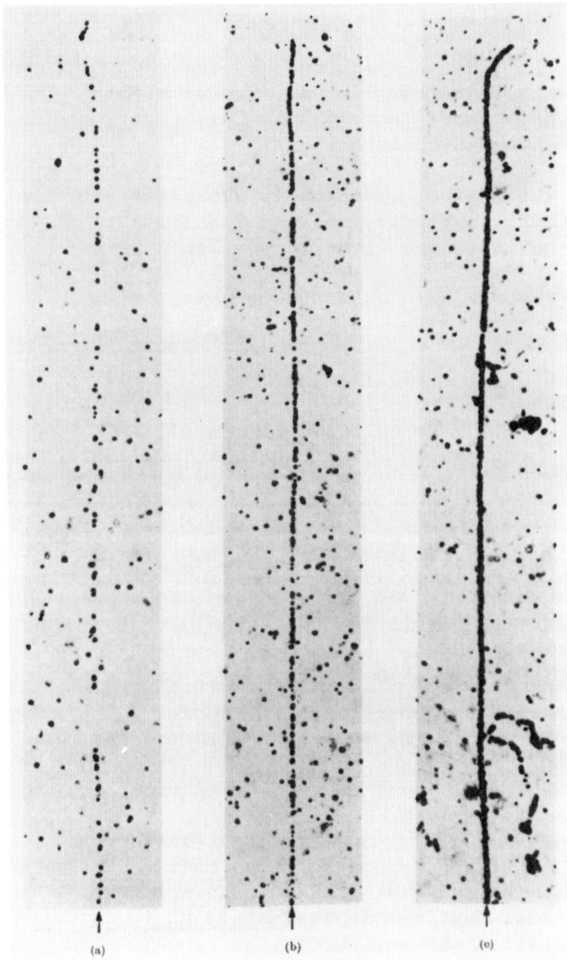

Bristol

Fig. 2. Tracks in different emulsions of protons approaching end of their range.

(a) Ilford 'half-tone' emulsion; 1939; this emulsion was produced for general photographic work. (b) Ilford C2 emulsion; 1946, produced by Glassett & Waller. The greatly increased grain-density in the track, compared with (a) is clearly visible. Tracks in this emulsion value for particles with the elementary electronic charge. (c) Kodak NT4. 1949, produced by Berriman. This was the first 'electron-sensitive' emulsion, and gave recognisable tracks of charged particles of all velocities. It will be seen that the track of the proton is almost continuous. The track of a slow electron, not associated with the proton, is seen near the bottom of the photograph. The proton has produced several slow δ-rays, which appear as short spurs on the track. The increased grain-size as compared with (b) is made manifest by the slight increase in the thickness of the track.

At the same time Don Perkins published a beautiful 'σ' meson of 3 prongs in an Ilford B1 emulsion. At that time the natural interpretation of the σ was that it was caused by a μ^- coming to rest and being captured by Ag or Br. The μ-meson experiments had then showed that μ^- captured in nuclei with $Z \geq 13$ did not decay.

The two double mesons listed in Table 2 are shown in Figs. 5 and 6. These are the discovery pictures of the π. The paper was published in May 1947. The first is not quite perfect. The parent is a clear meson and comes to rest at A. A second clear meson is emitted from the end and passes out of the emulsion after ∿ 550 μm. It shows unmistakable signs of slowing down and the grain density has more than doubled.

The second double meson event is perfect, and it clinched the matter. It was found just a few days after the first. It is complete. The parent is unmistakably coming in to the point A and appears to come to rest there where the second track originates. This has a range of 610 μm and is clearly a meson - it slows down rapidly. This range is consistent with that estimated for the first, so the conclusion was:-

The first meson gives rise to the emission of the second in a 2 body process

$$\pi \text{ (for primary)} \rightarrow \mu + ?$$
$$\sim 4 \text{ MeV}.$$

Is this interpretation justified? Could the primary meson be a μ, say, and be rejuvenated? F.C. Frank looked into this question carefully and concluded there was no likely source of 4 MeV in the emulsion.

So if X is a γ-ray or neutrino one can determine the mass of the parent from conservation of momentum and energy. This gives $\sim 1.32 \times m_\mu \sim 270\ m_e$.

Clearly it was imperative to obtain more examples to refine the range of the secondary meson, and to check if the ranges indeed formed a single peak.

With the need to obtain good new exposures at high altitude at the forefront of his mind, here I will digress. Guilio Lattes went to the Geography Department here, and found there was a Meteorological Station at Chacaltaya only 20 km from La Paz at 5,500 m altitude. He proposed to Powell and Occhialini that he go there to expose some more emulsions. Professor Tyndall's recommendation was that he should fly BOAC to Rio. However, so Guilio tells me, the Brazilian airline, Varig, had a new aeroplane, a Constellation, and was serving steaks so thick, so he flew Varig, which, as he says, was wise, because the British 'plane crashed in Dakar with no survivors.

Back to the Science. Additional emulsions were also exposed on the Pic du Midi, and after the second set from Chacaltaya had been scanned the numbers of mesons observed had risen to over 600. Forty of these were double meson events, of which a further 10 provide complete secondaries. The details are in Table 3. The clear conclusion from the data on double mesons given in Table 3 was that they indeed represent two-body decay of the parent meson at rest.

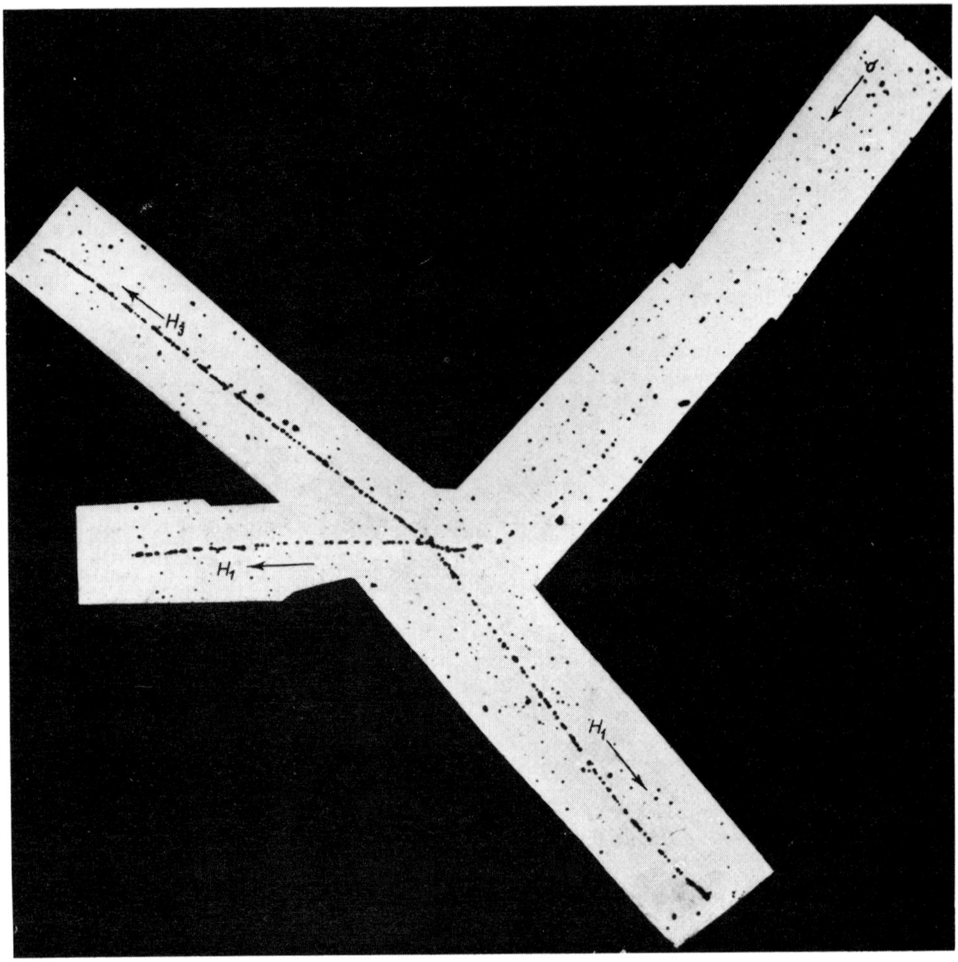

Ilford B1 emulsion Perkins (1947)

Fig. 3. First observation of a disintegration by a negative meson

The change in grain-density as the meson approaches the end of its range, and the increased scattering, are clearly displayed. These features provided the decisive evidence that the particle had indeed approached the centre of disintegration and had not receded from it; and that the particle, which must have produced the disintegration, was moving very slowly when it interacted with the nucleus. It followed that the energy of disintegration was provided by the disappearance of rest-mass of the incident particle,a process which had not previously been observed.

The good discrimination of this early emulsion is well displayed, the differences in the rate of change of grain-density along the tracks of the meson and of the secondary hydrogen nuclei being clearly apparent. The attribution of one of the tracks to a triton depends upon a comparison of the rates of change of grain-density along the tracks.

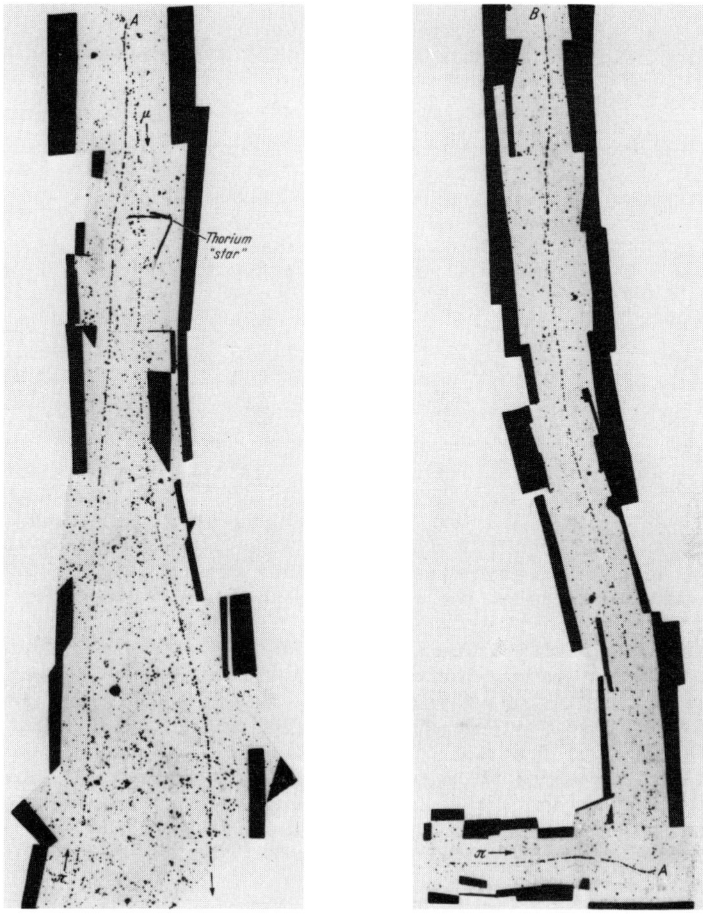

Ilford C2 emulsion Lattes, Muirhead, Occhialini and Powell; Nature 159,694(1947).

Fig. 4. First observations of the decay of a π-meson

(a) The parent particle π reached the end of its range at the point A and the secondary μ-meson was ejected nearly backwards along the line of motion of the π-meson. The variation in the scattering and grain-density along the tracks made it evident that both particles were much less massive than a proton, and allowed their directions of motion to be established without ambiguity. A comparison of the grain-density in the two tracks showed that, at the point where the μ-meson left the emulsion, its residual range was small. The black edges were retained to show the extent of the individual photographs of which the mosaic is constructed.

(b) This second example of the decay of a π-meson was found shortly after the first. Although the track of the parent particle is short, it gives unambiguous evidence for the direction of motion of the particle and for its low mass. The secondary μ-meson came to rest in the emulsion, its range being 610 .

These events have been reprinted from the original negatives to the same magnification to facilitate the comparison of the two historic events.

TABLE 3

Details of the first 11 complete 'double mesons'
Lattes, Occhialini and Powell, October 1947

Primary range μm	Secondary range μm	
133	613	
84	565	
1040	621	RMS
133	591	deviation
117	638	of
49	595	range
460	616	
900	610	= ± 31 μm
239	666	cf 26 μm
256	637	for straggling
81	590	

Mean range 614 μm ± 9

= 4.1 MeV if μ-meson.

The clear conclusion from the data on double mesons given in Table 3 was that they indeed represent the two-body decay of the parent meson at <u>rest</u>.

The spread in range around the mean is consistent with a unique energy at emission for the μ-meson. Their estimate of the μ-meson KE was 4.1 MeV based on an extrapolation of the directly measured Range-Energy, relations for protons. Analysis of the data presented in Table 3 allowed the authors to evaluate the true fraction of the meson events that were πμ decay. With an emulsion thickness of 50 μm and a μ range of 614μ only ∿ 4% of μ-mesons so produced would stop in the same emulsion to form a complete πμ. So the 11 complete πμ decays represented 270 ± 80 parents in the sample of 660 mesons - a very sizeable fraction ∿ 40% of all mesons found.

3. FURTHER MEASUREMENTS ON MESONS

The next question to receive attention was the mass of the parent meson. A comparison of grain density versus range on parent and secondary enables the comparison to be made. Positions where the grain density are the same have equal velocity and hence a residual range from that point is proportional to the mass. One must restrict the data to those parents whose range is at least 600 μm and only two were suitable. The ratio obtained was m_π/m_μ ∿ 1.8:1. From this figure they concluded with caution that X in the equation

$$\pi \rightarrow \mu + X$$

had to be of mass ∿ 200 m_c.

On the other hand measurements of Multiple Coulomb Scattering, while of less apparent accuracy, always indicated a lower mass ratio (Muirhead, Occhialini) ∿ 1.35, which supports the rest mass of X being zero.

We now know that a ratio of 1.32 is correct value. I believe with hindsight that the errors in grain density measurements were underestimated, especially that from the change in sensitivity with depth. There was also the fact that the parents tended to have a less favourable geometry, and that the difficulties in correcting for dip with respect to the emulsion plan resulted in some systematic error.

By 1948 Goldschmidt-Clermont, King, Muirhead and Ritson published their mass measurements by multiple scattering of stopping mesons.

π μ parents	260 ± 30
π μ secondaries	205 ± 20
σ-mesons	275 ± 15
ρ-mesons	200 ± 10

4. ORIGIN OF THE π-MESONS

Where did the π-mesons come from? Were they like the μ-meson indirectly produced or did they come direct from high energy nuclear collisions? This question soon had an answer.

σ-mesons - those that produce multi-prong stars at the end of their range can be produced in nuclear collisions. In this example the parent had an energy less than 2 MeV and interacted at rest, very probably with oxygen in the following manner:

$$\pi^- + {}_8O^{16} \rightarrow 3\ {}_2He^4 + {}_1H^1 + 3\ {}_0n^1 + 95\ \text{MeV}.$$

Since the πμ's represented ∿ 40% of all the mesons detected, and the parent π's must be π^+, otherwise how could they decay at rest into a monoenergetic secondary? So the σ-mesons must surely represent the π^--mesons. Only π^--mesons were seen emitted with very low energy. The Coulomb barrier repels the π^+-meson strongly, this inhibits low energy emission, and at the same time enhances π^--emission.

In the next 18 months, the number of mesons detected increased considerably, and many of the questions were answered. Thicker emulsions were generally employed → ∿ 200 μm which increased the information received substantially.

At this stage, however, the main work on π's shifted to Berkeley. This followed Lattes's move there in 1948. With the emulsion and microscope techniques he applied, meson tracks were found in great abundance, in contrast cosmic ray physicists had a source of best strength only ∿ $2\pi^{\pm}/cm^3$ day! It was at Berkeley that the neutral π was discovered by Bjorklund, Crandall, Mayer and York in 1950. It decays normally into 2 photons in ∿ 10^{-15} sec.

$$\pi^o \rightarrow 2\gamma$$

The photon pairs could be identified as coming from a π^o and its mass determined if both the γ-rays were detected, with the momentum of each being determined with reasonable precision with a pair spectrometer. The mass of the π^o was shown to be slightly lower than that for the $\pi^{\pm}$-meson at 265 m_e.

The π^o meson has a less common mode of decay predicted by Dalitz in 1951, where an $e^{\pm}$ pair replaces one γ-ray.

$$\pi^o \rightarrow \gamma + e^+ + e^-$$

The effects of these two processes can be seen in emulsion and are of great importance to the propagation of the cosmic radiation through the atmosphere.

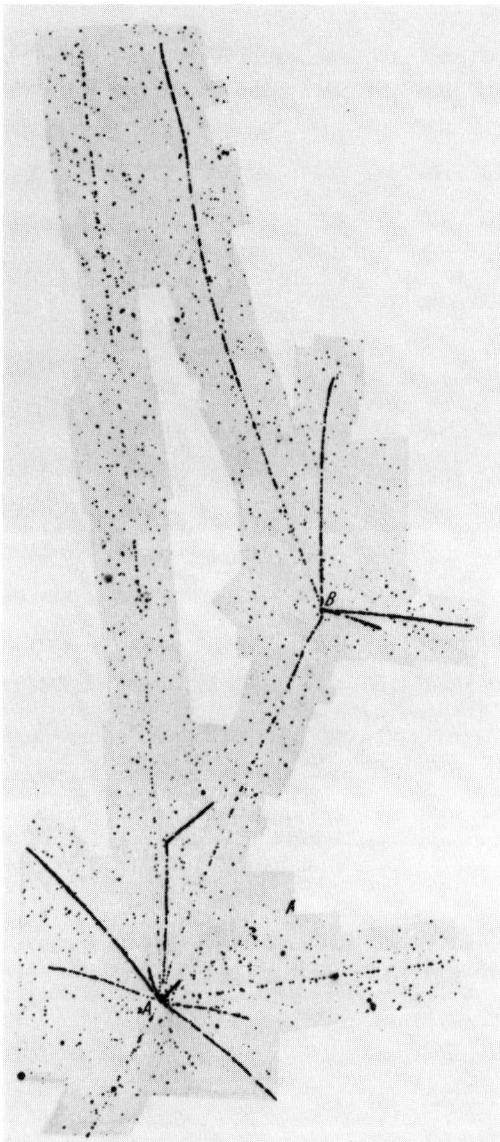

Ilford C2 emulsion Lattes, Occhialini and Powell; Nature 160,486(1947).

Fig. 5. Creation of a negative π-meson in a nuclear disintegration

One of the first two observations of the creation of a π-meson. The π-meson leaves the disintegration at A, and reaches the end of its range at B, where it is captured by a light element and disintegrates it. The distribution of grains along the track may be compared with that in the tracks of the two hydrogen nuclei which leave the disintegration B moving in an upward direction.

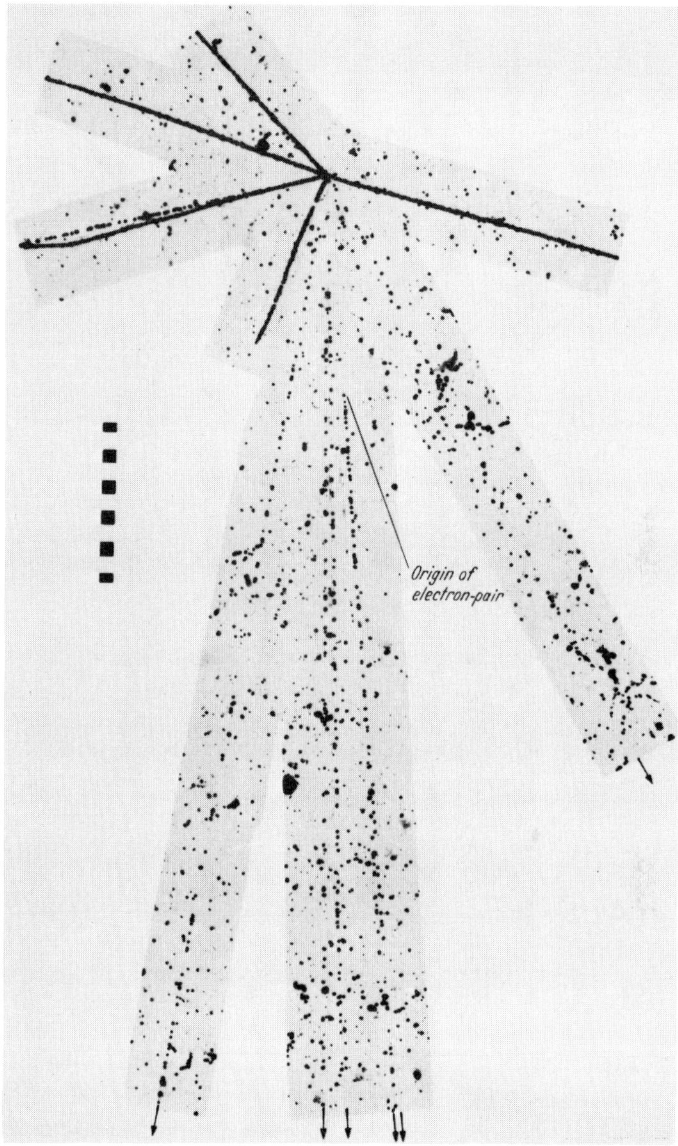

Ilford G5 emulsion Carlson, Hooper and King; Phil. Mag. 41, 701(1950).

Fig. 6. Electron pairs due to γ-rays produced in the decay of a π^{0}-particle

The bisector of the tracks of the electron pairs intersects the centre of disintegration and makes it almost certain that the two events are associated. The disintegration was produced by a neutral particle, probably a neutron. Such 'pairs' are referred to as 'associated pairs' to distinguish them from those, due to the alternative mode of decay, which appear to originate in the disintegration; see Fig. 7.

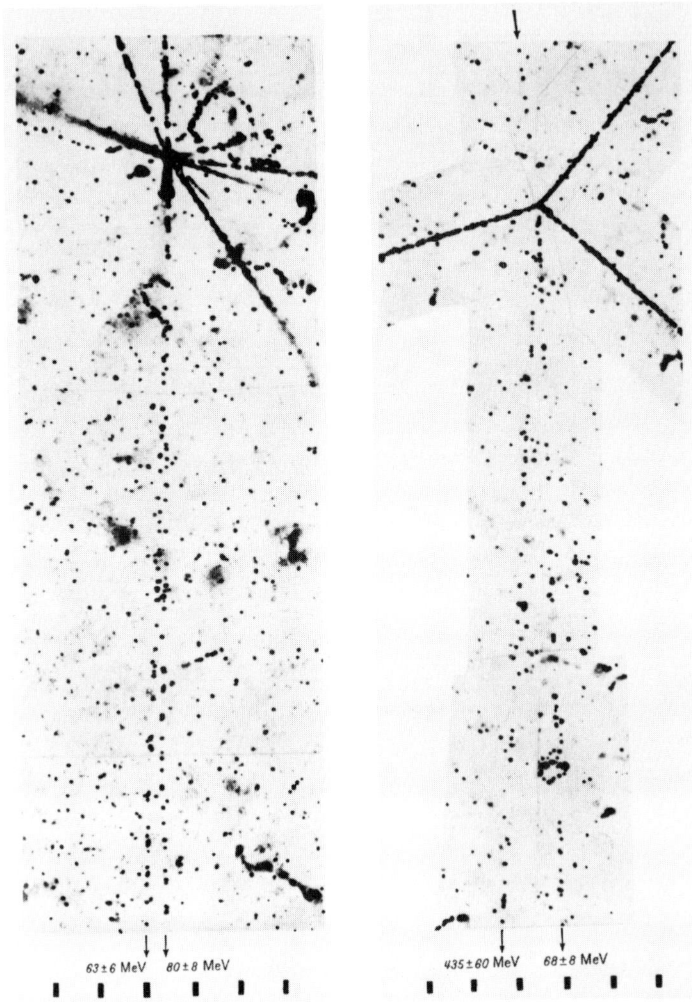

Ilford G5 emulsion   Anand (1953).

Fig. 7. Electron pairs, apparently emerging directly from nuclear disintegrations, attributed to the rare alternative mode of decay of the neutral π-meson

In the left-hand photograph, the first grains in the track of the pair cannot be distinguished because of the dense track due to a slow evaporation fragment. The individual electrons were identified by observations of scattering and grain-density and their energies are indicated in the photograph.

In the right-hand photograph only one of the two tracks believed to constitute an electron pair can be identified with certainty. The other, of energy 430 MeV, produces a track with 'plateau' grain density g_p, whereas a π- or μ-meson with the same scattering would have a value about 0.95 g_p.

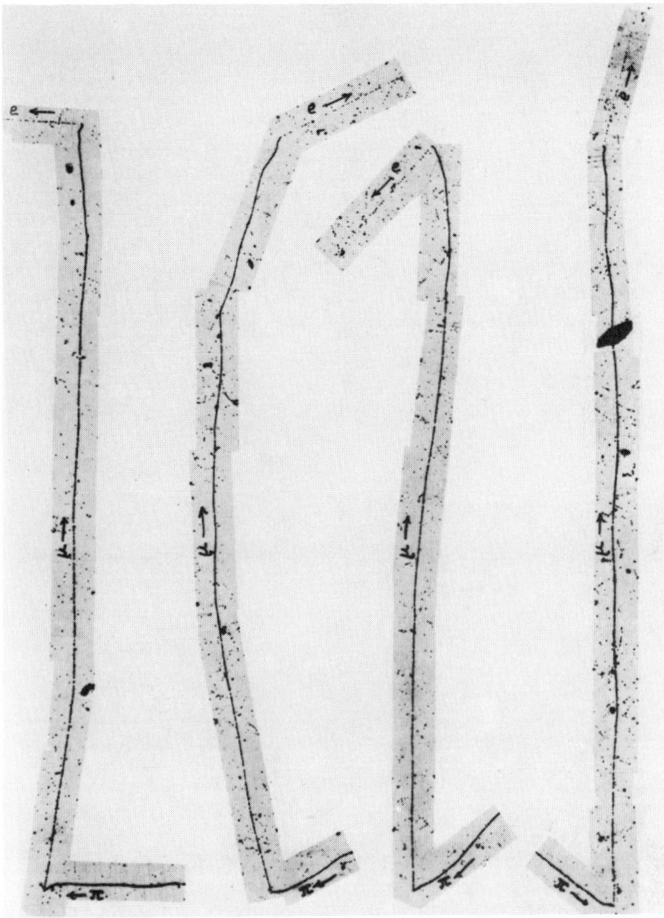

Kodak NT4 emulsion Powell; Rpts. Prog. Phys.13,384(1950).

Fig. 8. Successive decay of the π-meson, $\pi \rightarrow \mu \rightarrow \beta$

With the development of the electron-sensitive emulsions (BERRIMAN 1948), it became possible to observe the tracks of electrons moving at relativistic velocities. Among the effects observed shortly afterwards were examples of the β-decay of the μ-meson (BROWN et al.1949). Fig. 8 shows four examples of the process $\pi \rightarrow \mu \rightarrow \beta$. In favourable cases, it was possible to measure the energy of the electrons by the scattering method (DAVIES et al.1949), and the values obtained were similar to those for cosmic-ray mesons as measured in Wilson chambers operated in a magnetic field. These observations proved that the penetrating particles in the cosmic radiation are identical with the μ-mesons, and that they are commonly produced by the decay 'in flight' of fast positive and negative π-mesons. The latter result directly from nuclear collisions occurring in the atmosphere.

The four μ-mesons shown have nearly equal ranges; the variations in range due to 'straggling' are commonly much greater than is suggested by these photographs.

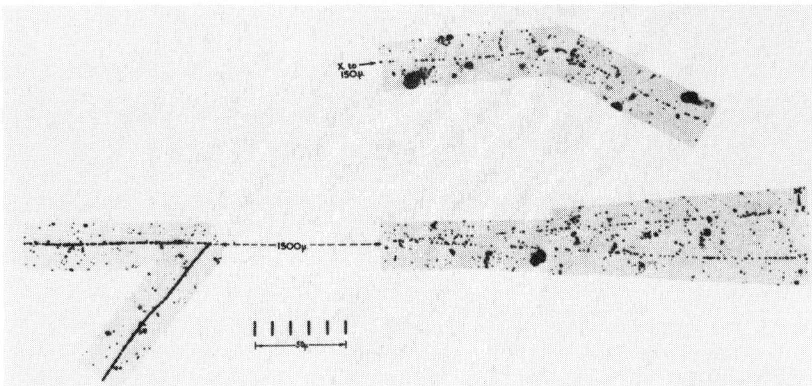

Ilford G5 emulsion O'Ceallaigh; Phil.Mag.41,838(1950).

Fig. 9. Pair of charged particles apparently associated with the decay of a π-meson

This event was found by O'Ceallaigh in a search for electron pairs in a volume of emulsion around the line of recoil of the neutral particle in πμ-decay. The event appeared to originate at a point close to the line of motion of the neutral particle. Further, the 'bisector' of the pair was close to the expected line of motion of the neutral particle. It was very difficult to admit that this pair was due to two electrons formed by the materialisation of a γ-ray of energy 30 MeV, because:

a) The grain-densities in the two tracks were substantially greater than the minimum value; and b), the particles producing the two tracks appeared to stop suddenly in the emulsion without undergoing the characteristic increase in ionisation of an electrically charged particle at the end of its range. No similar observation has been recorded, and it now appears certain that the two tracks were due to two heavily ionising particles emitted from a disintegration which occurred whilst the plate was swollen; either during drying after manufacture, or just before or during development. When due to heavily ionising particles, such tracks, sometimes referred to as 'prehistoric', have grain-densities in the finished emulsion similar to the values observed. Such an explanation is consistent with the observed scattering and the absence of any increase in ionisation at the end of the range. The grain-densities observed correspond to an emulsion swollen with with water to about five times its normal thickness.

A complete transformation occurred in the summer of 1948 with the first new 'Electron Sensitive' emulsion. Particles of all β could now be detected; there should be β^+ particles on all πμ decay chains, if the interpretation was correct. In October 1948 we received just 45 cm^2 of 200 μm emulsions of the new emulsion Kodak NT4A. These were taken by David Ritson to the Jungfraujoch, exposed for just a few days close to a magnet and returned, processed at once and within a few days were on the microscope! There were only a dozen π-mesons in the whole set! The first πμ seen is shown on the left-hand side of Figure 8.
Rosemary, now my wife, found this. The decay chain is indeed complete. The three others are shown also. The μ^+ on πμ decay surely is the first clear demonstration of recoil from a neutrino. I have this event under the microscope for you to view if you so desire. The event spans about six fields of view.

There was one further significant piece of work that was undertaken in 1949/50 by O'Ceallaigh. This involved scanning in the direction opposite to the μ^+ in complete πμe events to see if there was any sign of a γ-ray converting into an electron pair. No electron pair was seen, so X in the equation for πμ decay was not a γ-ray, and the neutrino hypothesis was confirmed. But the event shown in Fig. 9 was found - that superficially appears as a pair of tracks - that are neither electrons or protons. The interpretation of this event is that it was a two prong star that occurred when the emulsion was fully expanded - either during manufacture or in the presoak at the start of processing. This expansion ensures that the number of AgBr crystals penetrated by the tracks is reduced substantially from normal - so the tracks are a set of well separated grains. This affair held us up for quite a time.

To conclude this talk I must give some personal reminiscences of this most exciting period that started in 1946. I was a final year undergraduate at the time, but was well aware of the happenings on the 4th floor and came up as often as I could. Also I worked on the microscopes in all the vacations. A considerable team of microscope observers was acquired who provided the material for Cecil Powell and his team of young scientists from all over the world to study. One never knew what might turn up! The excitement only really started with the discovery of the π meson in 1946, and it continued for many years.

REFERENCES

Kinoshita and Ikeuti 1915 Phil. Mag. 29 420
Taylor 1935 Proc. Roy. Soc. (A) 150 382
Lattes, Muirhead, Occhialini and Powell 1947 Nature 159, 694
Perkins 1947 Nature 159 126
Frank 1947 Nature 160 525
Lattes, Occhialini and Powell 1947 Nature 160 453 486
Goldschmidt-Clermont, King, Muirhead and Ritson 1948
Proc. Roy. Soc. 61 183
Bjorklund, Crandall, Moyer and York 1950 Phys. Rev. 77 213
Carlson, Hooper and King 1950 41 701
Dalitz 1951 Proc. Roy. Soc. (A) 64 667
Anand 1953 Proc. Roy. Soc. (A) 220 183
Powell 1950 Rpts. Prog. Phys. 13 384

Berriman 1948 Nature 162 792
Brown, Camerini, Fowler, Muirhead and Ritson 1949 Nature 47 82
Davies, Lock and Muirhead 1949 Phil. Mag. 40 1250
O'Ceallaigh 1950 Phil. Mag. 41 838
Powell, Fowler and Perkins 1959 The Study of Elementary Particles by the Photographic Method (Oxford:Pergamon) for general references.

Discussion

Goldschmidt-Clermont

As you know, I arrived in Bristol after the discovery of the π - μ decay. You mentioned that the second example was found by Mrs. Roberts. I think the first example was found by Marietta Kurz.

Fowler

That is correct.

Goldschmidt-Clermont

I think she had it under the microscope for a day or two before she convinced people that actually a meson decayed into another and convinced them to look into her microscope. So the 30 seconds you mentioned only occurred after an initial delay!

Fowler

Well, I was in the lab, but probably working at lectures or something.

Lattes

I can confirm the version with Marietta Kurz, but it didn't take long until we looked.

Fowler

Good. I would have been surprised if you didn't! But in your statement in your festschrift, Guilio, you have the first one discovered as the complete one. I think it's the other way round?

Lattes

Yes, that is true.

Fowler

The first one was the one which just left and the second one, a few days later, was the first complete one. The pair together makes the case.

Marshak

I'd like to ask Lattes a question which I've been curious about. When you went to Berkeley and showed how to develop the emulsion and started seeing all those pions, there was a period of several months when Berkeley also thought they were producing the muons directly. Did you have any sense of what was going on? Later it turned out that they were decaying and the magnetic field was making them go back. What was going on then?

Lattes

After hearing so many times and reading it that Gardner didn't know how to process emulsion, maybe you've convinced me, but I must tell you that I didn't change anything in the processing of the emulsion. I changed the

© 1988 IOP Publishing Ltd

rate of the scanning. Professor Gardner was a sick man and had no scanners and couldn't look at the microscope very much. There was no change in the developing technique. I also heard or read that Gardner had Eastman plates and that I took or sent Ilford plates. That's not true. Ilford plates had been available for over two years already and Gardner was quite an alert physicist. Now, as for the muons coming directly, I think this is a question of semantics. I told MacMillan that the muons were also being made, and Bethe understood that to mean "knocked straight off". But "made" means that we were detecting them and they came from pions stopping in the targets.

Marshak

It seems very simple the way you express it, but actually, at the second Shelter Island Conference in April '48, Serber was there because the Shelter Island Conferences were only open to theorists, which was changed at the Rochester Conferences. Serber reported on the work at Berkeley and claimed that the muons were being produced directly; it was not a question of just having seen them. If you read the Feynman report of that conference, published in Physics Today, he says that a strange phenomenon was going on of direct production of muons. At the MIT Conference in June 1948, Alvarez reported that and Fermi believed it. It took a little time to straighten that out.

Lattes

I never stated at the American Physical Society meeting that they were produced directly. At that meeting MacMillan was saying "directly". I explained to Bethe what I meant by "being produced". So, I am not responsible for what theoretical people said!

Fowler

I think it is very interesting that I must have got it wrong about the processing at Berkeley. It's new to me. You're saying that he had done all the right things and the tracks were good?

Lattes

That's right. We had the impression that he had underdeveloped because we had plates exposed directly to the α beam and these he had underdeveloped. There was no special type of development needed for his plates.

Rochester

I think, Peter, that one ought to put the question of deflection of particles in a magnetic field in perspective. A team consisting of a group from Brussels and Manchester, among whom the prominent people were Connie Dillworth, Steven Goldsack, Goldschmidt-Clermont and others, showed quite clearly when we had exposures on the Pic that one could get magnetic curvature far in excess of the scattering. The problem with the emulsion was that of course one had to use enormously high fields. The magnet was coned to give a field of order 40,000 Gauss so that one could only use small volumes of emulsion. So it was extremely inefficient. Later, with very high fields which were produced at CERN of, for example, 200,000 Gauss, the method did find some application in the measurement of the magnetic moment of the hyperon.

Fowler

Connie, can I ask you, didn't you have a gap between the plates?

Dillworth

Yes, it was a sandwich, a magnetic field between the two plates. Yves Goldschmidt-Clermont can tell you about that - he used it. The other technique was inside the emulsion. The point is, you need a high field, but it's also possible because the curvature, due to scattering, increases as the square root of the length, whereas the magnetic curvature as the length itself. So if you have enough track length, it's simple.

Fowler

I was talking about practicalities, but I was more dogmatic than I should have been!

Conversi

If I may add, I think we used this method, at least at the qualitative level, at the WA17 experiment at CERN using the fringe field of BEBC, which was 28 kG.

Perkins

While we are on the subject of the charge of mesons in cosmic rays, I think the first definite experiment was done by Barbour at Chicago on a balloon flight using sandwiches of emulsion and permanent magnets. He showed that the $\pi - \mu$ decay of the pion were due to positives, the σ's were negative and the ρ's in C2 emulsion were a mixture of positive and negative. This was the Swan Song of the C2 emulsion, I think, in 1950.

Williamson

Professor Fowler, you told us about the crucial role of the emulsion technique and how important it was to meson physics. Was there any parallel improvement in viewing microscopes, and was that carried out here?

Fowler

No, over the whole period that I am recording there has been no improvement in either the Vickers-Cooke microscope or the Leitz microscopes, the new ones and the old ones. In fact, they might have gone backwards. The optics were very good in those early days.

Ballam

Just for the sake of completeness, the bubble chamber technique has gone down to a resolution of 20μm. The size of the bubbles are down quite a bit from the 1 mm you showed.

Pinkau

The question was on the improvement of the technique. I just wanted to remind you of the glass prism to improve the scattering measurement : when you had the objective resting against the glass prism so you could have an absolute zero and the noise in scattering was reduced. That is perhaps the only technical improvement that happened.

British patent 580,504 and Ilford nuclear emulsions

C Waller

Formerly Ilford Limited
Present address 23 Denehurst Gardens, Woodford Green, Essex IG8 0PA

ABSTRACT: By a new technique disclosed in British Patent 580,504 gelatin with silver halide is precipitated from emulsion by adding a surface active agent. This technique was used from 1945 to produce emulsions with about eight times the normal ratio of silver halide to gelatin. The technique also facilitated the combined use of production and laboratory resources for their reliable manufacture on a fairly large scale.

I was gratified to be invited to this Conference, but when I agreed to speak at it I was aware of at least two handicaps I would carry. First I would probably be the only chemist among physicists, and second I might be the one industrial scientist among academics, skilled and long-practised in expressing themselves before an audience. Even my presumed expertise could now be suspect, since for 15 years I have been absent from photographic technology and living happily with a self-imposed ban on even thinking about it.

At least this was true until just a year ago, when Professor Connie Dilworth, writing now as a historian of science, asked me "if there existed any scrap of paper with information on the how, when and why of the development of concentrated nuclear emulsions". Such a "scrap of paper" exists and has been available for public examination since the end of 1946. It is the British Patent 580,504 (Waller and Woosley, 1944), and now I have Ilford Limited's permission to draw attention to it, and to speak about its connection with Ilford nuclear emulsions.

Example III of this patent is so informative in very few words that I hardly need to refer to anything else. It was first printed in the Provisional Specification dated 3 July 1944 but I have reproduced the duplicate on p.7 within the Complete Specification (Fig.1).

Half-way through the preparation of an emulsion the silver halide crystals have grown to the size that is desired and are still in the environment of silver halide solvent plus gelatin that has brought about this growth. This is the emulsion at the start of Example III, which then describes how to precipitate the gelatin carrying the silver halide with it, by adding the right quantity of a "detergent", a surface active sodium alkyl sulphate or sodium alkylaryl sulphonate, the alkyl groups having around 12 or 14 carbon atoms. The precipitate forms a soft, dense, cohesive layer, and when it has been separated by decanting the

© 1988 IOP Publishing Ltd

liquid phase, and redispersed in more gelatin solution, nearly all the water-soluble electrolyte has been discarded, and the subsequent stages that are a normal part of emulsion preparation may continue with little change from normal. Most importantly this invention provides a new method of concentrating emulsions so that they may be used efficiently in film manufacture. The method is simple and convenient, and compared with the technique it replaced much more economical in plant and materials. In fact it is hard to imagine a simpler and more satisfying concentrating procedure than just to pour off all the water.

To 1 litre of an aqueous photographic emulsion at 125 F. containing suspended silver halides 11.5% by weight, ammonium nitrate 0.6 N, ammonium bromide 0.06 N, ammonia 0.5 N, and gelatin 0.4% by weight is added 8 ccs of Teepol (a commercially available anion soap believed to be a 20% by weight aqueous solution of a mixture of secondary alkyl sodium sulphates mainly dodecyl and tetradecyl), stirring thoroughly for 30 seconds. After 10 minutes the clear supernatant liquid is decanted and the residue dispersed in the following solution at 125 °F.:-

Gelatin	16 gms.
Water	110 ccs.
Ethyl alcohol	25 ccs.

To the completely dispersed liquid an addition of 500 ccs. of ethyl alcohol (64 O.P.) is made which results in the precipitation of the gelatin bearing the silver halides. The supernatant aqueous spirit containing the wetting agent is easily decanted from the precipitate, which is then soaked in water, and redispersed in gelatin in known manner to form a photographic emulsion substantially free from the salts present in the original emulsion.

Fig. 1. Example III from British Patent 580,504

It will be clear that the high concentration of emulsion for film coating just referred to is a matter only of water content. This of course makes no difference to the composition of the dried layer which typically is (or was) for films and plates around 250 g gelatin per mol of silver halide (108 g silver), about 13% of silver halide by volume.

It must have been in September 1945 that Powell started to get the photographic industry again interested in making emulsions for nuclear physics research. In the 1930s we had made two special emulsions of very low fog background and low sensitivity, but the commercial emulsion of Ilford's Half-Tone plate seems to have been as good as any other. We made it available as plates with thick emulsion layers, and as layers thicker still without glass or other support. This was a very limited undertaking, a small part of the work of a small section. I was not involved with it, and I had not even thought about how the photographic technique was made to work in nuclear physics and why thick emulsion layers were sometimes a good idea.

Coming on to the autumn of 1945, we had to bring ourselves back to peace-time conditions of work. Our research projects were geared to the needs of war and we had to plan and undertake new ones, aimed to bring profit to the Company in the peace-time future. Briefly we needed for the time being to cultivate our own research and not be too eager to help others with theirs. So when Powell told the Company that he would like to get Kodak as well as ourselves interested in making nuclear emulsions we could not object, nor would we have wanted to. All the same we heard from Bristol soon afterwards that much more concentrated emulsions were needed.

By this time the new process for concentrating and washing emulsions was in routine use in the production of most film emulsions, and naturally it had become part of the techniques used in emulsion research. It came into my mind that when the process has reached the point of the first redispersion in gelatin solution (represented by line 45 of Fig 1) the product of this redispersion is an emulsion which (compared with a conventional liquid emulsion) has about one-eighth the gelatin content in about one-eighth the volume, with all the silver halide. So it is "conventional" in regard to gelatin concentration and has a silver halide concentration about eight times as great. Probably only the gelatin concentration would be significant in the techniques for producing plates or emulsion layers, the silver halide content having little or no effect. The emulsion evolved in this way had still to be set, shredded, washed in water and sensitised. Without any new experimental work plates were made from an emulsion of "Half-Tone" type adapted for precipitation by surface active agent. On our tests by light exposure and ordinary processing they appeared to have survived and we sent them to Bristol. This was probably in November 1945 and this prototype emulsion was warmly approved to us verbally by Powell and Occhialini within a few days.

A great deal was known at Ilford about "Half-Tone" and related emulsions and how their crystal size could be varied by controlling the conditions of silver halide formation and crystal growth. I do not think that at this time, 40 years later, I could usefully report on just how C2 came to be judged generally acceptable so that it became the first of two "standard" Ilford emulsions. Certainly things moved fast, since Occhialini and Powell (1947) exposed C2 plates for 5 weeks on the Pic du Midi and still had time to send a letter to Nature on their results before the end of 1946. C2 had a mean linear crystal size of about 0.15 μ, with the maximum sensitivity that was obtainable by sulphur sensitisation, and with low background fog.

It is strange that at this time (with C2 launched) those of us at Ilford who were concerned with nuclear emulsions were unfamiliar with the ideas of "electron sensitivity" and "minimum ionisation". Perhaps the composition of the team was faulty in one respect (the lack of a physicist), perhaps it worked too much in isolation, but I am not going to offer very belated excuses. We certainly learnt about both ideas in 1948 from the work of R.W. Berriman and R.H. Herz and from Bristol accounts of the NT4 emulsion they made. They were a balanced team of emulsion chemist and radiation physicist; they showed what was possible, and being Kodak provided the strongest possible incentives for our own immediate efforts in response.

We continued just as before - but with the new objective - sensitivity to minimum ionisation. In our case a sensitivity increase to about 4 times that of C2 was needed, and it seemed likely that, as in Berriman's case, about equal contributions could come from increasing the crystal size, probably to 0.3μ, and from a more efficient sensitising process. Such a process may use traces of gold in addition to labile sulphur, but we knew already that this would give heavy fog with any emulsion of C2 type, including one with the required larger crystals, and no means of avoiding this had been discovered.

My brief is still the interaction between British Patent 580,504 and Ilford nuclear emulsions. The concentrating process is "friendly" (to borrow the adjective from computer jargon) and never did an emulsion any harm! This meant that we could have the whole of Ilford's emulsion technology in view, fairly confident that an emulsion that had the right crystal size distribution, and was **uncomplicated** in its **response** to efficient sensitisation, would convert simply to nuclear composition without degradation of its qualities. Fortunately we did not have to look very far or for too long. The starting point for G5 was found within an experimental series of emulsions made at Ilford which until then had hardly been exploited, mainly because they needed extremely precise control of the environment in which the silver halide was first formed. Bristol seemed to have deserted us, but when G5 was introduced in 1949 we made warm acknowledgement to Dilworth and Occhialini, by then in Brussels, and to E.P. George and A.C. Jason much closer at hand at Birkbeck, "for advice and for criticism of experimental materials". There was plenty to criticise at first; I seem to remember looking for three or four grains in line so as to claim them as a primitive track of a relativistic particle!

By the time we were selling G5 we had made one final call on the benificence of the British Patent. Since in making the basic G emulsion we needed to control meticulously the rates of addition of reactants and the simultaneity of twin additions, there was a big advantage in operating on a large scale - in fact the production scale of the Plate Factory emulsion department. Then when the stages of precipitation, redispersion and water washing had been passed (still within the production unit), the volume had been reduced to about one-eighth, a volume (and weight) of concentrated emulsion which was now convenient for the laboratory to take over for sensitisation. Production of G5 emulsion was thus a joint undertaking of production and laboratory staff, the former making a great contribution through qualities of experience, discipline and enthusiasm for the unusual task. We would need the production capacity that this organisation created as the demand for G5 expanded world-wide during the next decade.

REFERENCES

Occhialini G P S and Powell C F 1947 Nature 159 93
Waller C and Woosley D P 1944/5 British Patent 580,504

Discussion

Fowler

You said something about "printing out". What does that mean?

Waller

"Printing out" is the effect of light on silver bromide without any development - just going grey.

Payne

It was said in 1945 or 1946 the Canadians were on the point of producing emulsions. Can you say anything about that?

Waller

They were quite early in doing this work. I didn't hear anything about it at the time. There was Demers of course. He was a university worker and did this all on his own without knowing anything about industrial technology. Certainly his methods of working, particularly with respect to sensitization, were on too small a scale. He was making a beaker full of emulsion and it was quite difficult to make it reproducible. As is said in the Powell, Fowler and Perkins book, our emulsion was available on a large scale and his wasn't.

Marshak

I was in Rochester then and Broughton Peters had found the heavy primaries; I mention this because Isabel Powell provided the original emulsion which they examined. They got Kodak in Rochester very interested and pushed them to compete with Ilford. I had the impression that Kodak surrendered to Ilford after a while because they didn't think they'd make enough money. Did you make enough money?

Waller

We never make enough money.

Frank

Did you make some money?

Waller

Well, it's difficult to tell with overheads etc.

Rochester

I'd like to add one small comment about the Demers emulsion. As I understood it, he made very small quantitities for the purpose of radioactive studies in Montreal, and worked with what I would call "kitchen sink" methods which weren't obviously very suitable for manufacture on a large scale. I do recall, and Don Perkins may know this

© 1988 IOP Publishing Ltd

better than I, that on the emulsion committee, probably in 1947, Rotblat, the chairman, brought along some small squares of emulsion which had been given to him by Chadwick and which he had obtained from Demers, and on those we could definitely see the ends of electron tracks. Do you remember Don?

Perkins

I wasn't at that meeting. I certainly don't remember.

Dillworth

If I remember correctly, the Kodak NT4 was not as reproducible as the G5. In fact, people gradually gave it up as there wasn't such steady reproduction. The Kodak production was entirely from the laboratory. It didn't involve production lines and so you didn't get the repeatability. I think that is the big difference as Dr. Waller pointed out. The emulsion became a standard instrument because it was repeatable.

Waller

When I started writing this talk I found it at first quite impossible; the subject was too large. What saved me and made the job possible was that I decided that all I would write about was this patent and how it influenced the origin and development of nuclear emulsion. There is an awful lot more which could be said, but I don't want to say it. For instance, I haven't even acknowledged all my very essential staff in this. When I was writing I kept coming back to my scheme, and this is it.

Early work with electron-sensitive emulsions

W.O. Lock

CERN, Geneva, Switzerland

ABSTRACT: Electron-sensitive nuclear research emulsions, that is, emulsions sensitive to minimum ionising particles, became commercially available for the first time in 1948. A survey is given of the major results obtained at Bristol in the years 1948-1951 by C.F. Powell's research group from studies of both Kodak and Ilford emulsions exposed to the cosmic radiation at mountain altitudes and by means of free balloons. An attempt is made to try to capture the international spirit of the research group by reference to some of the personalities involved.

1. INTRODUCTION

In the summer of 1947 I had just completed my second year as an under graduate here at Bristol, studying physics. I had the good fortune to spend a month that summer on the 'Fourth Floor' not studying cosmic rays, but helping Hans Heitler with the analysis of some of the emulsions exposed earlier in a scattering camera experiment at the Liverpool cyclotron. However, I got to know the international group then forming around Cecil Powell - Ugo Camerini, Connie Dilworth, Giulio Lattes, Hugh Muirhead, Ron Payne and Beppo Occhialini and I was attracted by the exciting work then being carried out and just described to you by Peter Fowler. The following year, 1948, I graduated together with Fowler, who had returned from the Royal Air Force to complete his degree. He was already working with Powell while I was given the choice of continuing for a Ph.D. either with L.C. Jackson on low temperature physics in the basement or with Powell higher up in the laboratory. I chose the latter and spent the next four years as an active member of the growing research team.

2. FIRST WORK WITH ELECTRON SENSITIVE EMULSIONS

The summer of 1948 saw the availability of the first electron-sensitive Kodak NT4 emulsions (Berriman 1948). Ilford followed the next year with their G5 emulsion which was soon adopted for general use in preference to the NT4. The new Kodak emulsions only 2" x 2" in size and 200 μ, later 400 μ, in thickness were quickly exposed and processed on the Jungfraujoch by David Ritson, where other cosmic ray studies were in progress (Camerini et al 1948) and he brought them back to Bristol in the autumn of 1948, just as John Davies and I became Hugh Muirhead's first research students.

© 1988 IOP Publishing Ltd

In this short talk I shall try to mention a few of the significant results obtained in the following three years until particle beams from accelerators gradually took over from cosmic rays for the study of elementary particles. For a description of the overall context of this work in relation to other cosmic ray studies, particularly with cloud chambers, reference should be made to the excellent talk given by C. Peyrou (1982) at the International Colloquium on the History of Particle Physics held in Paris in July 1982. For more details of much of the work which I will describe see Powell (1950).

A first examination of the plates from the Jungfraujoch immediately showed many examples of the creation and subsequent capture of slow π^--mesons, confirming the earlier observations of the capture of slow mesons in less sensitive emulsion by Perkins (1947) and by Occhialini and Powell (1947). In addition for the first time the complete decay sequence of a pion to a muon and then of the muon to an electron could be clearly seen, thus giving further evidence for the correctness of the π-μ decay hypothesis. First results were published in Nature early in 1949 (Brown et al 1949a) and in more detail later (Brown et al 1949b) in the first of series of papers published in the Philosophical Magazine under the general title "Nuclear Transmutations Produced by Cosmic Ray Particles of Great Energy". Figure 1 shows the creation and subsequent capture of a slow negative pion while Figure 2 shows the complete π-μ-electron decay sequence. The handwriting is Powell's and the date of October 1948 is

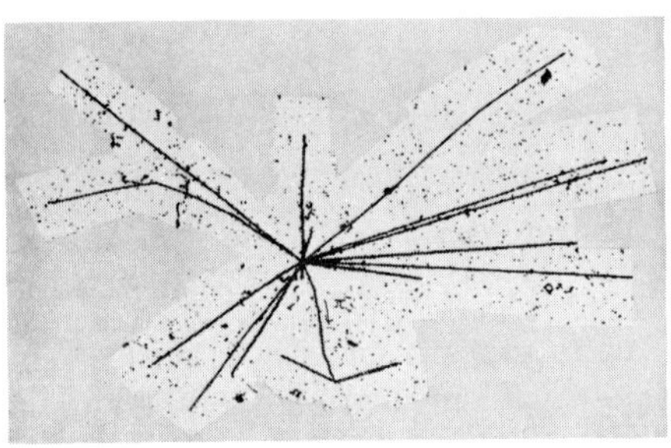

Fig. 1. A neutron induced star with the emission of a π^- meson of short range which produces a second disintegration. (From Brown et al 1949b)

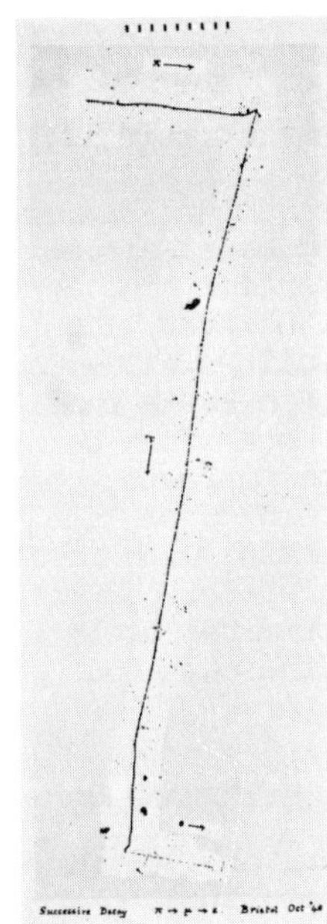

Fig. 2. A complete $\pi^+ \rightarrow \mu^+ \rightarrow e^+$ decay sequence, labelled by C.F. Powell in October 1948 (Photograph University of Bristol; the event is the same as that subsequently published by Brown et al (1949a), but with a different labelling)

given; it is interesting to note that when this photograph was published in the Nature article of January 1949 he replaced the 'e' for the electron by the symbol 'η' and carefully wrote in the caption "The particle, η, arising from the decay of the μ-meson, is commonly assumed to be an electron, but this assumption remains to be finally established" (Brown et al 1949a).

An example of the production of a slow positive pion subsequently showing the $\pi^+ \rightarrow \mu^+ \rightarrow e^+$ decay sequence is shown in Figure 3.

In the same emulsion a few examples were also seen of high energy nuclear disintegrations with the production of several "shower particles" of minimum ionisation, presumably the same as the production of showers of penetrating particles obtained in counter and cloud chamber experiments. Figure 4 shows a typical example; note Powell's habit of always acknowledging the scanning girl who found the event shown. The analysis of events of this type will be discussed in more detail in Section 5.

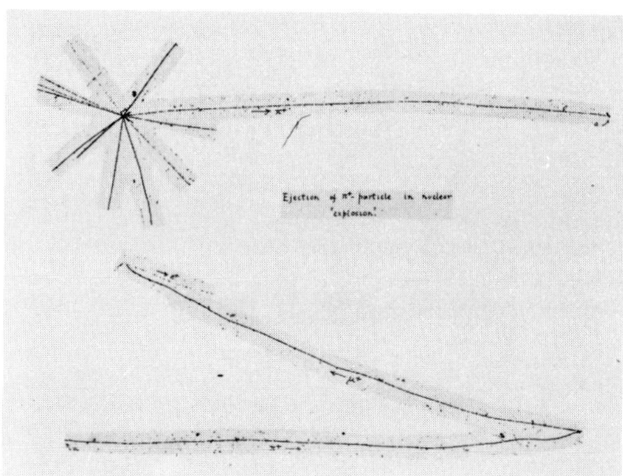

Fig. 3. Production of a slow π^+-meson which stops in the emulsion and decays to a μ^+ meson which also stops and decays to an electron. The pion track (top right) is continued as from bottom left (From Powell 1950)

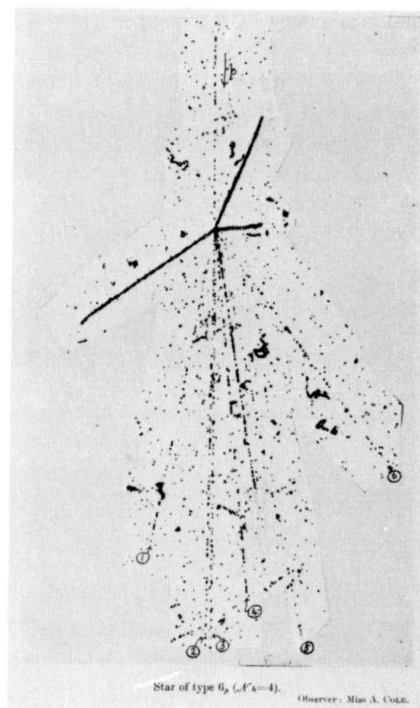

Fig. 4. A typical high energy nuclear disintegration produced by a proton with the production of six "shower particles". (From Brown et al 1949b)

3. EVIDENCE FOR HEAVY MESONS

However, the most exciting discovery in the new emulsions in the autumn of 1948 was that of a new event - a heavy meson apparently decaying to three pions (Figure 5). Hugh Muirhead (1984) has recalled that the first emulsions sent back from the Jungfraujoch by Ritson arrived in Bristol on a Thursday and the new event was found on the Saturday. It was immediately plotted out on paper using a high power projection microscope, called the "telepanto" developed by Occhialini. Together with John Davies and Hugh Muirhead it was one of my first tasks to measure the multiple Coulomb scattering of those particles of sufficient length in the emulsion, using the method developed by Goldschmidt-Clemont et al (1948). The angles were measured with a very simple device - a normal protractor as used with a draughtsman's drawing board as illustrated in Figure 6.

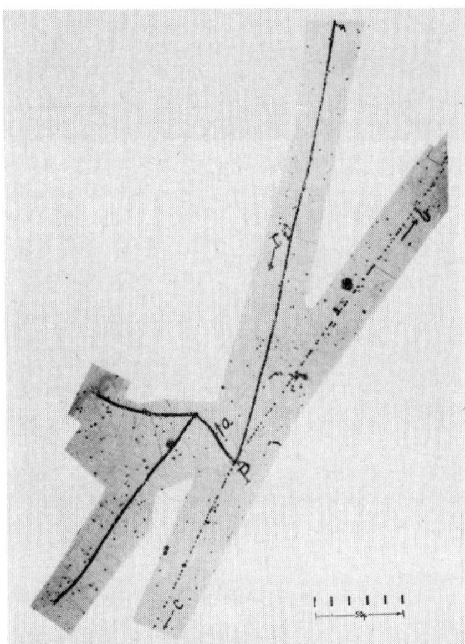

A heavy meson, τ, comes to the end of its range in the emulsion and decays into 3 charged particles, one of which is a π^--particle of short range which produces a nuclear disintegration. There is strong evidence that the tracks *b* and *c* were also produced by π-mesons. The direction of motion of the three secondary particles are co-planar. Observer Mrs. Van der Merwe. Photograph by A. R. Gattiker and Miss P. Dyer.

To face page 1048

Fig. 5. The original τ-meson microphotograph as re-published by Fowler et al (1951), with their caption

Fig. 6. The apparatus used by Davies et al (1949) to measure the multiple scattering of electron tracks from μ-meson decays (From Lock, 1951)

Detailed analysis of momentum balance and coplanarity together with mass measurements on the primary particle and one of the secondary particles strongly suggested that the event was that of a heavy meson of mass around $1000m_e$ decaying into three pions. The details were published in January 1949 as the second part of the Nature paper of Brown et al (1949a). In this paper the heavy meson was called a K-particle; it was only later that Powell (1950)+ introduced the terminology of τ-meson for particles of mass intermediate between that of the pion and proton, which was subsequently restricted to the three charged pion decay mode (Fowler et al 1951), the other heavy meson decaying to a single charged particle being called a K-meson (O'Ceallaigh 1951). The photograph of the first τ-meson event shown in Figure 6 is taken from the paper of Fowler et al (1951) and it can be seen that the primary meson track has been re-labelled τ by Powell.

The possible existence of heavy mesons of about half the proton mass had been suggested a few years earlier by the observation of Leprince-Ringuet and L'Heritier (1944) of the collision of a fast cosmic ray particle with an electron in a cloud chamber picture and by the two cloud chamber events of Rochester and Butler (1947), which were interpreted as the decay of a charged particle of mass around 1000 m_e and of a neutral particle of mass not less than 800m_e respectively. However, the event of Brown et al (1949a) was perhaps the first convincing evidence for the existence of heavy mesons, aided by the fact that the decay was to three charged particles and by the high resolving power of the emulsion compared to that of the cloud chamber. As is well known, no further event of this type was found for nearly two years, until Harding (1950) observed two more examples followed by a further event at Bristol (Fowler et al 1951) and at Imperial College (Hodgson 1951). (See also Section 6.)

4. THE DECAY OF THE MUON

Although isolated examples of meson decays (now known to be muons) to presumed electrons had been seen in cloud chambers since the work of Williams and Roberts (1940), in 1949 it was not established whether the μ → e decay process was a 2-body decay or a 3-body decay. Since a large number of μ → e decays were observed in the plates which had been exposed on the Jungfraujoch, Davies, Muirhead and I embarked on a systematic study of measuring the multiple scattering and hence the energy of the long electron tracks in the emulsion, using the projection method mentioned above. Figure 7 shows typical plots of (a) a shower particle and (b) and (c) decay electrons of 40 MeV and 9 MeV respectively.

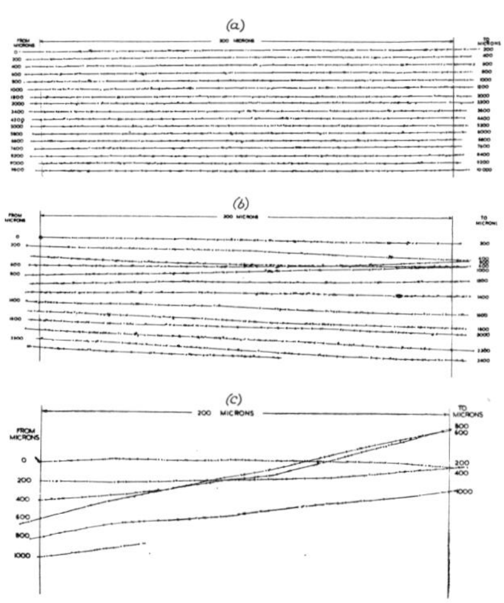

Fig. 7 Facsimile tracing of three particles at minimum ionization. The secondary particle of a shower is given in (a), plots of 200 μ length have been made at 800 μ intervals illustrating the negligible distortion which can be achieved in the emulsion. Plots (b) and (c) show decay electrons of 40 MeV and 9 MeV, respectively; the latter is from the μ-meson of a π-μ decay. (From Davies et al. 1949)

+ It was probably Leprince-Ringuet (1949) who first called heavy mesons τ-mesons; see Pais (1986)

The results of measurements on 81 decay electrons (Davies et al 1949) are given in Figure 8(b) where the dotted line shows the distribution expected for a 2-body decay. They clearly indicate a 3-body decay, probably $\mu \rightarrow e + 2\nu$ from the average electron energy of $1/3m_\mu$. Very similar results were obtained by Leighton et al. (1949) from cloud chamber measurements on 75 electrons (Figure 8(a)) and by Steinberger (1949) using counter techniques. Later in more extensive emulsion studies of muon decay O'Ceallaigh discovered the first $K \rightarrow \mu$ decay process which I shall mention briefly at the end of this talk.

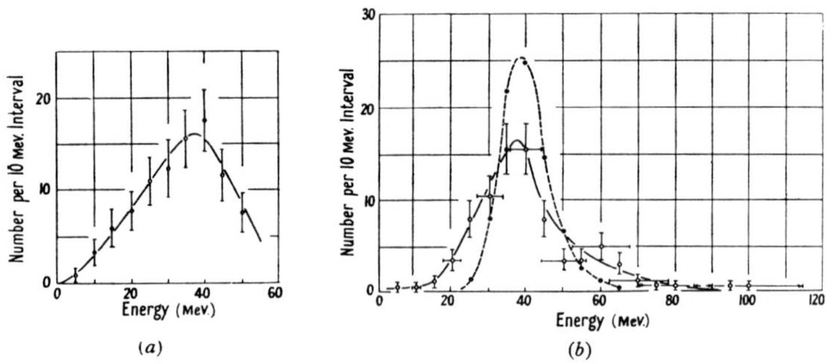

Fig. 8 Energy spectra of the electrons from μ-decay. Figure (a) Leighton et al (1949); figure (b) Davies et al (1949) (From Powell 1950)

5. EARLY BALLOON FLIGHTS: STUDIES OF HIGH ENERGY COLLISIONS

It was soon realized that many more events of all kinds would be obtained by exposing emulsions at higher altitudes. Largely due to the persistence of Hugh Muirhead and Ugo Camerini and with the assistance of Tom Coor, then US Scientific Attaché from the Office of Naval Research in London, early in 1949 we started a programme of flying emulsions to high altitudes by means of free balloons. In the first attempts small rubber meteorological-type ballons were used, which expanded as they gained altitude. They were launched from a bomb-site near to the Physics Laboratory at the top of St. Michael's Hill, where the Bristol Maternity Hospital now stands. One of those who helped us was Kevin Tindall, who retires next week after many years as Laboratory Superintendent of the Wills Physics Laboratory here at Bristol. Figures 9 and 10 show balloons being prepared for launching. The balloons were inflated such that when one burst the remainder would just balance the load and give a level flight, or one was detached by radio signal when a given altitude was reached. When more balloons burst the apparatus descended by parachute. The altitude was measured by a simple radio-sonde. Typical flight curves are shown in Figure 11 while the very simple apparatus for the exposure of the emulsion is shown in Figure 12. Technical expertise for these balloon flights was provided by John Davies, Hans Heitler and Max Roberts.

Fig. 9. Balloon flying in 1949 or 1950; holding the left-hand balloon from the left the late M.A. Roberts and W.O. Lock and holding the balloon in the foreground is U. Camerini (Photograph University of Bristol)

Fig. 10. Balloon flying 1949 or 1950; from the left W.O. Lock and the late M.A. Roberts (Photograph University of Bristol)

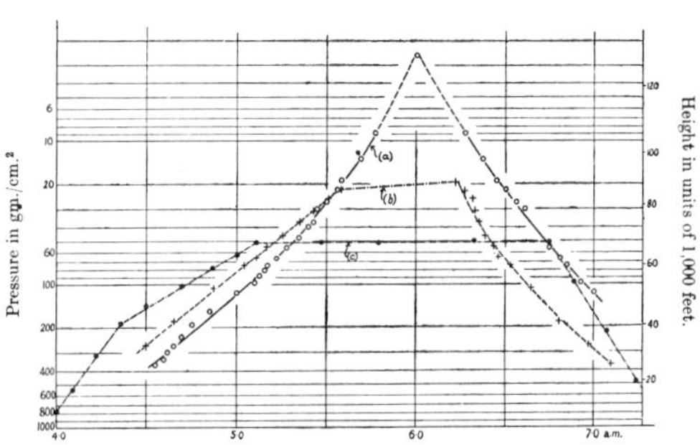

Fig. 11. Typical balloon flights showing the variation of altitude with time. Note that the launchings were made at dawn! (From Camerini et al 1949)

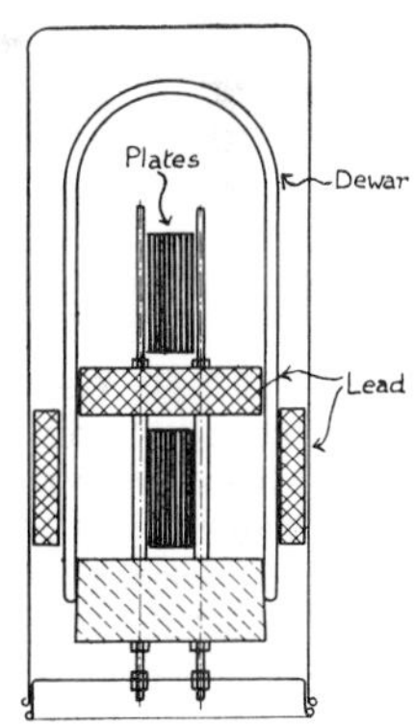

Fig. 12. Plate assembly in Dewar flask with lead blocks (From Camerini et al 1949)

The first results from flight (c) of about 1 1/2 hours at 70'000 feet with Ilford G5 emulsion 400 μ thickness, were already published as Part II of the "Nuclear Transmutations" series (Camerini et al 1949). Figure 13 shows the authors in the summer of 1949 together with other members of Powell's group, including the essential bevy of scanning girls.

Fig. 13. Cecil Powell's research group at Bristol in the summer of 1949 (Photograph University of Bristol)

Rear row, left to right: A.R. Gattiker, Mary Cole, Mrs. M.L. Andrews, Mary Merritt, Alex Cole, Peggy Ford, Miss P. Dyer, June Cowie, Grace Hussey, Mrs. B. Moore, Mrs. J. Van de Merwe, Mary Jones, Margaret Stott.
Second row: W.O. Lock[*], J.H. Davies[*], D.T. King[*], H. Heitler, S.O.C. Sorensen[*], Mrs. I. Powell[*], C.F. Powell.
Front row: T. Coor, H. Muirhead[*], U. Camerini, C. Franzinetti, N. Tobin, P.H. Fowler[*].
Those marked with an asterisk were present at the July 1987 Bristol Conference.

One of the first results of the study of the shower particles from the high energy nuclear disintegrations was that in a few cases where the track of the particle in the emulsion was long enough, it was possible to show by the combination of multiple scattering measurements (proportional to $p\beta$) and grain density counts (β) that the particles in question were π-mesons. However, the scattering measurements by the projection method described above were very laborious. Since the tracks were better defined in the new emulsions compared to those in the earlier less sensitive C2 emulsions, Powell suggested to Muirhead to try again to measure the multiple scattering directly on the microscope by measuring sagittas with respect to the stage movement (Muirhead 1987). They were quickly able to show that the method worked if precision microscopes with an accurate x-movement were used. The track under study was aligned along the x-axis and the y-coordinate read off at equal x intervals (say of 100 μ) using an eyepiece with a graduated scale. The mean value of the second difference $\bar{D}$ is proportional to the mean scattering angle $\bar{\alpha}$ which in turn gives $p\beta/Z$ where p is the momentum, β the velocity and Z the charge of the particle. Measurements were made by Fowler (1950) on several hundred tracks from high energy disintegrations observed in the emulsions exposed in flight (c) and he was able to show that the great majority of the slow mesons emitted were π-mesons and not μ-mesons. He was subsequently joined by Camerini, Muirhead and myself and we published a more extensive analysis in the summer of 1950 (Camerini et al 1950). The results of measurements of multiple scattering and grain density on about 700 tracks are shown in Figure 14.

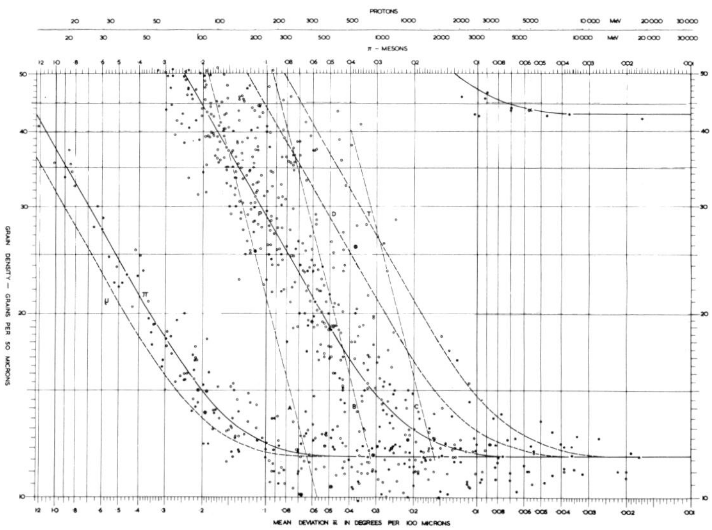

Fig. 14. Relation between grain-density and $\bar{\alpha}$ (100μ) for the tracks of 700 particles. "Primary" particles are shown thus ●; "Secondary" particles thus σ; and "interconnecting" particles thus -ϕ-.
A, B and C are lines of constant momentum and correspond respectively to 500, 1000 and 2000 MeV/c.
Kinetic energy scales for π-mesons and protons are included at the top of the figure. (From Camerini et al 1950)

It can be seen immediately that there is strong evidence that the low energy mesons are pions; the corresponding mass distribution for the particles in Figure 14 is given in Figure 15. The average value of the mass of the mesons with grain density more than 15 grains per 50μ was found to be 283 $\pm$ 7 m_e, in good agreement with the value found at Berkeley for artificially produced π-mesons, (e.g. Barkas et al 1950). For the mesons of higher energy, i.e. 150 to 1500 MeV, no conclusion could be drawn but very ingenious counter experiments by Piccioni (1950) at the same time, gave strong evidence for the production of π-mesons in high energy nuclear disintegrations produced by cosmic rays. No evidence was seen for any energy loss of the shower particles in traversing the emulsion which would have been seen if a substantial fraction of them were electrons.

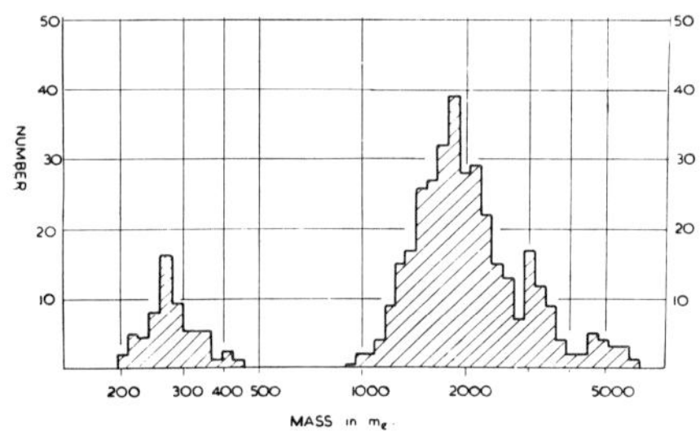

Fig. 15. Distribution in mass of the particles of charge /e/ corresponding to the measurements represented in Figure 14 (From Camerini et al 1950)

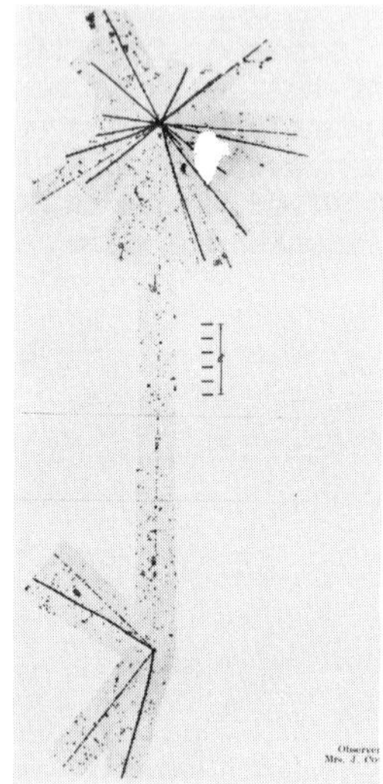

Fig. 16. The original caption reads "A shower particle ejected from a $18+5_n$ star is observed to produce a second nuclear disintegration of the type $5+0_p$; the interconnecting track is 10,600 μ long and can be identified as due to a meson ($\alpha = 0.049°$ per 100 μ; grain-density = 10.4 grains per 50 μ). (From Camerini et al 1950)

As well as providing evidence for the copious production of π-mesons from high energy collisions, Camerini et al. were also able to show that they were strongly interacting particles from a study of the average distance travelled in emulsion before they themselves caused a nuclear disintegration. Examples of such 'double stars' as they were called is shown in Figure 16. The interconnecting track is definitely identified as a meson. Although based on limited statistics (15 double stars) the result obtained was that the interaction mean free path was close or equal to the geometric cross section, a conclusion which was confirmed by later work (Lock and Yekutieli, 1952). From the scattering measurements it was also possible to derive the energy spectrum of the secondary mesons, protons and deuterons + tritons. In a continuation of this work it was possible to compare the measured energy of the primary particles with the observed energy in charged particles and deduce from the missing energy that neutral mesons should also be produced in the collisions studied (Camerini et al 1951).

6. STUDIES OF NEUTRAL MESON PRODUCTION

In parallel with the studies just described, another group in the Laboratory, Carlson, Hooper and King (1950) were studying the energy spectrum of electron pairs observed in the same emulsions exposed in the same balloon flight. (Note that Carlson subsequently changed his name to Ekspong.) When these were observed very near a disintegration as, for example, in Figure 17, they were easy to see, but to find a pair of minimum ionising tracks suddently starting in the emulsion required very careful scanning under high magnification. This was carried out and the energy of some 200 electron pairs was determined by multiple scattering measurements as described above.

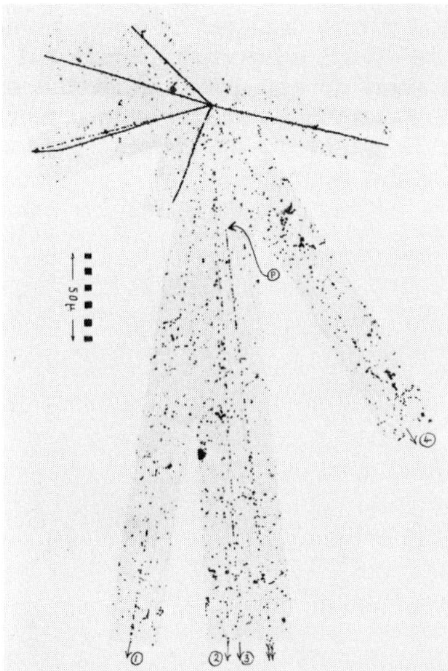

Fig. 17. A large shower of mesons (6+4p) showing the origin of a related pair very close to the star. The pair has been marked by an arrow p in the microphotograph.
Observer: Miss E. James (who subsequently became Mrs. Lock)
(From Carlson et al 1950)

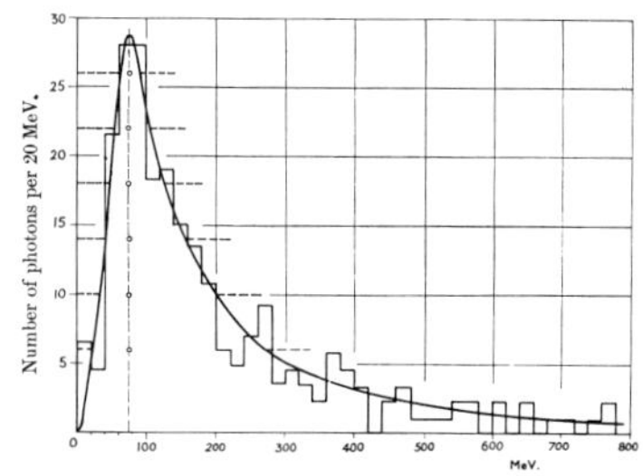

Fig. 18. Spectrum of the γ-radiation at 70,000 feet. The full line is that giving the best fit with the experimental results and the mass of the neutral particle was deduced from it
(From Carlson et al 1950)

The corresponding photon spectrum is shown in Figure 18; the full line gives the best fit to the data and gives a neutral meson mass of 295 ± 20 m_e. They were also able to obtain an estimate of the lifetime of the presumed neutral meson from an elegant analysis of the geometrical relationship of electron pairs found to be associated with stars (see Figure 19). The greater the value of the lifetime, the greater the average distance l and the corresponding value of r. For 15 pairs with $r < 40\ \mu$ (background expected 0.2) the lifetime was shown to be less than 5×10^{-14} seconds. In making this estimate the authors neglected the momentum imparted to the nucleus involved in the formation of the electron pair (Bethe 1934); it was for this reason that they obtained a lifetime limit some twenty times longer than the presently accepted value of $\tau \sim 10^{-16}$ seconds (Hooper 1987).

These observations of Carlson et al. provided convincing evidence that neutral π-mesons of mass approximately the same as that of the charged pions were produced in nuclear explosions with a frequency comparable to that of the positive and negative mesons. The existence of π°-mesons produced by accelerators had of course been demonstrated earlier in the same year by Bjorklund et al (1950) and Panofsky et al (1950). However, the observations of Carlson et al. clearly demonstrated that the main source of the "soft" component of the cosmic radiation was the production of π°-mesons in the same nuclear disintegrations which gave rise to charged π-mesons; the π° decayed to two γ-rays which then transformed to electron-positron pairs while the charged pions decayed to muons and electrons, the muons being the bulk of the 'hard component' observed at sea level. Further, extensive air showers seen at sea level were the end result of a complex cascade process throughout the atmosphere initiated by an incident cosmic ray particle (proton, α-particle or indeed heavier nuclei) of extremely high energy high up in the atmosphere.

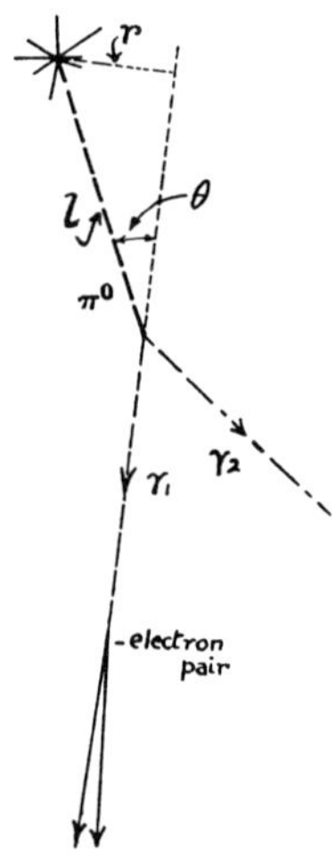

Fig. 19. Principle of the method for the determination of the lifetime of the neutral π-meson (From Carlson et al 1950)

7. MORE HEAVY MESONS

As this talk was to have been given by Cormac O'Ceallaigh, who is prevented by ill health from attending the Conference, I should like to make a brief mention of his discovery of the first clear examples of K-meson decay in 1951.

When studying the energy spectrum of the decay electrons from μ-decay Davies et al. (1949) had seen a few examples of electrons with much higher energy than allowed from the kinematics of the decay process (e_{max} ~ 55 MeV). However, the average mass of those parent particles for which the electron energy was > 60 MeV (6 cases) was measured to be 204 ± 19 m_e.

When thicker emulsions (400 μ) became available O'Ceallaigh continued the study of the decay spectrum in order to obtain much better statistics. During this study (using plates exposed under 30 cm of lead at the Jungfraujoch) he found what looked like a μ → e decay but with an apparent electron energy of 250 MeV (Figure 20). Scattering versus grain-density and residual range measurements on the primary particle, carried out by Menon and Rochat, gave a mass value of 1320 ± 170 m_e while the energy of the secondary if a μ-meson is 180 MeV (mass 200-300 m_e).

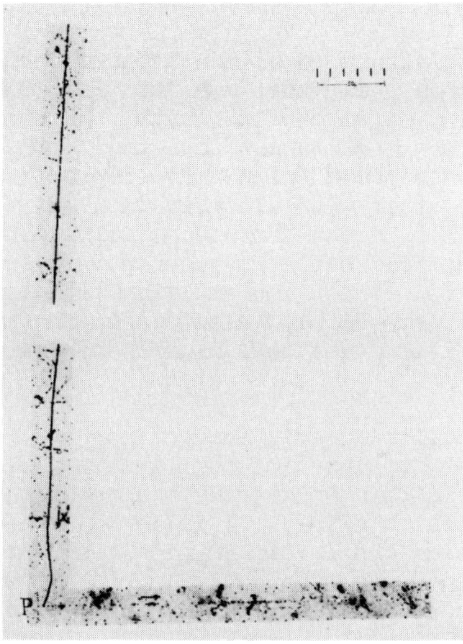

Fig. 20. A heavy meson, k_1, reaches the end of its range and decays with the emission of a single particle of charge /e/ . The track of k_1 was longer than 4000 microns and measurements of scattering and grain-density give a value for the mass of the particle $1320 \pm 170\ m_e$. The energy of the secondary particle, if a μ-meson, is 180 MeV; if an electron, 250 MeV (From O'Ceallaigh 1951)

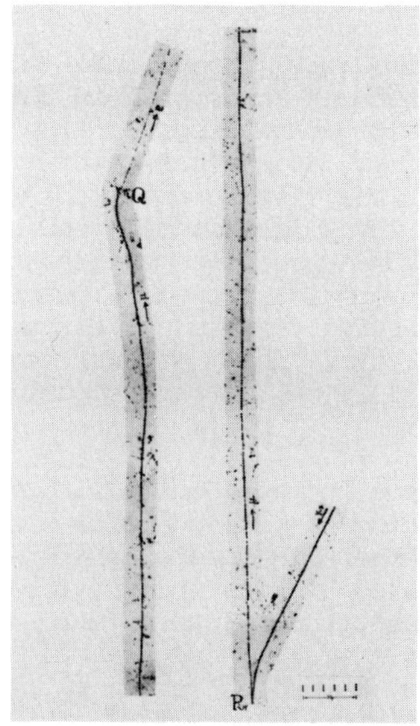

Fig. 21. A heavy meson, k_2, reaches the end of its range at a point P and decays with the emission of a μ-meson of range 1098μ. The μ-meson stops at the point Q and emits an electron. The track of the μ-meson is given in two parts. The mass of the particle as given by measurements of scattering and grain-density is $1125 \pm 140 m_e$
(From O'Ceallaigh 1951)

In the same batch of emulsions, O'Ceallaigh (1951) had the good fortune to find what superficially was a π-μ decay (Figure 21) but with a muon of nearly twice the range for such a decay, which stopped in the emulsion and emitted what appeared to be an electron, although the track was too short for measurements to be made. The mass of the primary particle, as given by measurements of scattering and grain-density was $1125 \pm 140\ m_e$.

It was also in the same batch of emulsions that at about the same time another τ-meson decay was found by Fowler et al (1951). O'Ceallaigh's paper and that of Fowler et al. were published as two papers entitled "Masses and Modes of Decay of Heavy Mesons".

This was the beginning of several years of confusion about the various kinds of heavy mesons and their different decay modes and which was only cleared up with the advent of the large accelerators which allowed them to be studied under controlled conditions.

8. CONCLUDING REMARKS

I have only been able to mention briefly some of the important work carried out at Bristol in the first few years of the use of electron-sensitive emulsions. In retrospect, in comparison with the large detector and accelerator facilities now available to the modern physicist, it is remarkable how much was achieved with very simple means (emulsions, microscopes and balloons) together with a combination of man and woman power (sometimes known as Cecil's Beauty Chorus) as illustrated in Figure 13. Powell's award of the Nobel Prize for Physics in 1950 was a fitting reward for his perseverance to perfect the emulsion technique ever since his pre-war exposures of emulsions, also made on the Jungfraujoch, at the suggestion of W. Heitler (Heitler, Powell and Fertel 1939).

Working on the 'Fourth Floor' under Powell's gentle direction was an exciting experience for us all, most of us in our twenties. The atmosphere in the period I have covered in this short talk has been well captured in a short article by O'Ceallaigh (1982). I will end with an unusual group picture taken in 1951 outside the Wills Physics Laboratory after a small hydrogen balloon had exploded after a faulty launch (it was blown into the side of the building) for a Festival of Britain Exhibition (Figure 22)!

FIG. 22. After the explosion of a balloon in the summer of 1951; from left to right: J.H. Mulvey, the late M.A. Roberts, W.O. Lock and J.E. Hooper (Photograph University of Bristol)

ACKNOWLEDGEMENTS

I should like to express my sincere thanks to: Professor Peter Fowler for kindly letting me see a copy of the draft manuscript of his talk at the Conference; Dr. A.J. Herz for his advice and encouragement and Dr. Hugh Muirhead for elucidating for me some of the early history of methods of making multiple scattering measurements in emulsions.

I am indebted to the various authors and publishers for their permission to reproduce figures or plates viz: the University of Bristol for a number of photographs, Adam Hilger (for the Institute of Physics) for figures from Volume 13 of Reports on Progress in Physics (1950) and Messrs Taylor and Francis for a number of figures and plates which appeared in papers published in the Philosophical Magazine.

Lastly, I am most grateful to Mr. M. Audria of CERN's Publications and Exhibitions Group for the preparation of the figures and slides and above all to Miss Helga Schmal of the Director-General's office at CERN for her hard work and careful preparation of the final text.

REFERENCES

Barkas W H, Bishop A S, Bradner H and Smith F M 1950 Phys. Rev. 77 462
Berriman R W 1948 Nature 162 992
Bethe H A 1934 Proc. Camb. Phil. Soc. 30 524
Bjorklund R, Crandall W E, Moyer B J and York H F 1950 Phys. Rev. 77 213
Brown R H, Camerini U, Fowler P H, Muirhead H, Powell C F and Ritson D M 1949a Nature 163 47, 82
Brown R H, Camerini U, Fowler P H, Heitler H, King D T and Powell C F 1949b Phil. Mag. 40 862

Camerini U, Muirhead H, Powell C F and Ritson D M 1948 Nature 162 433
Camerini U, Coor T, Davies J H, Fowler P H, Lock W O, Muirhead H and Tobin N 1949 Phil. Mag. 40 1073
Camerini U, Fowler P H, Lock W O and Muirhead H 1950 Phil. Mag. 41 413
Camerini U, Davies J H, Fowler P H, Franzinetti C, Muirhead H, Lock W O, Perkins D H and Yekutieli G 1951 Phil. Mag. 42 1241

Davies J H, Lock W O and Muirhead H 1949 Phil. Mag. 40 1250

Fowler P H 1950 Phil. Mag. 41 169
Fowler P H, Menon M G K, Powell C F and Rochat O 1951 Phil. Mag. 42 1040

Goldschmidt-Clermont Y, King D T, Muirhead H and Ritson D M 1948 Proc. Phys. Soc. 61 183

Harding J B 1950 Phil. Mag. 41 405
Heitler W, Powell C F and Fertel G E F 1939 Nature 144 283
Hodgson P E 1951 Phil. Mag. 42 1060
Hooper J E 1987 Private communication

Leighton R B, Anderson C D and Seriff A J 1949 Phys. Rev. 75 1432
Leprince-Ringuet L and L'Heritier M 1944 C.R. Acad. Sci. Paris 219 618
Leprince-Ringuet L 1949 Rev. Mod. Phys. 21 42
Lock W O 1951 Ph.D. Thesis, University of Bristol
Lock W O and Yekutieli G 1952 Phil. Mag. 43 231

Muirhead H 1984 Topics on Cosmic Rays (Campinas, Brazil) 1 14
Muirhead H 1987 Private communication

Occhialini G P S and Powell C F 1947 Nature 159 186
O'Ceallaigh C 1951 Phil. Mag. 42 1032
O'Ceallaigh C 1982 J. de Physique 43 Suppl. No.12 185

Pais A 1986 Inward Bound (Oxford University Press) p 512
Panofsky W K H, Aamodt R L and York H F 1950 Phys. Rev. 78 825
Perkins D H 1947 Nature 159 126
Peyrou C 1982 J. de Physique 43 Suppl. No.12 7
Piccioni O 1950 Phys. Rev. 77 1, 6
Powell C F 1950 Rep. Prog. Phys. 13 350

Rochester G D and Butler C C 1947 Nature 160 855

Steinberger J 1949 Phys. Rev. 75 1136

Williams E J and Roberts G E 1940 Nature 145 102

Discussion

Fowler

This explosion: I believe it was the Festival of Britain, or something. Those with beards looked dreadful afterwards because their beards were singed, but the following day they were happy because they'd shaved, but everyone else was bright pink and very sore!

Marshak

Professor Lock didn't have time to talk about the heavy primaries but because Mrs. Powell was involved I'd just like to tell this little story, because the discovery relates back to this laboratory. The point is that in November 1947 I visited Bristol to talk over the pions and muons and so on and it became so interesting that I asked Mrs. Powell if she could give me a plate with some of the $\pi - \mu$ decays. I have a letter from November 14th, 1947 in which she sends the plate and says that it took her five hours to find 14 events. When I received the plate I said to Broughton Peters, who was preparing for the synchro-cyclotron, "Let's see how long it will take you to find the events." She also sent me a sheet with the x-y coordinates of the events on. They came back in two weeks and said they had found the correct $\pi - \mu$ decays and what should they do now? I said that Frank Oppenheimer was flying a small cloud-chamber at 94,000 feet and maybe he would be willing to accommodate some Ilford plates. Of course he was willing and this led in the Spring of '48 to their paper, which I have here, but there isn't time to project it. Let me mention the name of the discoverers of the heavy primaries, all due to the interest of Broughton Peters - Freier, Lofgren, Nye, Oppenheimer, Bradt and Peters. In the paper they have six photographs, one of an done with the cloud chamber and the other five of heavier primaries going up to Iron all done with the nuclear emulsion. I mention this to show how powerful the nuclear emulsion was in the heavy primary game in the Spring of 1948.

© 1988 IOP Publishing Ltd

The use of emulsions from the π meson to beauty

D H Davis

Department of Physics and Astronomy, University College London.

A description is given of the contributions of the emulsion technique to particle physics in the accelerator age. These include determinations of particle properties, hypernuclear physics and more recently charm and beauty searches with hybrid detectors in which emulsion stacks form the high spatial resolution vertex detectors.

INTRODUCTION

The invention of the bubble chamber shortly after the advent of accelerators capable of producing strange particles, together with the further sophistication of electronic techniques diminished the prominent roles of cloud chambers and emulsions in particle physics which they had enjoyed during the cosmic ray era. However, for emulsions at least, a few tasks remained which required their high spatial resolution and resulting precision of measurement.

This talk will concentrate on three areas where emulsions have been used to advantage, namely in determining some properties of particles; hypernuclei; and charm and beauty searches.

1. Particle Properties

The particle discovery days were over, at least until the coming of charm, with one exception, that of the anti-lambda hyperon (Prowse and Baldo-Ceolin, 1958). As a glance at the Particle Data Group tables shows (Aguilar-Benitez et al 1986), emulsion measurements of particle properties such as masses and lifetimes were well to the fore in the early sixties but these have now been largely superseded by determinations using other techniques. A comparison of some sample emulsion results with those presently accepted is shown in TABLE 1.

As can be seen the emulsion results are remarkably accurate but today rarely precise. Among the precise measurements, that of the kaon mass (Barkov et al 1979) used well calibrated emulsions to measure ranges, hence kinetic energies, of kaons formed at an e^+e^- collider from the decay of ϕ mesons whereas Bohm et al (1972, 1973) used range measurements of the Σ^+ hyperon from K^- meson captures at rest on hydrogen and the proton from at rest Σ^+ hyperon decays to determine simultaneously the Σ^+ hyperon mass and the density of the emulsions at the time of the exposure.

The apparent discrepancy in the mass of the Σ^- hyperon determined from the ranges of the Σ^- hyperon and the π^+ meson following K^- meson capture on hydrogen led to the observation by Barkas et al (1963) that negative

© 1988 IOP Publishing Ltd

TABLE 1

Quantity	Emulsion	P.D.G.(86)
$m_{\pi^\pm}/\mathrm{MeV}/c^2$	139.68 ± 0.15 (1)	139.5685 ± 0.0010
$m_\pi - m_\mu$	33.890 ± 0.076 (1)	33.9093 ± 0.0011
τ_{π^o}/s	$(1.0 \pm 0.5) \times 10^{-16}$ (2)	$(0.869 \pm 0.038) \times 10^{-16}$
$m_{K^\pm}/\mathrm{MeV}/c^2$	493.670 ± 0.029 (3)	493.667 ± 0.014
$\tau_{K^\pm}/s$	$(1.25 \pm 0.22) \times 10^{-8}$ (4)	$(1.2371 \pm 0.0026) \times 10^{-8}$
$\mu_\Lambda/\ eh/4\pi m_p c$	$-\ 0.66 \pm 0.07$ (5)	$-\ 0.6130 \pm 0.0044$
$m_{\Sigma^+}/\mathrm{MeV}/c^2$	1189.33 ± 0.04 (6)	1189.365 ± 0.057
$\mu_{\Sigma^+}/\ eh/4\pi m_p c$	3.3 ± 1.0 (7,8,9)	2.379 ± 0.020
$m_{\Sigma^-} - m_{\Sigma^+}$	7.91 ± 0.23 (6)	7.97 ± 0.07
$\Sigma^+ \rightarrow n\pi^+/\mathrm{all}$	0.484 ± 0.015 (10)	0.4836 ± 0.0030
τ_{Σ^-}/s	$(1.41 \pm 0.09) \times 10^{-10}$ (10)	$(1.482 \pm 0.011) \times 10^{-10}$
τ_{Σ^o}/s	$< 1 \times 10^{-14}$ (11)	$(0.58 \pm 0.13) \times 10^{-19}$

(1) Barkas et al (1956); (2) Stamer et al (1966); (3) Barkov et al (197); (4) Barkas et al (1961); (5) Dahl-Jensen et al (1971); (6) Bohn et al (1972, 1973); (7) Kotelchuck et al (1967); (8) Sullivan et al (1967); (9) Coombe et al (1968); (10) Tovee et al (1971); (11) Davis et al (1962).

particles lose energy in a medium at a lower rate than do positive particles of the same mass and velocity. The need of a further term in the Bethe-Block formula dependent upon the sign of the charge was confirmed by Heckman et al (1969) and Bohm et al (1972).

There exist some sparse results for K^- - proton processes at energies too low to be accessible to the hydrogen bubble chamber technique and in addition a study of K^- meson captures on hydrogen in emulsion produced a value of the ratio $\gamma = \Sigma^-\pi^+/\Sigma^+\pi^-$ of 2.34 ± 0.08 (Tovee et al, 1971) strongly in disagreement with that of a hydrogen bubble chamber study of Kim (1966), namely 2.06 ± 0.06. This conflict was resolved by Nowak et al (1978) who found $\gamma = 2.38 \pm 0.04$ using a track sensitive target hydrogen/neon bubble chamber.

Studies of K^- meson captures on complex nuclei in emulsion first showed a significant contribution ($\sim$ 20%) of multinucleon processes

$$K^- + N + N \rightarrow Y + N$$

often seeming to occur on α-like structures in light nuclei (Evans et al 1961) whereas a measurement of the ratio

$$\left({}^{\Sigma^-\pi^o} / {}_{\Sigma^\pm\pi^\mp} \right)_{\mathrm{Auger}} \Big/ \left({}^{\Sigma^-\pi^o} / {}_{\Sigma^\pm\pi^\mp} \right)_{\mathrm{recoil}} = 5 \pm 1$$

by Davis et al (1967) was cited by Burhop (1967) as evidence for a neutron-rich region in the extreme nuclear periphery of heavy nuclei.

2. Hypernuclei

The first hypernucleus was observed by Danysz and Pniewski (1953) and this was followed by a few more isolated examples from other cosmic ray experiments. However their systematic study began with the introduction of separated K^- beams, studies in which the ability of emulsions to separate closely spaced disintegrations was essential.

Among these are the determinations of the B_Λ values (Λ separation energies) for a variety of hypernuclei with A ≤ 15, a few of which are presented in TABLE II (Juriĉ et al 1974) (Cantwell et al 1974).

Hypernucleus	B_Λ/MeV	Hypernucleus	B_Λ/MeV
${}^{3}_{\Lambda}H$	0.13 ± 0.05	${}^{8}_{\Lambda}Li$	6.80 ± 0.03
${}^{4}_{\Lambda}H$	2.04 ± 0.04	${}^{9}_{\Lambda}Be$	6.71 ± 0.04
${}^{4}_{\Lambda}He$	2.39 ± 0.03	${}^{11}_{\Lambda}B$	10.24 ± 0.05
${}^{5}_{\Lambda}He$	3.12 ± 0.02	${}^{12}_{\Lambda}B$	11.37 ± 0.06
${}^{7}_{\Lambda}Li$	5.58 ± 0.03	${}^{13}_{\Lambda}C$	11.69 ± 0.12

It is to be noted that there are no Λp or Λn bound states, ${}^{3}_{\Lambda}H$ is only just bound and that there is a monotonic rise in B_Λ of ∿ 1 MeV/ nucleon with no wild fluctuations as is the case for extra nucleons, the Λ hyperon being allowed to descend to the 1s state. Whereas charge symmetry requires the Λp and Λn interactions to be equal a small but real difference in the B_Λ values of the mirror pair ${}^{4}_{\Lambda}He$, ${}^{4}_{\Lambda}H$ of 0.35 ± 0.05 MeV is observed. This charge symmetry breaking has been ascribed by Dalitz and von Hippel (1964) predominantly to electromagnetic mixing between nominally pure isopsin states such as (Λ, Σ^o) and (π^o, η). The B_Λ values do not continue to increase by 1 MeV/nucleon since the Λ-nucleus interaction saturates. This has been demonstrated by determining the B_Λ values of heavy spallation hypernuclei and from a sample of such formed by K^- meson capture on Ag, Br nuclei, Lemonne et al (1965) deduced a value for the Λ-nucleus potential well depth D_Λ of 27.2 ± 1.3 MeV, a value considerably smaller that the corresponding nucleon-nucleus one.

Spin values have been determined for ${}^{3}_{\Lambda}H(\frac{1}{2}^+)$, ${}^{4}_{\Lambda}H(0^+)$, ${}^{8}_{\Lambda}Li(1^-)$, ${}^{11}_{\Lambda}B(5/2^+)$ and ${}^{12}_{\Lambda}B(1^-)$ and in all of these cases $J_{HF} = J_{core} - \frac{1}{2}$. It is thus concluded that the Λ-N interaction is spin dependent, being stronger in the ${}^{1}S_{o}$ configuration than the ${}^{3}S_{1}$, opposite to the NN case. Figure 1 depicts the variation of the ratio $R_4 = {}^{4}_{\Lambda}H \to \pi^-\ {}^{4}He/{}^{4}_{\Lambda}H \to \pi^-$(all) with that of the p and s wave amplitudes, $p^2/(p^2 + s^2)$ in free Λ hyperon decay to $\pi^- p$, as calculated by Dalitz and Liu (1959) for the two spin possibilities 0 and 1 for ${}^{4}_{\Lambda}H$. Angular momentum conservation demands that $R_4 = 0$ when p = 1 and J = 0 and also when p = 0 and J = 1. Since $p^2/(p^2 + s^2)$ is known to be small, 0.126 ± 0.006, (Overseth and Roth, 1967) and R_4 large, 0.69 ± 0.02 (Bertrand et al 1970) it is concluded that $J({}^{4}_{\Lambda}H) = 0$. This, together with the observation of the K^- meson capture reaction

$$K^- + {}^{4}He \to \pi^o + {}^{4}_{\Lambda}H$$

in a helium bubble chamber determined the kaon parity to be odd. Since all particles involved are spinless it follows that $P(K^-) = P(\pi^0) = -1$.

Hypernuclei with $A \geqslant 5$ decay π^- - mesically dominantly to well defined states of the residual nucleus and the observed branching ratios and angular distributions have enabled the hypernuclear spins to be determined As an example, Figure 2 shows that the decay ${}^{8}_{\Lambda}Li \rightarrow \pi^- \; {}^{4}He \; {}^{4}He$ proceeds almost exclusively through well defined states of the intermediate nucleus ${}^{8}Be$ (Vilain et al, 1969).

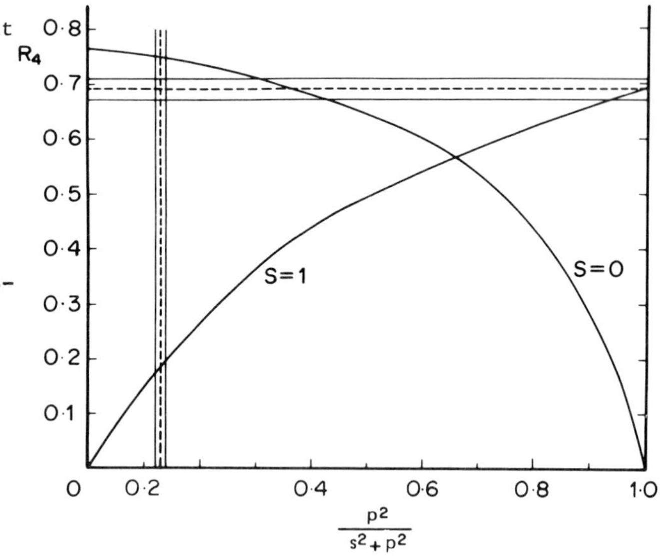

Figure 1.

Two examples of double hypernuclei, systems containing two bound hyperons, have been reported, both produced following the capture of Ξ^- hyperons in light emulsion nuclei. The first (Danysz et al, 1963) was identified as an example of ${}^{10}_{\Lambda\Lambda}Be$ with the decay sequence

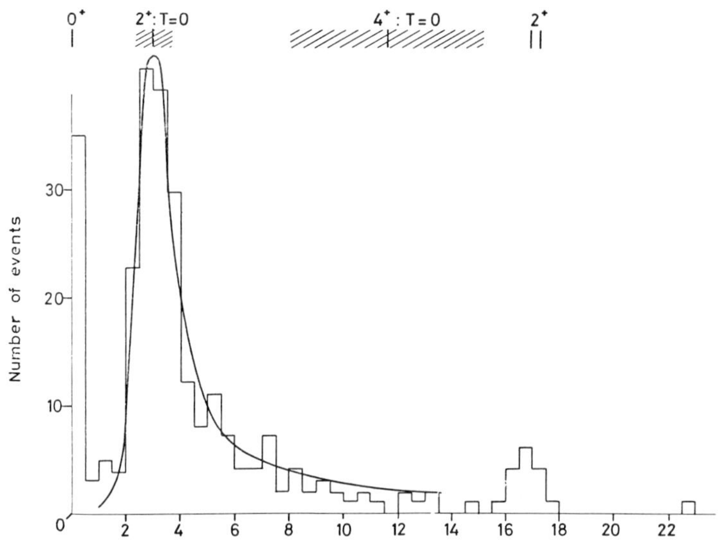

Figure 2

$${}^{10}_{\Lambda\Lambda}Be \rightarrow \pi^- + p + {}^{9}_{\Lambda}Be$$
$$\hookrightarrow \pi^- \; p \; {}^{4}He \; {}^{4}He$$

the second (Prowse, 1966) as ${}^{6}_{\Lambda\Lambda}He \rightarrow \pi^- \; p \; {}^{5}_{\Lambda}He$

$$\hookrightarrow \pi^- \; p \; {}^{4}He \; .$$

Both permitted measurements of the $\Lambda\Lambda$ separation energies, $B_{\Lambda\Lambda}$, to be made and it was concluded from these that the ΛΛ interaction in the $^1S_0$ state was attractive. However, with recent speculation on the existence of quark-gluon plasma states, the very existence of double hypernuclei takes on a new significance since it would seem to deny that of the H, a deeply bound six quark state, (uu dd ss), postulated by Jaffe (1977).

While most emulsion studies refer to hypernuclei in their ground states, it was shown, by adopting an iterative momentum-energy balance procedure (Bohm et al 1970), Jurič et al 1972) that the π^- meson energy spectrum resulting from the process

$$K^- + {}^{12}C \rightarrow \pi^- + p + {}^{11}_{\Lambda}B$$

exhibited a sharp peak near to the phase space limit which suggested the formation and subsequent decay of an intermediate excited state ${}^{12}_{\Lambda}C^*$. This was attributed by Kerman and Lipkin (1971) and others to a strangeness analogue state formed by the coherent substitutions of a Λ hyperon for a neutron. The coherence was postulated over all the neutrons in a nucleus, but because of the large difference between the Λ and n - nucleus potentials Hüfner et al (1974) pointed out that such substitutions would likely be coherent only over a given shell. Since the B_Λ value of the observed ${}^{12}_{\Lambda}C^* \sim 0$, it was suggested that the Λ hyperon was in the p-shell.

This discovery led to the (K^-, π^-) spectroscopy experiments at CERN and Brookhaven. These were performed using a variety of targets including ^{12}C and most were restricted to small momentum transfers to the struck nucleus such that $\theta_{K^-\pi^-}$ was near to 0° and hence to final hypernuclear states of spin equal to that of the target. Results for a ^{12}C target, which are typical of these counter experiments in that the energy resolution is $\sim 2 \div 3$ MeV FWHM, are shown in Figure 3 (Brückner et al 1978) for $\theta_{K\pi} \sim 0^o$ and Figure 4 (Chrien et al 1979) for $\theta_{K\pi} = 15^o$. Both distributions show prominent states of ${}^{12}_{\Lambda}C$ with $B_\Lambda \sim 0$ MeV but whereas in the $\theta_{K\pi} \sim 0^o$ data this would be expected to be a 0^+ state, only higher spin states would be excited strongly at $\theta_{K\pi} = 15^o$. The ground state of ${}^{12}_{\Lambda}C$ (1^-) with $B_\Lambda \sim$ 11 MeV is seen in both distributions but is far more prominent at the larger Kπ angle.

In order to resolve the question of whether $B_\Lambda({}^{12}_{\Lambda}C^*)$ was positive or negative a reanalysis of old emulsion results, including those of non-mesonically decaying hypernuclei (Montwill et al, 1974)

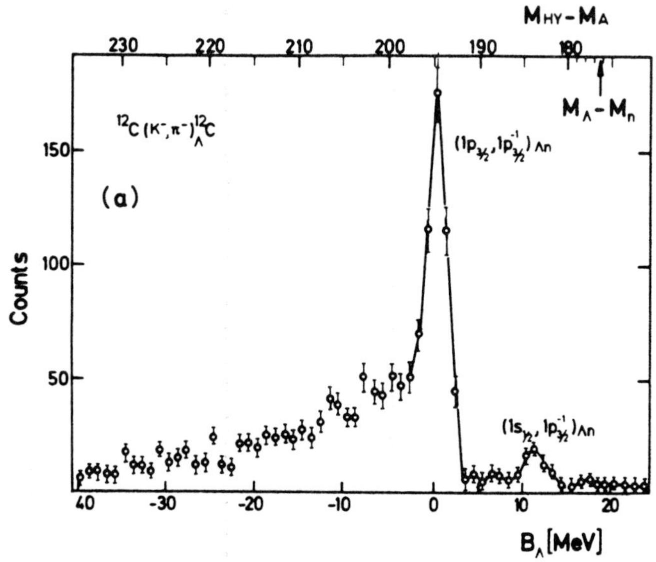

Figure 3. The ${}^{12}_{\Lambda}C^*$ spectrum obtained with p_{K^-} = 715 MeV/c and $\theta_{K\pi} \sim 0^o$.

was made using a consistent set of nuclear and particle masses (Dalitz et al, 1986). In Figure 5 the resulting T_π distributions are shown separately for π^- - mesic decays (known to be $^{11}_{\Lambda}B$) and non-mesic ones (assumed to be $^{11}_{\Lambda}B$). The coincident peaks and the consistent ratio of non mesic/mesic decays suggests that all events with T_π > 156.3 MeV are examples of $^{11}_{\Lambda}B$. While the statistics are poor, it is obvious that the data are not well represented by one, or even two. Breit-Wigner distributions, a good fit requires three.

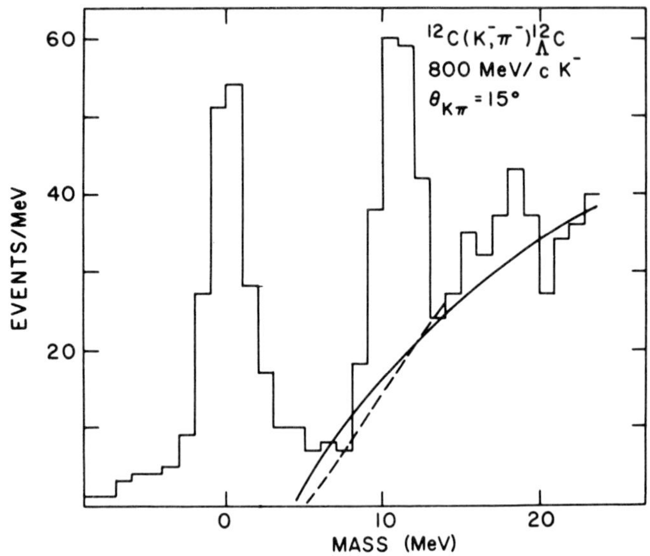

Figure 4. (The mass of $^{12}_{\Lambda}C$ ground state is zero on this scale).

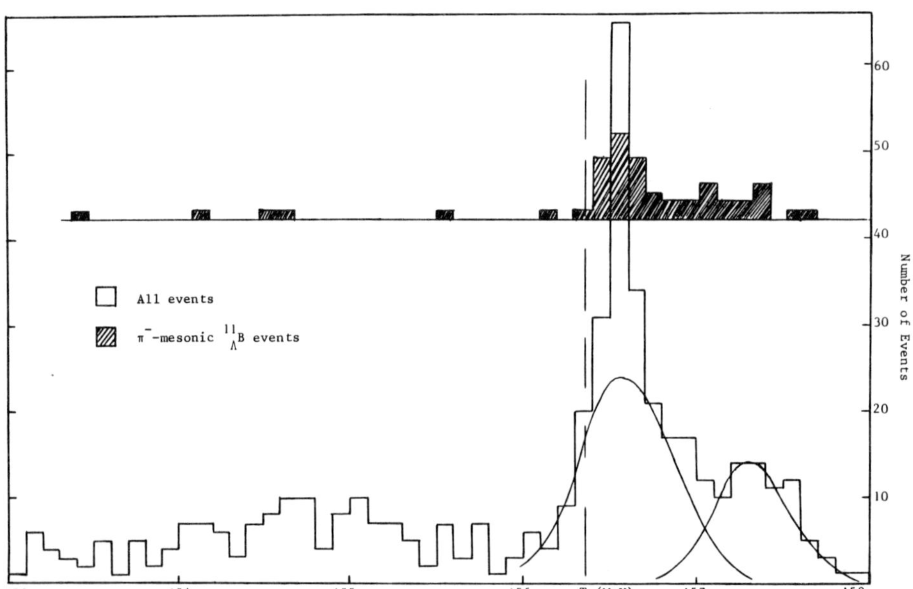

The T_π distribution for all events interpreted as $K^-+^{12}C \rightarrow \pi^-+p+^{11}_{\Lambda}B$ for K^- stopping in nuclear emulsion.

Figure 5.

The parameters of these are given in TABLE III

TABLE III

T_π/MeV	σ/MeV	Weight
156.50	0.054	55
156.52	0.222	181
157.27	0.219	67

Since other than 1s, it is the $1p_{3/2}$ sub-shell which is dominantly populated in ^{12}C the only Λ subsitution states in p-wave needed to be considered are $(1p_{3/2})_n^{-1}\ (1p_{3/2})_\Lambda$ and $(1p_{3/2})_n^{-1}\ (1p_{1/2})_\Lambda$. The first gives rise to $^{12}_{\Lambda}C^*$ states with $J^P = 0^+, 1^+, 2^+, 3^+$ and the second, $J^P = 1^+, 2^+$. The situation is simplified since the $K^-n \rightarrow \Lambda\ \pi^-$ capture is s wave with no spin-flip, so the $K^- \rightarrow \pi^-$ transition may only allow a transfer of orbital angular momentum L with parity change $(-1)^L$. Therefore as $J^P = 0^+$ for ^{12}C only the 0^+ and the two 2^+ states are expected to be produced. Now ^{10}B has $J_p = 3^+$ and the ground state of $^{11}_{\Lambda}B$, $5/2^+$. Thus to conserve both angular momentum and parity, the decay $^{12}_{\Lambda}C^*\ (0^+) \rightarrow p\ ^{11}_{\Lambda}B$ requires d wave proton emission but s wave for $^{12}_{\Lambda}C(2^+)$ decays. The Q values for these decays are low, being at most 1.4 MeV, so the expectation would be for d-wave proton emission to be highly suppressed.

The most natural interpretation of the observations is to ascribe the very narrow state to $^{12}_{\Lambda}C^*\ (0^+)$. [The calculated experimental error on T_π is $\sim$ 70 keV so the real width of this state may well be much smaller than observed] and the two wider states to the two expected 2^+ levels.

To conclude, the answer to the original question is that the Λ hyperon is bound in all three states, 0.14 ± 0.05 MeV in the 0^+ state and 0.17 and 0.92 MeV in the 2^+ states. The emulsion technique has resolved these states where counter experiments have failed. The widths, at least for the 2^+ states, have been measured for the first time and the small but real difference of 0.75 ± 0.04 MeV in the location of the 2^+ states may be ascribed to spin-orbit splitting but this effect is much smaller than in the NN case.

3. Charm and Beauty Searches

In the early seventies, the GIM mechanism to explain the non-observation of flavour-changing weak neutral currents, the discoveries of weak neutral currents and the J/ψ particle, the rise in the ratio R $(e^+e^- \rightarrow \text{hadrons}/\mu^+\mu^-)$, all suggested the existence of a fourth quark, charm, which, with a mass $\sim$ 1.5 GeV/c^2 would lead to charmed particle lifetimes of a few times 10^{-13}s, well within the compass of the emulsion technique. At the same time, dilepton events, often associated with strange particles were seen in high energy neutrino experiments in both bubble and spark chambers. These were attributed to charmed particle production at a rate of some 10% of high energy neutrino interactions. Such a favourable signal to noise ratio indicated the desirability of a neutrino experiment in emulsion. However, there remained the attendant problem that a worthwhile experiment would require a large emulsion volume ($\gtrsim$ 20 litres) whereas its exhaustive scanning would take a team of ten scanners about 100 years!

In 1965, well before the confident expectation of charmed particles, neutrino interactions were located in emulsion using an emulsion/optical spark chamber arrangement (Burhop et al, 1965). A similar experiment E247 (Burhop et al, 1976) was thus performed in the wide band neutrino beam at FNAL, in which interaction vertices were located in an emulsion stack within a search volume of $\sim 1\ cm^3$, defined by the point of convergence of tracks seen in an external wide gap spark chamber. 37 neutrino interactions were located and from one of them, a candidate for a charged charmed particle of lifetime a few times 10^{-13}s was seen to emerge. Thus encouraged, essentially the same team performed an emulsion/bubble chamber hybrid experiment in the wide band neutrino beam at CERN in which 30 litres of emulsion were exposed in front of the entrance window of BEBC. In this, measurements of bubble chamber tracks diverging from the entrance window determined, by backwards extrapolation, the coordinates of interaction vertices in emulsion, and for events of interest determined the signs of charge, momenta and sometimes identities of the particles producing them. In addition the external muon identifier enabled the charged current events to be selected. In all 169 neutrino interactions were located, and these gave rise to five charged and three neutral charmed particle decays, one of which was unambiguously identified as the charmed baryon $\Lambda_c^{\pm}$ decaying via the mode $pK^-\pi^+$ after a flight time of $(7.3 \pm 0.1) \times 10^{-13}$s (C Angelini et al, 1979). Figures 6a and 6b show a photomicrograph of this event and its accompanying bubble chamber photograph which depicts the elastic scattering of the proton in the liquid.

Three further hybrid emulsion neutrino experiments for charm were subsequently carried out at FNAL in the single horn focussed ν beam. The most successful of these, E531, was similar in concept to E247 but had a much increased analysis capability, including time-of-flight hodoscopes, magnetic spectrometer, lead glass counters and a muon identifier. They reported (Ushida et al, 1986 a and b).

$6\ D_s^{\pm}\ (F^{\pm})$	$\tau = (2.6\ {}^{+\ 1.6}_{-\ 0.9}) \times 10^{-13}$ s
$13\ \Lambda_c^+$	$= (2.0\ {}^{+\ 0.7}_{-\ 0.5}) \times 10^{-13}$ s
$28\ D^{\pm}$	$= (11.1\ {}^{+\ 4.4}_{-\ 2.9}) \times 10^{-13}$ s
$58\ D^o\ (\bar{D}^o)$	$= (4.3\ {}^{+\ 0.7}_{-\ 0.5}) \times 10^{-13}$ s

A different approach to the problem was made by the WA58 experiment at CERN in which frequently changed emulsion pellicles were coupled with the Omega Prime spectrometer and exposed in the tagged photon beam. Many examples of associated production of charmed particles were seen but the lifetime estimates tended to be lower than those quoted above (Adamovich et al 1984 a, b).

As happened with the strange particles, the results of these emulsion experiments have now been superseded by those of the silicon microsctrip, tagged photon spectrometer experiment of Anjos et al (1987) who quote $\tau_{D^+} = (10.6 \pm 0.5 \pm 0.3) \times 10^{-13}$ s, $\tau_{D^o} = (4.35 \pm 0.15 \pm 0.10) \times 10^{-13}$ s.

In the late seventies history repeated itself with the discovery of a further narrow resonant state, the Υ at 9.4 GeV/c^2 and a further rise in R at $E_{cm} \sim 10$ GeV, both indicating a further quark, b. For beauty particle

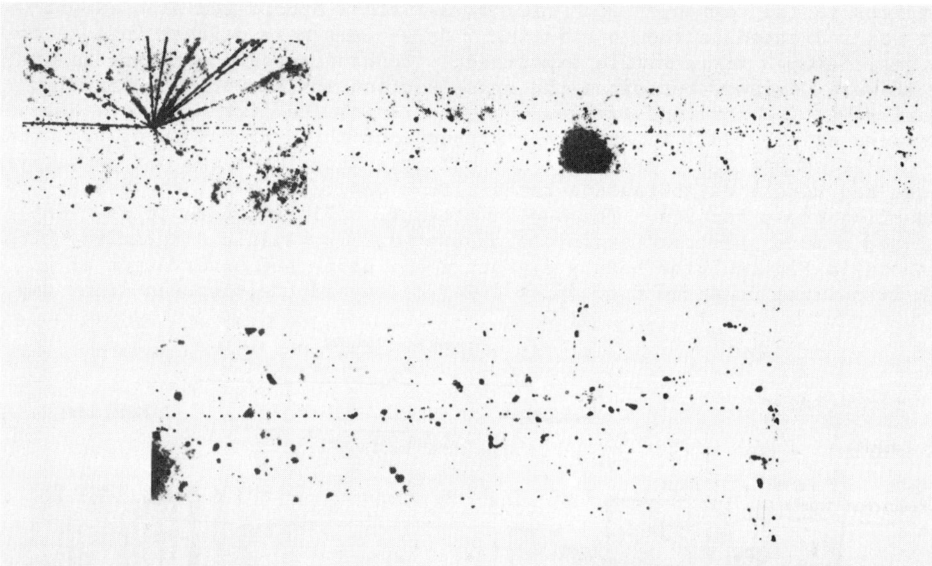

Figure 6a A photomicrograph of the decay $\Lambda_c^+ \rightarrow pK^-\pi^+$.

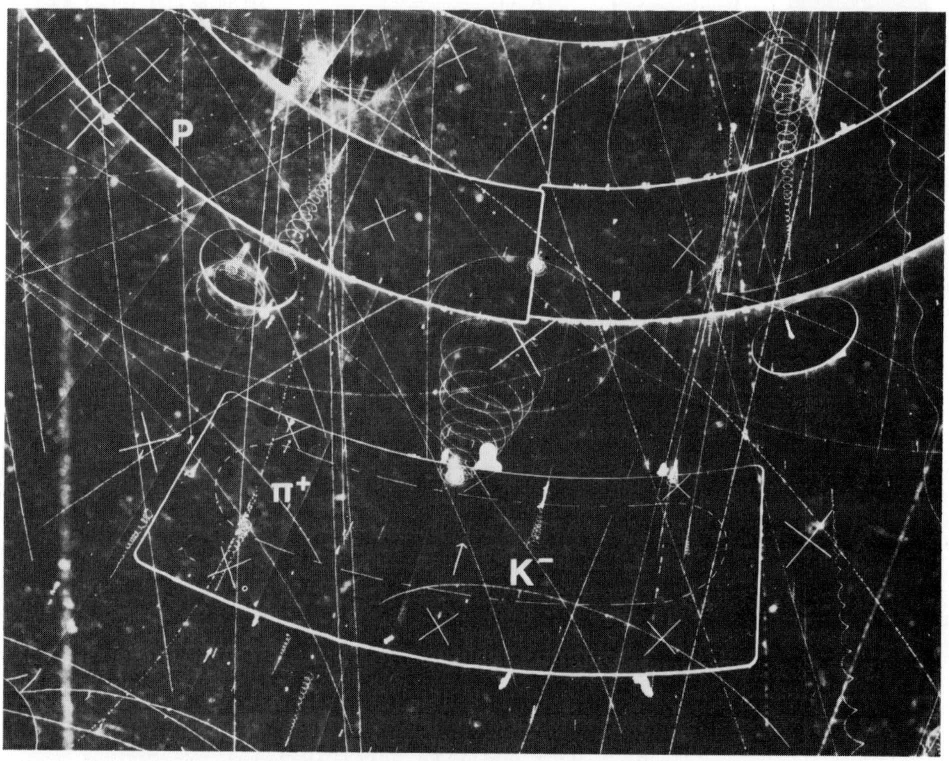

Figure 6b The associated bubble chamber photograph.

lifetimes in the region of 10^{-12} 10^{-14}s a further hybrid emulsion experiment was indicated, although existing ν or $\bar{\nu}$ beams were neither intense nor energetic enough for a viable experiment. So hadro-production experiments for $B\bar{B}$ were devised, but since the cross section was known to be very low a very selective external trigger was essential. The first attempt, NA19, (Albanese et al, 1983) required the presence of three prompt muons from the $B \rightarrow C \rightarrow S$ and $\bar{B} \rightarrow \bar{C} \rightarrow \bar{S}$ decay chains, but this experiment was not sensitive enough and merely established a rather large upper limit for the beauty production cross section. The second attempt, WA75 (Albanese et al, 1985) included a muon spectrometer in the apparatus, a schematic diagram of which is shown in Figure 7, the beauty signature now being the observation of a high transverse momentum muon (p_T > 1 GeV/c) beyond the tungsten-iron dump.

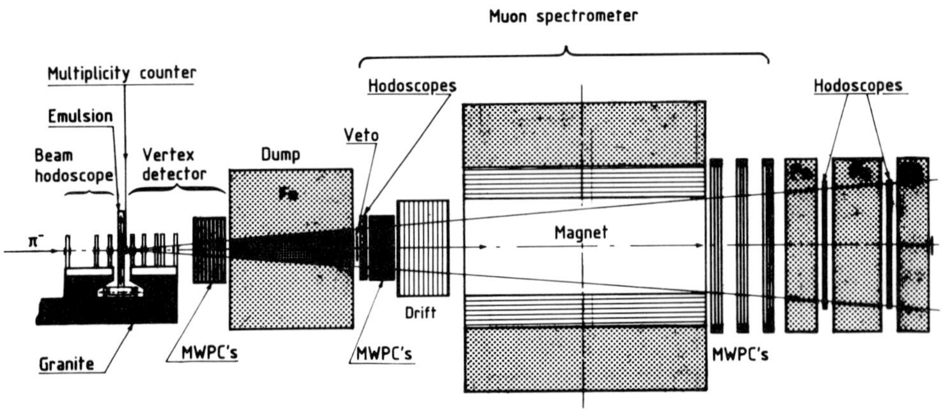

Figure 7. A schematic diagram of the apparatus of the beauty search experiment, WA75.

In all some 80 litres of Ilford and Fuji emulsion were exposed to 4×10^9 350 GeV/c π^- mesons producting 3×10^8 primary interactions in emulsion. The number of these to be considered was reduced successively to 1.5×10^6 on-line observing a muon beyond the dump and finally to 10^4 by the off-line requirement that its p_T > 1 GeV/c.

Over 200 examples of the associated production of charmed particles have been found, plus two double charmed events and one charm + J/ψ, and one, unfortunately still only one, beauty, anti-beauty event. In this two negative muons, one with high p_T (2.16 GeV/c), were observed in the muon spectrometer, 15 space tracks were reconstructed in the silicon microstrip vertex detector and 11 of these, including those of both muons were used to define the interaction volume. Figure 8 shows a schematic diagram of the event and Figure 9 successive photomicrographs of its various features. Since the direction of the tracks after the observed deflections matched those of the muons observed in the spectrometer the event was interpreted as the production of a $B^- \bar{B}^0$ pair with the observed decay sequences

$$B^- \rightarrow \mu^- + D^0 + \text{neutrals}$$
$$\quad\quad\quad\quad\quad\quad \hookrightarrow 4 \text{ charged} +$$
$$\bar{B}^0 \rightarrow D^- + \text{positive particle} + \ldots$$
$$\quad\quad \hookrightarrow \mu^- + \text{neutrals}$$

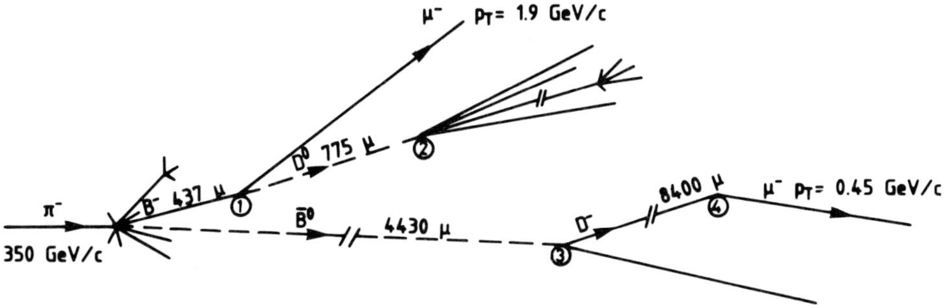

Figure 8. A schematic drawing of the beauty event.

The excellent matching of tracks from the decays of the neutral particles with those observed in the vertex detector established without doubt the overall correlation of the various features of the event observed in emulsion. This plus the presence of a high p_T muon rendered alternative interpretations of the event extremely improbable, at least $< 10^{-8}$.

Since, apart from the muons, the identities and energies of the particles are unknown, complete kinematic reconstruction of the decays is impossible. Nevertheless, from the observed angular distributions of the decay particles estimates of the energies, hence flight times, of the beauty particles have been made.

They are

$$t_{B^-} = (0.8 \pm 0.1) \times 10^{-13}\,\mathrm{s}$$

$$t_{\bar{B}^0} = (5 {}^{+2}_{-1}) \times 10^{-13}\,\mathrm{s}$$

which are similar to, although somewhat shorter than, those recently reported from e^+e^- collider experiments. The experiment is now nearing completion and the finding of one event corresponds to a beauty cross section of 1 or 2 nb.

CONCLUSIONS

Despite its often being labelled passé in the past thirty years, the emulsion technique has made a number of timely contributions to knowledge. That these, with the exception of some results in hypernuclear physics, have subsequently been bettered using more sophisticated techniques is only to have been expected. However, as recent charmed and beauty searches have demonstrated, when coupled with external detectors, emulsions have played, and may well continue to play, an important path-finding role in particle physics.

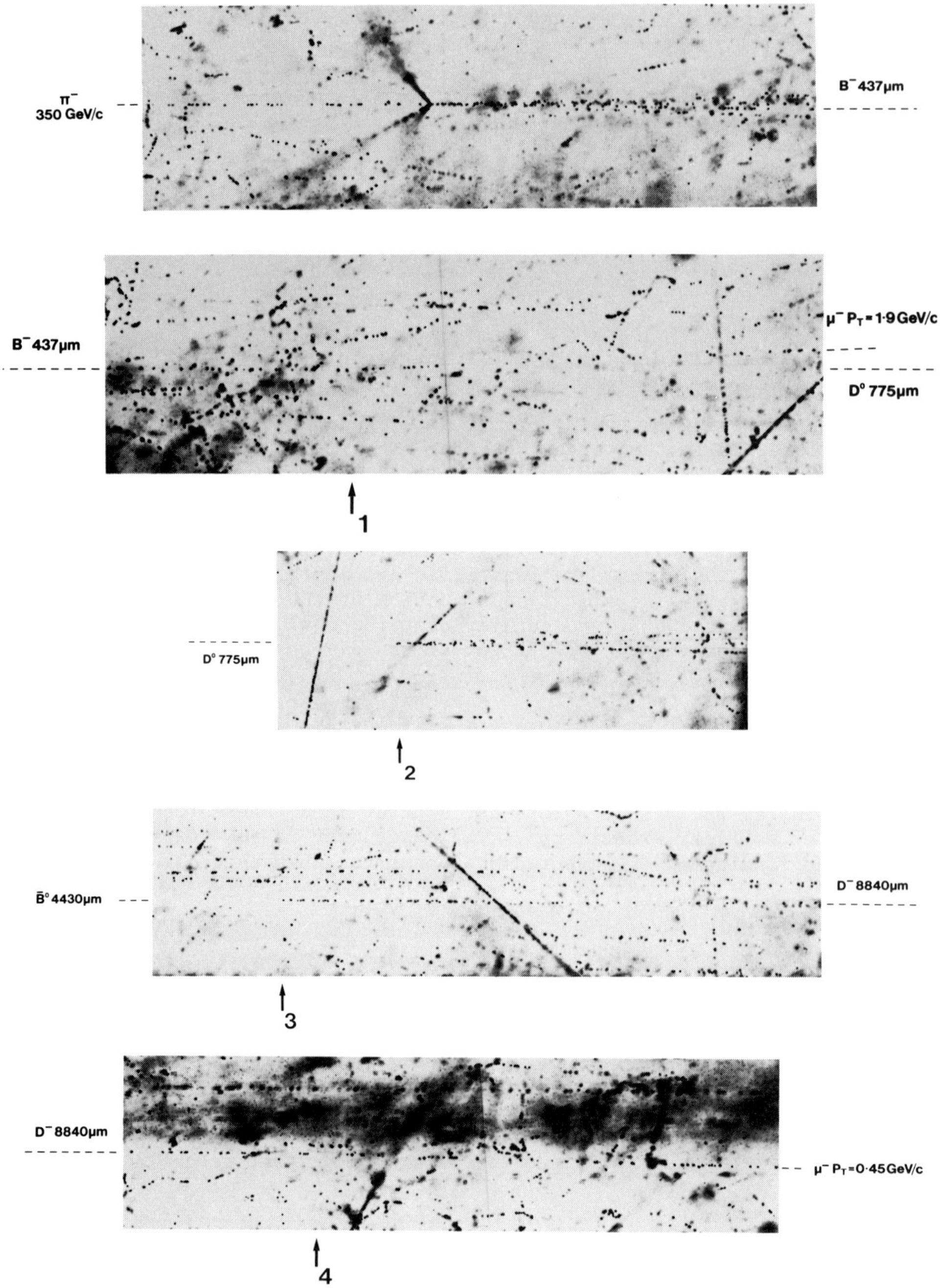

Fig. 9. A series of photomicrographs portraying the salient features of the beauty particle decays: 1) the B^- decay; 2) its associated D^0 decay; 3) the $\bar{B}^0$ decay; 4) its associated D^- decay.

Photography by D. Rooks.

References.

Adamovich M I 1984a Phys. Lett. 140B 119

Adamovich M I 1986b Phys. Lett. 140B 123

Aguilar-Benitez M et al 1986 Phys. Lett. 170B 1

Albanese J P et al 1983 Phys. Lett. 122B 197

Albanese J P et al 1985 Phys. Lett. 158B 186

Angelini C et al 1979 Phys. Lett. 84B 150

Anjos J C et al 1987 Phys. Rev. Lett. 58 311

Barkas W H, Birnbaum W and Smith F M 1956 Phys. Rev. 101 778

Barkas W H, Dyer J N and Heckman H H 1963 Phys. Rev. Lett. 11 26

Barkas W H et al 1961 Phys. Rev. 124 1209

Barkov L et al 1979 Nucl. Phys. B148 53

Bertrand D et al 1970 Nucl. Phys. B16 77

Bohm G et al 1970 Nucl. Phys. B24 248

Bohm G et al 1972 Nucl. Phys. B48 1

Bohm G et al 1973 Bull. Univ. Libre de Bruxelles 2

Brückner W et al 1978 Phys. Lett. 79B 157

Burhop E H S 1967 Nucl. Phys. B1 438

Burhop E H S et al 1965 Nuovo Cim. 39 1037

Burhop E H S et al 1976 Phys. Lett. 65B 299

Cantwell T et al 1974 Nucl. Phys. A236 445

Chrien R E et al 1979 Phys. Lett. 89B 31

Combe J et al 1968 Nuovo Cim. 57A 54

Dahl-Jensen E et al 1971 Nuovo Cim. 3A 1

Dalitz R H and Liu L 1959 Phys. Rev. 116 1312

Dalitz R H and von Hippel F 1964 Phys. Lett. 10 153

Dalitz R H Davis D H and Tovee D N 1986 Nucl. Phys. A450 311c

Danysz M and Pniewski J 1953 Phil. Mag. 44 348

Danysz M et al 1963 Phys. Rev. Lett. 11 29

Danysz M et al 1963 Phys. Rev. Lett. 49 121

Davis D H et al 1962 Phys. Rev. 127 605

Davis D H et al 1967 Nucl. Phys. B1 434

Evans D et al 1961 Proc. Roy. Soc. A262 73

Heckman H H and Lindstrom P J 1969 Phys. Rev. Lett. 22 871

Hüfner J, Lee S Y and Weidmuller H A 1974 Nucl. Phys. A234 429

Jaffe R L 1977 Phys. Rev. Lett. 38 195 and 617

Jurič M et al 1972 Nucl. Phys. B47 36

Jurič M et al 1973 Nucl. Phys. B47 1

Kerman A K and Lipkin H J 1971 Ann. Phys. 66 738

Kim J K 1966 Thesis (Columbia) unpublished.

Kotelchuck D et al 1967 Phys. Rev. Lett 18 1167

Lemonne J et al 1965 Phys. Lett. 18 354

Montwill A et al 1974 Nucl. Phys. A234 413

Nowak R et al 1978 Nucl. Phys. B139 61

Overseth O E and Roth R F 1967 Phys. Rev. Lett. 19 391

Prowse D J 1966 Phys. Rev. Lett. 17 782

Prowse D J and Baldo-Ceolin M 1958 Phys. Rev. Lett. 1 179

Stamer P et al 1966 Phys. Rev. 151 1108

Sullivan C R et al 1967 Phys. Rev. Lett. 18 1163

Tovee D N et al 1971 Nucl. Phys. B33 493

Ushida N et al 1986a Phys. Rev. Lett. 56 1767

Ushida N et al 1986b Phys. Rev. Lett. 56 1771

Vilain P et al 1969 Nucl. Phys. B13 451

Discussion

Pniewski

At the beginning I would say frankly that we were just lucky not to overlook the proper interpretation of the observed event. I think the reason was that at that time Marion Danysz and myself we were both still the nuclear and not yet the elementary particle physicists. The event we found represented an unstable fragment, as we would say unstable nuclear fragment, produced in high energy interactions. A rough estimate of the energy released in its decay was very high, but its lifetime, as estimated from its range in emulsion, had to be higher than a few picoseconds. We started a long exciting discussion. Nevertheless, being just nuclear physicists, we could not find any reasonable explanation of the event consistent with nuclear physics laws. We could exclude that it was an accidental overlapping of two independent events since the stack was no heavily irradiated and such a chance coincidence would be very improbable. We were continuing our discussion for several weeks and finally, just being in a coffee house, we noticed that the energy released in the decay of the fragment seemed to be similar to the energy corresponding to the rest mass of the pion. We decided that this coincidence could not be accidental and we started to look for the role the pion could play in this case. At first we were thinking about a pionic mesic atom. However, in a high energy collision one could not expect a very low energy pion to be captured on a Coulomb orbit of the emitted fragment. This interpretation, attractive as it was, we could not seriously consider as a reasonable one and the next idea which came to our mind was that of a bound V_1^o particle, i.e. bound with the nucleons of the fragment. We decided to send letters to some prominent physicists and among them Powell, Heisenberg and Skobeltzyn. Powell was not happy about our interpretation and advised us not to present this idea in our publication, but Heisenberg was rather encouraging us and Skobeltzyn answered after one year, sending us sincere congratulations when six events were already found. In fact, at that time, not knowing anything about strangeness, one might have an objection that V_1^o particles produced in a fast process of high energy interaction of two nucleons could not be bound with the nucleons.

© 1988 IOP Publishing Ltd

M DANYSZ & J. PNIEWSKI Phil. Mag. Ser. 7, Vol. 44, Pl. 13.

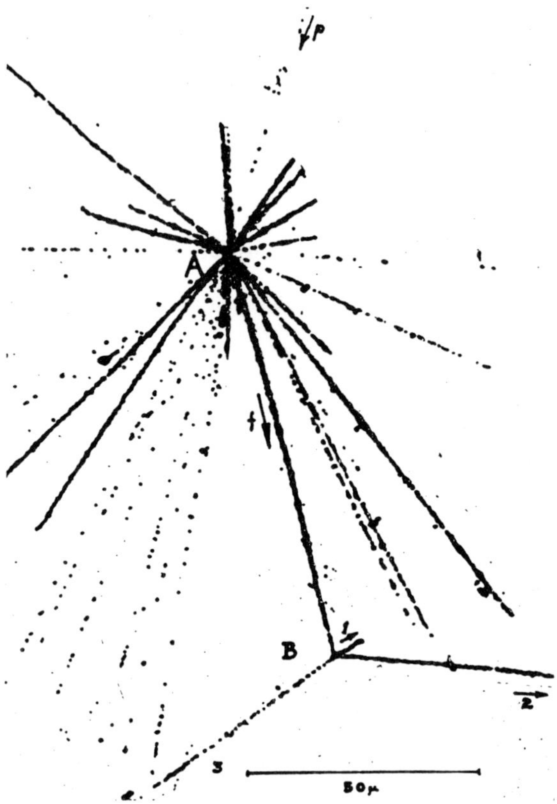

Figure 1. The first hypernucleus event

UNIVERSITY OF BRISTOL

C. F. POWELL, M.A., F.R.S.
Melville Wills Professor of Physics.
Tel. No: Bristol 24161 Ext. 110.
Night Service 25027.

H. H. WILLS PHYSICAL LABORATORY,
ROYAL FORT,
BRISTOL 8.

November 19th, 1952.

Dear Danysz,

Thank you very much for your letter of the 26th October. The event is certainly most striking, but I feel that you would be well advised not to publish it at this stage. Inspite of the most remarkable precision with which the heavy particle ends its range at the point of origin of the second star, you still have to meet ~~to~~ the objection that you are dealing with a chance juxtaposition of unrelated events. Because of this, I think it would be best , either to wait until a second example of the same phenomenon is found, or, to publish a photograph of it with a minimum of descriptive material. There seems to be no point, for example, in giving a detailed description of the big star from which the heavy particle was emitted.

With kind regards,

Yours sincerely,

C.F. Powell

Figure 2. Letter from C.F. Powell to M. Danysz, November, 1952

MAX PLANCK - INSTITUT FÜR PHYSIK
GÖTTINGEN BÖTTINGERSTRASSE 4

GÖTTINGEN, 10th November, 1952
Tel.: 3653

Dr. M. Danysz

Instytut Fizyki
Uniwersytetu Warszawskiego
Fizyka Doświadczalna

Warszawa / Polen

ul. Hoża 69

Dear Danysz,

Thank you very much for your letter of October 26th and the preprint of your paper which, I think, is very interesting. Prof. Heisenberg has also read it with great interest. He agrees that the event cannot be explained as the delayed disintegration of an excited nucleus since the time 10^{-11} sec is much too long. On the other hand, the probability for the event being a "delayed σ-star" is extremely small, too: The binding-energy of a π^--meson in the K-shell is of the order of 1 MeV whereas the average energies of the mesons ejected in the disintegration at A are apparently much greater. So it is unlikely that a π-meson would have been captured by the fragment. But even if it had been captured, it would have to be expected to interact with the nucleus within a time much shorter than 10^{-11} sec.

Your suggestion that the event might be explained in connection with the V_1^0 or a similar particle seems to be very reasonable, however. The V_c^1 particles appear to be different from the nuclear force mesons in that they may be created and annihilated only in conjunction with their anti-particles. In your event the V_1^0 may have been separated from its anti-particle, which flew off in a different direction, and was left within the fragment where it decayed after its life-time had elapsed.

I wonder what the future will teach us about all these funny particles.

With best wishes,
Yours sincerely,

K. Gottstein.

Figure 3. Letter from K. Gottstein to M. Danysz, November, 1952

Falla

You referred to work, Professor Davis, done by yourself and Professor Burhop at University College, in which you showed through observing K -capture that the nuclear surface was neutron rich. I should like to point out that I came to the same conclusion in my thesis here in Bristol eight years before that.

Round table

Chair: Sir Charles Frank. Members: R. Marshak, H. Fröhlich

Frank

Unfortunately Professor Lattes is feeling ill and has returned to his hotel. Perhaps we might begin with Professor Kemmer who has something to add to his remarks of this morning.

Kemmer

I suddenly remembered something I had included in my manuscript but not mentioned. I said there were the three papers, by Yukawa, Bhabha and our own which came out at the same time and were virtually identical. This is true, but many years later I had a letter from Pauli saying "if only you had remembered what your right hand was doing when you were using your left hand you might have written a different paper". I had published a separate paper about charge independence which was not taken into account in the three papers. If I had put a charge independent term in, there would have been a (τproton)(τproton) factor which for the deuteron ground-state is -3. If this factor had been in one might not have chosen the vector meson, and might even have chosen the pseudoscalar!

Fröhlich

I was never directly connected with this work but I know a few things about how the photographic plate method was introduced to Bristol which I'm sure no-one else knows. It started out with Walter Heitler being interested in Cosmic Rays and hearing about the two Viennese girls Blau and Wambacher working on the properties of photographic plates and finding stars. So Heitler thought that since they were at high altitude this must be of use in investigating cosmic rays and so he asked Cecil Powell whether he thought he could develop a systematic method. He said yes and did it. Now the strange thing is that there was absolutely no competition, especially from Rutherford's pupils. The reason is as follows: some years ago at some formal dinner the man beside me said: "Oh yes, you are from Bristol. I am connected with that investigation. I worked under Rutherford and my problem was to investigate whether the photographic plate method would be of any use. The result was that we decided not". This was I think in 1925, and this was the reason why no-one tried to follow it up. So Cecil Powell had it to himself until the technique became better known.

The second thing I should mention is the question of the π-μ decay. Occhialini was a specialist in investigating caves. After the war he was in Bristol and he had a friend in Brussels and they decided to investigate a cave below a very high mountain in the Pyrenees. When he left, Occhialini said "Well, I may as well take a few plates and leave them on

© 1988 IOP Publishing Ltd

top". When they came back and developed them there were these tracks! Two years ago I was visiting Milan and Occhialini came in and said "Ah, you were the first to tell us that we had found a new particle". They were not interested in new particles at the time.

Frank

Peter Fowler was signalling when you mentioned 1925.

Fowler

I believe it was 1935.

Fröhlich

No, certainly not as late as that.

Perkins

There was a paper by Taylor in 1935 on the emulsion technique.

Marshak

Since this is at Bristol and the forty years celebration of the pion, I'd like to read a couple of sentences from the Rochester Conference number 3, in December 1952.

"Dr. Oppenheimer next called on LePrince-Ringuet for some comments on K-particles, 'so lucid that all will know it'".

Le Prince got up and said - my French is terrible - I can only say that when I was Professor de Change in Paris in '53-54 I was trying to be nice and gave my lectures on Meson Physics in French. They said "Do it in English!"

LePrince said

"Je voudrais reprendre simplement quelques uns des passages du superbe exposé de Perkins. En Europe, il y a pour les émulsions, Bristol, le grand soleil, et puis un tout petit nombre de petits satellites dont la dimension même en faisant la somme, reste très inferieure a celle de Bristol".

I'd like to offer that into the record for your celebration.

Fowler

Would you like to hear about this thing attributed to Taylor in 1935? The reference says that the photographic emulsion technique is no good. Here is the quote.

Fröhlich

No, no, but I met the man who did the work!

Fowler

And it isn't Taylor?

Fröhlich

No, he's director of an electrical industry or something.

Friedlander

I was at Bristol. There may be one or two points on the emulsion technique which might be of interest. More should be said on the production of a decent range-energy relation. As we found faster and faster secondaries, which came to rest in large emulsions, the real problem was how to estimate the energies even when you know the masses. This illustrates how innocent Bristol was of theory to a large extent. No-one thought of taking the Bethe-Bloch relation and integrating it. Credit goes to the Rome group, Castagnoli, Baroni and others who in the spring and summer of 1954 produced the first really high energy range-energy relation. There was one point at 340 MeV which came out of Berkeley, stopping in copper; the rest was a Heath-Robinson kind of apparatus to try to put together a range-energy relation for long-range things. Most of the stuff talked about today were relatively low energy particles; the real problem was how to extrapolate to mesons which were really fast. The production of good values first for the K - π-3 and then the K - π-2 did hinge on that and credit should be given to the Rome group. I think at around the same time the CERN group produced another range-energy relation. Then one could begin to have some confidence.

I think another couple of points on the emulsion are worth commenting on. Dr. Waller was very modest; one of the major technical feats was to produce pellicles which could be re-attached to glass, which is no easy trick.There were a lot of magical recipies for how you re-attached pellicles onto glass of increasing size. One of the first attempts at this was in the Bristol lab., and since it was very valuable Cecil Powell insisted on handling the emulsion himself. When it came to plate 22 it was dropped and it had to be taken off the glass and stretched. We had a stack of pellicles and had to jump 22 because it was distorted.

There was also a major technical feat in the production of long working range objectives. The high magnification x100 objective typically had a very short working distance between the end of the lens and the thing in focus. The typical depth of field was of order 1/2 - 1 micron and Leitz produced objectives which could have a working distance of close to 300 microns or 120 microns if you included the glass cover-slide. This was a major feature to have this high resolution and high magnification. The important point is that one was often very concerned with accidental spatial coincidences. Professor Pniewski mentioned that he has a fascinating letter which Powell wrote to him when the first hyperfragment was found in which Powell cautioned that it perhaps should not be published because of the possibility that the track leading to the disintegration was actually a random coincidence, that they were unrelated events. One of the fascinating double-stars which travelled the labs of Europe came out of LePrince-Ringuet's lab; it was published in two or three places, of a double star connected to a highly ionising track which was claimed at one stage to be a K^- ejected from one star and going to the next. For many years this was thought to be the explanation. However, I think in late 1952 or 1953 Beppo Occhialini brought this particular square centimeter of valuable information to Bristol so that we could look at it. It was Beppo's idea to re-soak the emulsion and swell it up whereupon the track and the star separated and weren't connected. The published report says that they were; it looked that way but you needed great resolution in depth to be sure.

Finally, an experiment which also ought to be mentioned is the G-stack experiment which was a desperate attempt to try to get more statistics. At one stage Powell was almost like a travelling salesman going around

Europe inducing other groups to join in for more manpower. It ended up with 36 people on the paper, which was enough to drive some of us out of particle physics because we thought the number was too large! At a stage of desperation in the analysis afterwards I have a clear recollection of Mischco Merlin grumbling about the size of the collaboration and how unwieldy it had become with 36 authors and that if he had known about it beforehand he would have not called it the G-stack but the M-stack. You can use your imagination for what the M is.

Fowler

Italian or French?

Friedlander

French

Waller

Dr. Friedlander said that I hadn't mentioned pellicles of emulsion. As I explained, I had to limit my talk to precipitation in order to be able to say anything at all. However, we were proud at Ilford of what was done in coating and drying emulsion and producing the pellicles. Also the equipment was developed for other things than emulsion, filters for example. I would like to take this opportunity to say to you how much you all owe to a man named Vincent, who was a remarkable chap at handling any kind of material. I can illustrate the type of man he was by telling you that he reconstructed harpsichords and made a lute still used by one of the leading lutinists in England. Unfortunately he didn't live long after retirement but I am glad to have this opportunity of mentioning him to you.

Dalitz

(See following manuscript.)

Frank

I'd like to fill in the story of the muonic atom a little bit. I never really belonged to the fourth floor, I belonged to Mott's Solid State Group, but I used to wander up to the fourth floor to talk to Beppo Occhialini to discuss the "art of the film", because in those days I could remember particular sequences from Pudovkin films in detail and he could remember likewise from W.W. Griffiths' films, and at the same time I would pick up what was going on on the fourth floor. On one of these occasions Beppo said to me, when three but not four π-μ decays had been observed, "Blackett says, 'Are you sure that the second meson is really different from the first; is it not possible that the meson stops, collects energy from the environment and kicks off again?'" I said, "Well, what kind of energy could there be in the environment?" You know, there are some amounts of energy of the right order of magnitude here and there in sundry elements, trapped by a double-β process where if there wasn't a severe prohibition against double-β processes, there would be a significant amount of energy released. Suppose the negative meson could interact with the nucleus and turn into a positive meson, then it could jump two charge steps and would be equivalent to a double-β decay. How could we get the meson to interact with the nucleus? I should remark that Sulphur and Tin were two β steps away. This meson, once it gets near a proton will form a mesonic atom which will look like a neutron and screen away the electrons from around it and be able to wander freely through matter and snuggle up with another nucleus. I did a calculation in which at first I omitted a factor of 2π from the exponent thereby making it look much more promising than it really was. Fortunately I corrected that before I published it. Beppo made me publish it. So I published the letter and Peter was absolutely right not to pay too much attention to it. However, Sakharov read it and Sakharov said that this is a way not only of getting the meson close to another nucleus but also a way of getting the hydrogen nucleus close to another nucleus. Supposing those two hydrogen nuclei were not protons but deuterons then we could get interesting fusion results. He went on to say that what's more the meson sticking onto a proton if it met a deuteron would transfer as it has a slightly lower energy. He quasi-published it; S.E. Jones' article in Nature of about 18 months ago says it was published as an article by Sakharov and Lebedev. In fact it's an internal report of Sakharov to the Lebedev Institute. Then a couple of years later Sakharov and Zel'dovitch published and they proceeded to calculate the rates of various processes. Their outcome was that you would get about five fusions per meson. They didn't know, and no-one could, that the chemistry of that particular heavy hydrogen molecule which is one deuteron, one proton, one muon and one electron happens to have an accidental energy level which makes it go a couple of hundred times faster than Sakharov and Zel'dovitch thought. They had also overestimated the probability that the fast α would carry away the muon with it. Therefore it is a much more promising process. How promising it is I don't know because I haven't read the latest "Scientific American". But anyway, it greatly delights me, whether or not this produces a fusion process without vast pressure and fields etc., to have instigated Sakharov's first open scientific publication.

I have one other question, why do you say pion? I had one other interaction with the fourth floor group, which is in my memory a coffee table discussion, when I argued, and found myself strongly supported by Cecil Powell, that they must not use nomenclature which implied they understood what they were observing. They must above all if possible make it a purely objective nomenclature related to the features seen in the photographic emulsion. Powell agreed with this strongly. Lock told us earlier about Cecil not saying "electron" when he wasn't sure it was an electron. I have a very clear memory of Cecil Powell repeatedly saying "a neutral particle of zero or very small mass" rather than say neutrino. I remember recommending them to use something as objective as letters of the Greek alphabet. I am under the impression that μ and ν being already adopted for the meson and neutrino there followed π ρ σ as purely objective labels for something seen in the emulsion. My memory is that for a week or two the labels π ρ σ purely stood for features in the emulsion and then presently the letter π got transferred to mean the particle at about the same time that they announced theoretically that it really was a particular kind of meson. If anyone has a better explanation, could it be contributed?

Fröhlich

The name meson came up very late; we called it mesotron, but were attacked in "Nature" by Greek scholars saying that mesotron must be an instrument, and then we introduced meson. This was rather late.

Payne

I have a sense of guilt about this. Peter, do you remember your Table 2 in your talk in which you talked about the primary and secondary meson? I asked Occhialini what these two columns represented. He said, "On the left hand side is the primary meson; the one on the left is the new meson, it has just been born". So, having been brought up by a mathematically inclined headmaster I put at the top of the left-hand column a π because it was a Primary and for the secondary column I put a μ, because ν was already in use for the neutrino. I think that's how it started.

Amaldi

I wanted to go back to a remark by Professor Friedlander; it connects also to the information referred to by Dr. Lock. He referred to the balloon flight made here from Bristol. Immediately after that Cecil Powell started to talk with some of us in Italy to see if it was possible to organise some expedition in the Mediterranean area with balloon flights in order to expose the emulsions to Cosmic Rays in the upper atmosphere at lower latitude where the primaries of low energy are cut away so that the spectrum is harder and there is a higher probability to produce some of these new particles. There were three expeditions; the one mentioned by Friedlander was the third, the G-stack. In this expedition the rôle of Cecil Powell and all the Bristol Group was very important. Many Italian universities were involved because the first expedition was launched, I believe in '52, from Naples and was in June-July, lasting about one month. The balloons carried the package of plates; when they fell into the sea they had a parachute and a small radio calling. We had a boat from the Italian Navy and an aeroplane from the Italian Air Force to find the package. The first expedition was not very successful because we didn't really know very much especially about the winds at 30 km. The second expedition one year later was very successful; the base was Cagliari in Sardinia. About twenty balloons were launched and many recovered. There were nineteen universities involved, fifteen from Europe, and four from other parts of the world, one from Australia and several from the U.S. Then there was a further expedition, the G-stack, one year later, which was carried out from Novi Ligure. "G" meant gigantic! There were a lot of results obtained from that expedition. Some of the most important from the second were discussed at a conference in Padua, attended by Marshak. At that time the measurement of the masses of the various heavy mesons was sufficiently precise to see that they were all close to each other and the errors were rather small. At that time we thought of different mesons, rather than the same mesons with different decay modes, which came later, so different names were given. The fact that the masses were close together was important. This point was cleared after the G-stack since there was no glass and it was possible to follow tracks to the end of their range and therefore get good mass measurements. I believe that the results of these expeditions together with measurements made by machines in the United States were the results which really prepared the material that was used by Lee and Yang for saying that all the decay modes were actually from a single particle rather than different ones. The last two expeditions were done under the auspices of CERN because the only work going on at CERN in this period was constructing machines and it was very important to introduce a spirit of research. In agreement with Powell we said that it was under the auspices of CERN even though CERN did not provide any money.

Marshak

You are absolutely right about the Padua conference. There was a convergence of several decay modes with different names, but the accuracy was still not great so what I presented on behalf of Maurice Levy and myself was using Pais' multiplicative quantum number, that you might have a triplet P_0 state and a singlet S_0 state just slightly separated so parity wasn't violated. It took another year or more before the θ-τ dilemma hit you in the eye.

Amaldi

That was '54 and parity came in '58!

Fowler

From the G-stack in '54, we got several, like four to five, examples of K-π-2 where the π stopped and we had four or five even of the K-μ-2 where the μ stopped. We then did have a range-energy relation, I take issue with Mike Friedlander over some of his statements, and they did give the same mass. Now the fact that the particles can decay into two πs and three πs if you prove they have got the same mass, does produce a paradox, but if you have an uncertainty in the mass then you can get out of it.

Marshak

That was just the point. We already knew in '54 that they were within 10-20 MeV.

Fowler

Oh, we already knew much better than that. We had brought them to rest.

Friedlander

There were two Italian expeditions in '52 and '53, and at the Padua conference the results of the '53 Sardinian expedition were reported. Later that year in the early fall was the flight from Novi Ligure for the G-stack and these results came out with a really high quality measurement at Pisa in '55. At this stage the masses were pinned down well enough. There was a sequence of papers by Lee and Yang, exploring all sorts of possibilities to try to wriggle out of it. Parity non-conservation was the last desperate paper to find some explanation. Many of these things were triggered by the experimental results and Dick Dalitz's analyses of these things were driving it.

Waller

I thought that Peters in India was probably the first one to send up an emulsion stack?

Marshak

Well, that's another story!

Historical remark on $\pi^0 \rightarrow \gamma e^+ e^-$ decay

R H Dalitz

Department of Theoretical Physics, Oxford University

I was very glad to hear Peter Fowler draw attention to the work of Carlson, Hooper and King (1950) in his talk this afternoon. I have always thought this paper to be a most remarkable piece of work, a real classic of the nuclear emulsion technique, in an area where one might not have expected this technique to be so successful. The analysis they carried out did depend on one assumption, of course, that the neutral meson sought would decay dominantly to two photons, a photon being identified by its electron-positron pair conversion in the emulsion, not far from its parent star. No attempt was made to fit the data to alternative hypotheses, such as neutral meson decay to three or more photons, possibilities which they might have been able to exclude. But no matter, they showed that the $\pi^0 \rightarrow \gamma\gamma$ decay mode did fit the data remarkably well, and their analysis led them to good estimates for the π^0 mass value, for the total rate of π^0 production in cosmic ray interactions in emulsion and for the resulting π^0 energy spectrum, the last two items being consistent with one-half of the net charged pion production. They also obtained by far the strongest upper limit available at that time for $\tau(\pi^0)$, the lifetime of the neutral pion. All of this information was obtained from the one emulsion exposure to the cosmic radiation at high altitude!

Unfortunately for Carlson, Hooper and King, the Berkeley groups had just come into action with their proton cyclotron and their electron synchrotron. Bjorklund et al. (1950) had measured single γ-ray spectra from proton-nucleus collisions at the end of 1949, and by the end of April 1950 Steinberger et al. (1950) were able to measure γ-γ coincidences following photo-excitation of nuclei by the γ-ray beam from their electron synchrotron, which they attributed to neutral meson production and decay, their results being published in June 1950. After a preliminary Letter (Panofsky et al. 1950a) in the same issue, Panofsky et al. (1950b) were soon able, by August 1950, to establish the decay process $\pi^0 \rightarrow \gamma\gamma$ very directly and quite unambiguously, by the study of the decay of π^0 mesons of rather low momentum resulting from the at-rest capture reaction $\pi^- p \rightarrow \pi^0 n$, and so to determine the π^0 mass with quite a high accuracy. The Berkeley groups got in first, and the fine paper of Carlson, Hooper and King published in July 1950 was scooped, and did not receive the attention it deserved. If it had been only several months earlier, the story would have been quite different.

However, what I want to mention here is a different outcome from this work. Early in 1951, when I was holding a research fellowship at Birmingham University, I returned to Bristol one week-end to visit old

© 1988 IOP Publishing Ltd

friends on the fourth floor of this laboratory. Hooper and King were then working particularly on tridents, electron pair production by high energy electrons or positrons which were themselves the result of electron pair conversion following cosmic ray stars in the emulsion. They showed me some of the emulsion events they were working on, and I admired the characteristic tracks for the electron and the positron, and their measurement of the electron and positron momenta. Then they said "Here are two peculiar pairs". In each of these cases, the outgoing pair of tracks were cleanly identified as electronic. What was peculiar about them was that the origin of each pair could not be seen as separate from the centre of the cosmic ray star from which it had emerged. What could give rise to these two pairs?

On my way back to Birmingham, I suddenly realized that the π^0 itself could give rise to them by a direct decay to (γe^+e^-), through the process of <u>internal</u> electron pair conversion of one of its product photons,

$$\pi^0 \rightarrow \gamma + \text{"}\gamma\text{"} \rightarrow \gamma + e^+e^- \qquad \text{(i)}$$

In those days, I think that most people were accustomed to associate electron-positron conversion of a photon with the presence of a strong external electric field, such as that due to a nucleus of charged Ze for large Z. Certainly, that was true for external pair conversion, as given by the Bethe-Heitler formula, but the formulae given in the books for internal pair conversion also depended strongly on the atomic number Z. The observations made of internal pair creation associated with transitions between nuclear levels were generally made for transition energies not greatly above 1 MeV and in heavy nuclei, and its branching ratio did depend strongly on the value of Z. A few cases of internal pair conversion were then well known for light nuclei, where the branching ratio $(e^+e^-)/(\gamma)$ was most significant for high transition energies, of order 5 to 10 MeV. These processes had been well studied theoretically but were well known only to specialists.

It happened that I had written my doctoral dissertation for Cambridge University the previous summer, on the topic "Zero-zero transitions in nuclei". It was concerned primarily with the case of ^{16}O, which has an excited state with spin-parity (0+) at energy 6.05 MeV. This state was known to decay to the ground state of ^{16}O dominantly by electron pair emission,

$$^{16}O(0+, 6.05\ \text{MeV})^* \rightarrow {}^{16}O(0+, \text{g.s.})e^+e^-, \qquad \text{(ii)}$$

since single photon emission is strictly forbidden for a zero-zero transition. My interest in this particular process had been aroused by the measurements which had been made on it by Sam Devons (1949) at the Cavendish Laboratory, Cambridge. Its zero-zero character gives it some special features which are not relevant for the present discussion. What mattered was that this is a case where Z is small and I knew from calculation that its internal pair conversion rate depended very little on Z, being little changed if Z were reduced to zero in the calculation, and certainly being non-zero for Z=0 (Dalitz 1951a).

So I calculated the rate for the internal pair conversion of one of the γ-rays in $\pi^0 \rightarrow \gamma\gamma$ decay for a free π^0 meson (Dalitz 1951b). This did <u>not</u> require detailed knowledge of the structure of the π^0 meson, because the transition from a high-energy photon (energy $k = m(\pi^0)/2 \approx 70$ MeV, here) to an electron-positron pair occurs dominantly with a four-momentum transfer of order only $2m_e$, a quantity negligible relative to $m(\pi^0)$. My result was that about 1.185% of π^0 mesons should undergo the γe^+e^- decay mode (Dalitz 1951b), increased to 1.195% by radiative corrections (Joseph 1960); the branching ratio is known empirically today to be (1.198±0.030)%. This branching ratio was one of the most reliable predictions a theoretician could make, since the dominant part of its rate involves the $\pi^0 \to \gamma\gamma$ matrix-element for one of the photons so slightly "off-mass shell". The branching ratio $(\gamma e^+e^-)/(\gamma\gamma)$ therefore depended very little on the details of the $\pi^0 \to \gamma\gamma$ decay amplitude.

I then wrote to David King to let the group know about this additional source of electron-positron pairs. How close to the star centre these pairs originated would depend on the magnitude of $\tau(\pi^0)$, the π^0 lifetime, and on the momentum distribution of the π^0 mesons from the star, which they had just determined. I pointed out that the study of these pairs might well allow them to make a determination of the π^0 lifetime in due course, which would be most interesting. I asked them to write a Letter for publication, describing their two events, to accompany my Letter describing my calculation, its result and its possible use for determining the π^0 lifetime. The answer which came back was surprising: "No, we don't want to do that. We have only two of those events. Your formula predicts many more than that, and we don't see them. We don't know whether or not these two events are due to your decay process." So I sent off my Letter alone, for publication in the Proceedings of the Physical Society. I didn't know either, whether or not their two events were examples of $\pi^0 \to \gamma e^+e^-$ decay. All I knew was that this decay process had to happen and that the rate predicted was a thoroughly reliable estimate. I could only conclude that they were missing a substantial fraction of these events, for some reason I could not specify. [I know now from Powell et al. (1959) that David King (1949) had suggested that such a loss of electron-positron pairs with a small angle of separation might occur because of coherence between the effects of the electron and the positron on the emulsion, the ionization being suppressed because the net charge of the pair is zero - even their electric dipole moment is then small - but I did not hear this at the time. King's suggestion was subsequently found to be correct, both by experiment (Perkins 1955) and by theoretical calculation (e.g. Burkhardt 1958).]

I wish that Carlson, Hooper and King had seen fit to publish a short Letter at that time, describing their events without prejudice, but I can understand their desire not to appear to claim the first observation of this decay process when they knew that they could not demonstrate that their events were examples of it. In those days, experimenters had rather high standards for their work; you don't see many experimenters around today with such high standards for publication! But Carlson, Hooper and King always had the highest standards possible for the work they did publish, and I respect them for that.

A year or two later, work at Bristol by Daniel et al. (1952) showed very clear evidence for the presence of the $\pi^0 \to \gamma e^+e^-$ decay mode, from more extensive emulsion studies of electron-positron pairs from cosmic ray interactions in the emulsion, which also gave a stronger limit on the π^0 lifetime, $\tau(\pi^0) < 10^{-14}$ sec. Much later, studies were made by Harris, Orear and Taylor (1959) of π^0 decay following the $K^+ \to \pi^+\pi^0$ decay mode for K^+ mesons coming to rest in emulsion. Since these π^0 mesons have a unique momentum at production, and much higher statistics were available with emulsions exposed to a stopping K^+ beam from the Bevatron, it was possible to strengthen this limit to $\tau(\pi^0) < 5 \times 10^{-16}$ sec. Of course, we know today that the π^0 lifetime is $(0.87 \pm 0.04) \times 10^{-16}$ sec, considerably shorter than this limit.

REFERENCES

Bjorklund R, Crandall W E, Moyer B J and York H F 1950 Phys. Rev. 77 213
Brisbout F A, Dahanayake C, Engler A, Fujimoto Y and Perkins D H 1956 Phil. Mag. 1 605
Burkhardt H G 1958 Nouvo Cim. 9 375
Carlson A G, Hooper J E and King D T 1950 Phil. Mag. 41 701
Dalitz R H 1951a Proc. R. Soc. A 206 521
—— 1951b Proc. Phys. Soc. (London) 64 667
Daniel R R, Davies J H, Mulvey J H and Perkins D H 1952 Phil. Mag. 43 753
Devons S and Lindsey G R, 1949 Nature 164 539
Devons S, Hereward H.G. and Lindsey G R 1949 Nature 164 586
Harris G, Orear J and Taylor S 1957 Phys. Rev. 106 327
Joseph D 1960 Nuovo Cim. 16 997
King D T 1949 Quoted on p.193 of Powell et al. (1959)
Panofsky W K H, Aamodt L, and York H 1950a Phys. Rev. 78 825
Panofsky W K H, Aamodt L, Hadley J and Phillips R 1950b Phys. Rev. 80 94
Perkins D H 1955 Phil. Mag. 46 1146
Powell C F, Fowler P H and Perkins D H 1959 The Study of Elementary Particles by the Photographic Method (Pergamon Press, London) p.192
Steinberger J, Panofsky W K H and Steller J 1950 Phys. Rev. 78 802

The beginning of particle physics: from cosmic rays to CERN accelerators

Edoardo Amaldi
Department of Physics, University of Rome "La Sapienza", Rome, Italy.

ABSTRACT: The author summarizes the most relevant events that brought to the creation of CERN as a permanent Organization the October 7, 1954. The first initiatives started to appear after the middle of the 1940's and lead, the February 15, 1952, to the creation of a provisional Organization which accomplished its task in the successive two and a half years. At the end the author reminds the glorious tradition of Great Britain in particle physics and concludes with an appeal for avoiding that in the near future it may be voluntarily interrupted.

1. INTRODUCTION

The Second World War was ended, also in the Pacific, before the end of August 1945 and already on 24 October of the same year the United Nations Organization (UN) was born during a conference held in San Francisco. The decision to create the United Nations Educational, Scientific and Cultural Organization (UNESCO) was taken in London in November 1945, by the delegates of forty-five nations. The constitution of this worl-wide Organization took place in Paris in 1946. Its main inspirers and organizers originated from European countries and Europe was the geographic region where these ideas, shortly afterwards, found the most fertile ground for the generation and development of regional organizations not closed in itself but open towards the rest of the world.

In this climate a number of mission oriented organizations were set up to tackle problems belonging to specific scientific or technological areas by an effort made in common by many European countries.

The European Organization for Nuclear Research (CERN) was the first European organization of this type to be created. The first official step took place at the General Assembly of UNESCO in Florence in 1950, a provisional Organization came into being in 1952, and the final permanent Organization - still in activity and evolution today - started to operate in 1954.

The history of CERN is of considerable interest not only from the point of view of high-energy physics, but also in general as the first example of a research laboratory of intergovernamental nature, created in Europe, that has been operated successfully for slightly more than thirty years, thus providing a remarkable model for the creation of other international organizations.
North-Holland has recently published the first volume of the "History of CERN" prepared by Armin Hermann, professor of History of Science at the University of Stuttgard, helped by a few younger historians. This 600 pages volume covers the years from 1949 to 1954. A second volume is scheduled for publication at the end of 1988. It will cover the period 1955-1965.

From the extensive presentation contained in the already published volume, John Krige, one of the members of Hermann's group, keeping in contact with his colleagues, will extract a more concise CERN history, contained in a single volume of 300 pages and addressed to a wider public.

© 1988 IOP Publishing Ltd

My speech, of course, will be of a completely different nature. I did not consult the archives of the Foreign Ministries or of the Research Councils of the Members States of CERN or of other intergovernamental organizations. What I say is based on a few well known documents, on my personal diary of that period and on a few reports and lectures, prepared years ago by Kowarsky or by me.

Everybody agrees that the early history of CERN can be divided into three periods. The first one includes the first initiatives and extends from the middle of the 1940s to 15 February, 1952. On this date the representatives of eleven European governments signed in Geneva the so-called Agreement, establishing a provisional organization with the aim of planning an International Laboratory and organizing other forms of co-operation in nuclear research.

The second period, the so-called planning stage, extends from February 1952 to 1 July, 1953 when the Convention establishing the final organization was signed by the representatives of twelve European Governments.

Finally, the third period, usually called interim stage, runs from July 1953 to 29 September, 1954, when a prescribed point of the ratification procedure by the Parliaments of the Member States was reached and the Convention entered into force.

I limit my presentation to these three early phases because of their peculiar nature and also because I was directly involved in them.

Niels Bohr (left) and G.P.Thomson in Geneva, 12-15 February 1952.

Pierre Auger in Geneva, 12-15 February 1952.

2. THE FIRST INITIATIVES

It is not easy to fix a date for the beginning of the first stage, the stage of the first initiatives.

I may start by mentioning that already in October 1946 a proposal was submitted by the French Delegation to the "UN Economic and Social Council" of creating "United Nations Research Laboratories" and that the "UN Atomic Energy Commission" created also in 1946 with the aim of bringing atomic weapons under some form of international control, was one typical circle were the problem of international collaboration, especially in nuclear physics, was raised and discussed.

These, as well as another initiative appeared in 1949 and involving the Netherlands, Belgium and the Scandinavian countries, refer mainly to low energy physics and its applications and, therefore, should be considered, in my opinion, more as precursors of the idea of Euratom than of CERN.

In those years in many European countries, in particular in France, Italy, West Germany and Belgium, the idea of moving towards some form of economic and/or political unification of at least a considerable part of the old continent, was considered of primary importance by many authoritative politicians who adopted it as a guiding principle of their immediate and long range program of action.

Another favourable element was the scientific, technical and administrative experience gained in a few countries, during and immediately after the war, of wide and complex organizations operating in the field of the nuclear sciences and their applications. This experience had brought about the creation in United States of a few big research laboratories such as the Argonne National Laboratory and the Brookhaven National Laboratory. In particular the Brookhaven National Laboratory had been created and run - very successfully, - by the Associated Universities Inc. Furthermore the dimensions of the geographic region involved and of the laboratory were very similar to those of a possible future European research establishment. For Europe, however, a program of research had to be chosen completely free from any limitation or restriction originating from military, political or even industrial secrecy.

At the same time in many European and overseas places, scientists were becoming aware of the continuously increasing gap between the means available in Europe for research in the field of nuclear physics and elementary particles and the means available in the United States, where a few high-energy accelerators had started to produce results, while others had already reached advanced stages of construction or design. It was becoming more and more evident that such a situation would be changed only by a considerable effort made in common by many European nations.

To these one should add a further important element: immediately after the Second World War cosmic ray research had a very high level in Western Europe and was in part carried out through successful international collaborations.

The discovery, in 1946, by Conversi, Pancini and Piccioni that the cosmic ray mesotron is a weak interacting particle and not a Yukawa particle, the discoveries in 1947 by Lattes, Muirhead, Occhialini and Powell of the π-meson and of its decay into the muon, and of the first two strange particle by Rochester and Butler (a Λ° and a θ°) are the most brilliant results of an extensive and rich production obtained in Western Europe.

As I have pointed out in a lecture I gave at the Varenna School in summer 1972, the mountain laboratories in Switzerland (Jungfraujoch), France (Pic du Mid) and Italy (Testa Grigia) and even more the nuclear emulsions laboratories of the Universities of Bristol and Bruxelles, led respectively by Powell and by Occhialini, had become in those years points of encounter of young physicists originating from many different countries.

The life in common in mountain huts and the co-ordination of the experiments planned by different groups paved the way to the idea of wider and more ambitious collaborations which were popping out in various places in Europe.

Cosmic ray physicists, of course, were interested in a European laboratory only if devoted to high-energy physics. I remember that in the years 1948-1950 the various aspects of the problem, including energy and cost of machines, were examined in Rome in frequent discussions between Bruno Ferretti and myself and in letters exchanged with Gilberto Bernardini, who in those years was at Columbia University where he had been invited by I.I.Rabi.

I remember that I became aware that similar problems were discussed in other European countries, in particular in France, when I heard of the European Cultural Conference held in Lausanne in December 1949. At the meeting a message from Louis de Broglie was read by Raoul Dautry, Administrator of the French Commissariat à l'Energie Atomique. In the message the proposal was made to create in Europe an international research institution without mentioning, however, nuclear physics or fundamental particles. To the message of de Broglie Dautry added the proposal that the Conference study the ways of strengthening collaboration in two fields, in astronomy and astrophysics by building powerful telescopes and all necessary auxiliary equipment, and in the field of atomic energy, by setting up a center with all the required modern apparatus.

As we see from these examples, at that time the opinions were still rather confused about the type of research to be tackled by the new organization and even more about the nature of the collaboration to be established.

In June 1950 the General Assembly of UNESCO was held in Florence and Isidor I. Rabi, who was a member of the delegation from the United States, made a very important speech about "the urgency of creating regional centres and laboratories in order to increase and make more fruitful the international collaboration of scientists in fields where the effort of anyone country in the region was insufficient for the task". In the official statement approved unanimously by the General Assembly along the same lines, neither Europe nor high-energy physics were mentioned. But this specific case was clearly intended by many people, in particular by Rabi himself and by Pierre Auger, who was Director of the Department of Natural Sciences of UNESCO.

I knew Rabi very well since 1936, when I had visited Columbia University for the first time. Since then I had revisited him in his laboratory a few times, before and after the Second World War. In June 1950, on occasion of his trip to Italy, I had with him a thorough conversation about the future European laboratory.

A further endorsement of this idea came from IUPAP which was at that time under the presidency of Kramers. I was one of the vice-presidents and I had asked Kramers, at the beginning of the summer 1950, to include the discussion of Rabi's proposal, with specific reference to Europe and high-energy physics, in the agenda of the meeting of the Executive Committee of IUPAP that was to take place at the beginning of September of the same year in Cambridge, Mass. Although Kramers could not preside at the meeting because of bad health, the problem was discussed at length under the chairmanship of Sir Darwin assisted by the Secretary General Fleury. As a conclusion I was asked to get in contact with Rabi and with physicists from various European countries in order to clarify the aims and structure of the new organization and to help in the co-ordination of the different efforts. My first step was to write to Auger who in the meantime had presented the problem to the conference on nuclear physics held in Oxford during the month of September.

After Rabi's proposal Auger had the authority to act but there was no money appropriated on the scale required for a detailed expert study of such a project.

In December 1950, Denis de Rougemont, director of the European Cultural Centre (which was founded at the already mentioned Lausanne meeting of 1949) in agreement with Auger called at Geneva a commission for scientific co-operation for discussing Rabi proposal. The participants, in addition to a few people of ECC, were: P.Auger, P.Capron (B), B.Ferretti (I), H.A.Kramers (NL), P.Preiswerk (CH), G.Randers (N), M.Rollier (I), and Verhaege (B).

The Commission concluded its works with a series of recommendations, the most important of which was "the creation of an international laboratory centred on the construction of an accelerator capable of producing particles of an energy superior to that forseen for any other accelerator already under construction", i.e. the cosmotron of Brookhaven (3 GeV) and the Bevatron of Berkeley (6 GeV).

The Commission also discussed and endorsed the estimates brought by Ferretti of the cost of such a machine obtained by comparison with the cost of the two American machines mentioned above.

Immediately after this meeting G.Colonnetti, President of the Consiglio Nazionale delle Ricerche of Italy, R.Dautry, Administrator of the French Commissariat à l'Energie Atomique and J.Willelms, Director of the Fond National de la Recherche Scientifique in Belgium made available to Auger some funds which, all together, amounted to about $ 10 000. This sum, although very modest, was sufficient for Auger to initiate the first steps for arriving at the planning and construction of a large particle accelerator.

W.Heisenberg (left) and A.Hocker, delegates from the Federal Republic of Germany. Geneva, 12-15 February 1952.

Alfen from Sweden in Geneva, 12-15 February 1952.

At the beginning of 1951 Auger established a small office at UNESCO and invited me to Paris at the end of April to discuss the constitution and composition of a working group of European physicists interested in the problem.

The first meeting of this "Board of Consultants" was held at UNESCO, in Paris, at the end of May 1951. The members of the "Board were: E.Amaldi (I), P.Capron (B), O.Dahl (N), F.Goward (UK), F.A.Heyn (NL), L.Kowarski and F.Perrin (F), P.Preiswerk (CH), and Alfven (S) in place of I.Waller.

Two goals were immediately established: a long-range, very ambitious, project of an accelerator second to none in the world, and in addition the construction of a less powerful and more standard machine which could allow at an earlier date experimentation in high-energy physics by European teams.

In this and a few successive meetings of the Board of Consultants a few other features of the new organization were also examined, as preparatory work for a conference of delegates of governments that was convened by UNESCO in Paris, in December 1951. This conference took place under the chairmanship of Franois De Rose, a French diplomat who had been involved in discussions with French scientists (P.Auger, F.Perrin, L.Kowarski) about the future European laboratory from the time of the UN Commissions mentioned above, and was elected, a few years later, President of the Council of CERN. It led to the signing of the Agreement, which took place in Geneva in February 1952.

3. THE PLANNING STAGE

The first problem tackled by the Council of the provisional organization created by the Agreement was the nomination of the officers responsible for the appointment of the remainder of the staff and for planning the laboratory.

Bakker from the Netherlands was nominated Director of the Synchro-Cyclotron Group, Dahl from Norway (with Goward from the U.K. as deputy) Director of the Proton-Synchrotron Group, Kowarski from France (with Preiswerk from Switzerland as deputy) Director of the Laboratory Group, which had to take care of site, buildings, workshops, administrative forms, financial rules, etc., Niels Bohr Director of the Theoretical Group and finally E.Amaldi as Secretary General with the task of maintaining cohesion between the four groups of which the provisional organization was composed.

Almost all these people had worked on the Board of Consultants, nominated by UNESCO, during the first stage and contributed later to the creation of the new organization and to the development of its activities.

In July of the same year, 1952, an international nuclear physics conference was held in Copenhagen; on that occasion the type of accelerator to be built as the main goal of the new European organization was amply discussed. A report of the conclusions reached by the participants was presented by Heisenberg to the Council which held its Second Meeting in Copenhagen immediately after the Conference. Thus the decision was taken that the Proton-Synchrotron Group should explore the possibility of constructing a 10 GeV proton-synchrotron which, at that time, represented the biggest machine in the world.

During the month of August Dahl and Goward went to Brookhaven in order to study in detail the Cosmotron (with a maximum energy of about 3 GeV) that was very close to completion. During their two-week visit, and in some way in connection with the discussions going on in relation to the European project, Courant, Livingstone and Snyder came out with the "strong-focusing principle".

This important discovery came soon enough to allow a change of the plans of the provisional organization: with the approval of the Council, the PS Group embarked on the study of a strong-focusing PS of 20-30 GeV instead of thc weak-focusing 10 GeV machine considered until then.

During the summer of 1952 four sites were offered for the construction of the new laboratory: one near Copenhagen, one near Paris, one in Arnhem in the Netherlands and one in Geneva.

After long and lively discussions the site in Meyrin near Geneva was unanimously selected.

Another point I would like to touch about those times concerns the participation of the European nations.

All the European members of UNESCO had been invited to the Conference opened in Paris in December 1951 and closed in Geneva on 15th February 1952, but no response came from the countries of Eastern Europe. Furthermore while the Agreement was signed at that date by the representatives of eleven countries, the Convention establishing the permanent organization was signed (the July 1st 1953) by twelve countries.

The difference was due to the U.K. which, at the beginning, was rather cautious in committing itself to take part in the new organization. The U.K. government preferred to remain in the formal position of an observer during the first two stages, while, by signing the Convention, it became a full-right member of the permanent organization.

I remember that in the autumn of 1952 it was decided that Bakker, Dahl and myself should go to Brookhaven to take part in the dedication of the cosmotrom that was foreseen for December 15.

Our trip was already arranged when I was called on the telephone from London by Sir John Cockcroft who had been from the start very much in favour of the participation of the U.K. in the new venture.

As a consequence of our conversation on the telephone I decided to leave earlier and to pass through London on my way to Brookhaven.

In London I went to the D.S.I.R. (Department of Scientific and Industrial Research) where I met its chairman Sir Ben Lockspeiser, who a few years later was elected president of the permanent CERN.

After rather long discussions about various organizational and financial aspects of the project as a whole, Sir John and Sir Ben brought me to Lord Cherwell who, at that time, was a member of the Churchill government.

Lord Cherwell appeared to be very clearly against the participation of the U.K. in the new organization. As soon as I was introduced in his office he said that the European laboratory was to be one more of the many international bodies consuming money and producing a lot of papers of no practical use. I was annoyed and answered rather sharply that it was a great pity that the U.K. was not ready to join such a venture which, without any doubt, was destined to full success, and I went on by explaining the reasons for my convictions. Lord Cherwell concluded the meeting by saying that the problem had to be reconsidered by His Majesty's Government.

When we left the Ministry of Defense, where the meeting had taken place, I was rather unhappy about my lack of self- control, but Sir John and Sir Ben were rather satisfied and tried to cheer me up.

A few weeks later Sir Ben Lockspeiser wrote me an official letter asking that the status of observer be given to the U.K. in the provisional organization, and the D.S.I.R. started to regularly pay "gifts", as they were called, corresponding exactly to the U.K. share calculated according to the scale adopted by the other eleven countries.

In spite of its very particular legal position, the U.K. gave in practice fundamental support to the provisional organization and was the first among the European countries to ratify the Convention.

4. THE FIRST SCIENTIFIC RESEARCHES UNDER THE AEGIS OF CERN IN ITS EARLY STAGES

The exploration of high altitude by means of balloons to study cosmic radiation had been tried extensively with success in the U.K., particularly by the Bristol Goup, around 1950. This group thus arrived at the discovery of the charged τ – and K-meson.

In 1951 it was suggested that a study of high energy events produced by cosmic radiation could be more conveniently carried out at lower latitudes. At a latitude of 40°, for example, only that part of the primary radiation which has an energy above 7 GeV per nucleon is allowed by the magnetic field of the Earth to enter the atmosphere. Thus, at these latitudes the magnetic field of the Earth removes a large number of primaries which are not very efficient in producing the new secondary particles because of their relatively low energy. The nearer one goes to the equator the stronger this effect is.

Thus a first international expedition was planned in 1952 under the sponsorship of CERN that recently had entered its "Planning Stage" (Sect.3). Naples and Cagliari were chosen as the most convenient of the available bases in the Mediterranean. Thirteen universities took part in the expedition which met some success. In particular was the first successful attempt to recover balloons at sea. Furthermore, the expedition made a survey of the winds at high altitudes which was of great importance for the expedition of the subsequent year. The results of this first expedition were briefly reported to the Third Session of the Council of CERN held on 4-7 October 1952 in Amsterdam.

A second expedition took place, also to Sardinia, in June-July 1953. This was much on the same lines but on a larger scale than the first one. Eighteen laboratories from European countries in addition to one from Australia took part in it. 25 balloons were launched and over 1000 emulsions were exposed 7 hours at altitudes between 25 and 30 km. This corresponds to 9.27 litres of nuclear emulsions weighting 37 kg. They were successfully recovered at sea on account of the employment of a seaplane and a corvette (Pomona) generously placed at disposal of the expedition by the Italian Airforce and the Italian Navy.

While in the 1952 expedition only glass-backed emulsions had been used, in 1953 expedition "stripped emulsions" were introduced. Thus, the tracks were followed almost always from one emulsion to the adjacent one, a point of great importance when high-energy events are studied. The development required special care and in particular the construction of special developing systems at the Universities of Bristol, Padua and Rome, where the emulsions of the whole expedition were processed.

In October 1953, a meeting was held at the Department of Physics of the University of Bern (Switzerland) for distributing the packages of exposed and developed emulsion among the participating universities and, in April 1954, an international conference was organized in Padua to discuss the first results and plan jointly the most efficient methods for the investigation.

On the 25 November 1959 J.Adams announced to the CERN Staff that on the evening before the PS has been put into operation.

Francis Perrin in Geneva, 12-15 February 1952.

Further results appeared in the normal scientific literature and in lectures given by various participants at the second Course of the Varenna International Summer School that took place during the summer 1954. At the same course, Fermi gave a series of "Lectures on pions and nucleons" which were his last contribution to the teaching of physics before his death in Chicago on November 29 of the same year.

Among the main results of the expeditions I can recall a number of determinations of the values of the masses of heavy mesons giving rise to different decays which brought suspicion that all these processes could be alternative decays of the same particle. Also the identification of the various hyperons and the determination of their masses and decay modes were considerably improved.

In July of the same year, 1953, a very important conference was held at Bagnéres de Bigorre in the French Pyrénées where the new heavy particles were discussed at length. At this meeting, for example, the Dalitz plot for the τ-meson was discussed by Dalitz, Michel presented the selection rules for decay processes and the Rome group introduced the logarithm of tgθ (later called rapidity) as a very appropriate variable for the description of high-energy events in which many secondary particles are emitted.

The speaker at the conference dinner was Cecil Powell who, in brilliant terms, pressed for the urgency of constructing big European accelerators for avoiding that in the old continent subnuclear particle research could be drawn by the mounting level of the sea of machine results already noticeable in the United States.

A third expedition was organized by the Universities of Bristol, Milan and Padua in October 1954 from Northern Italy (Novi Ligure). It consisted in the launghing of a single stack of 15 litres of nuclear emulsions (corresponding to a weight of 63 kg) which is indicated in the literature as "G-stack".

In the landing, which took place in the Apennines, the stack was partly damaged. The delicate operation of recovery of the damaged emulsions was made by the Milan group, while its processing, involving an exercise in small scale chemical engineering, was made at the Universities of Bristol and Padua.

The great advantage in employing a very large stack is evident for studying the different modes of decay of the heavy mesons produced in the collision of a high energy particle with a nucleus of the emulsions. A substantial fraction of the secondaries come to rest inside the stack so that their modes of decay can be observed and the energy determined with high accuracy from their range.

A great part of the results was presented at the International Conference on Elementary Particles held in Pisa in June 1955 to celebrate the Centenary of the Nuovo Cimento.

One of most important results of the G-stack study was the final recognition that the values of the masses of heavy mesons giving rise to different decay processes were identical within rather small experimental errors. Thus the interpretation of all these processes as alternative decays of the same particle was strengthened, contributing in an essential way to the general recognition of the so-called $\theta-\tau$ puzzle. This found its explanation only in 1956 with the discovery by Lee and Yang of the nonconservation of parity by weak interactions.

A striking fact that emerged in Pisa was that the time for important contributions to subnuclear particle physics from the study of cosmic rays was very close to an end. A few papers presented by physicists from the U.S.A. showed clearly the advantage for the study of these particles presented by the Cosmotron of the Brookhaven National Laboratory (3 GeV) but even more by the Bevatron of the Lawrence Radiation Laboratory in Berkeley (6.3 GeV).

5. THE INTERIM STAGE

After this detour on the cosmic rays researches carried out, mainly under the influence of the Bristol group, in a wide international frame and with a strong support from Italian Universities, I return to the development of CERN.

The four groups of the provisional organization at the beginning had started to work at the institutions of the corresponding directors. In October 1953 the PS staff was assembled in Geneva, partly at the Institute of Physics of the University and partly in temporary huts built in its vicinity. At that time the transition took place from mainly theoretical work to experimentation and technical designing.

John and Hildred Blewett of the Brookhaven National Laboratory arrived in Geneva in September 1953 and joined the PS group, which benefited of their experience in design and construction of accelerators for about one year.

In October an International Conference on proton synchrotrons was organized at the Institute of Physics by the PS staff. In March 1954 Goward died after a short and tragic illness. He was succeeded by John Adams (1920-1984) who became Director of the PS Division in 1955 when Dahl went back to Norway to direct the design and construction of the first Norwegian nuclear research reactor.

During October 1953 an administrative nucleus began to function in a temporary Geneva office and, from January 1954 on, in the Villa Cointrin at the Geneva airport.

First instrumentation workshops, then a library and a few laboratories were gradually set up also at the airport in the summer 1954. On 12 August 1954 France completed its ratification procedure of CERN Convention and the situation appeared sufficiently promising for daring to start, on 13 August, 1954, mayor excavation work on the Meyrin site for the foundations of the first permanent buildings, in particular that to house the Synchro-Cyclotron. I had taken this decision with the agreement of the Council, under the pressure of the Geneva weather months before all ratification formalities were in place. Sir Ben Lockspeiser was clearly pleased and commented: "Now we have another task - keeping Amaldi out of jail".

To arrive at the appointment of the Director General of the permanent Organization the Council of the provisional Organization set up (in June 1953) a Nomination Committee composed of Niels Bohr, Sir John Cockcroft, G.Colonnetti, W.Heisenberg, F.Perrin and P.Scherrer with Perrin as convener. The attention of the Nomination Committee was pretty soon focused on the name of Felix Bloch (1905-1983), proposed by Niels Bohr and supported with enthusiasm by all the other members.

On occasion of a short visit to Europe paid by Felix Bloch in March 1954, on invitation of the Nomination Committee, Bloch and I examined together the steps to be taken for creating in Geneva, as soon as possible, the right atmosphere of a research institution and agreed that a few of them had to be started before the beginning of Bloch's tenure of office.

The first point regarded the gradual displacement from Copenhagen to Geneva of the Theoretical Group (or Division as it was called by now). As first step B. d'Espagnat was engaged from 1 September 1954 and J.Prentki from 3 January 1955. Both of them were on leave of absence from the "Centre National de la Recherche Scientifique" in Paris. They worked together on meson production and scattering.

Picture taken on December 1975 when the CERN directorship passed from Willy Jentschke to John Adams and Leon Van Hove. From left to right: L. Van Hove, W.Jentschke, Paul Leveau, President of the Council, and J.Adams.

Niels Bohr presses the botton to put the PS into operation on occasion of the inauguration of this machine on Febraury 5, 1960.

The other step agreed with Bloch was the creation of a two-pronged cosmic-ray experimental programme at CERN. One project consisted in setting up, in collaboration with the Scientific and Technical Services Division directed by L.Kowarski, a nucleus for an experimental physics group. The first scientists involved in this group, starting in summer 1954, were Y.Goldschmidt- Clermont from Belgium and G. von Dardel from Sweden, with C.Peyrou, then at the University of Bern, as consultant. Until the accelerators in CERN could be taken into operation, this group could produce scientific results and gain experience for the future by performing experiments.

The second project consisted of taking over, in August 1955, the U.K. cloud chamber group working at Jungfraujoch Station, led by J.A. Newth, and which had followed Blackett, in his transfer from Manchester University to Imperial College in London, in 1953.

The Geneva group collaborated with this team in a new experiment for measuring the lifetime of K-mesons, using a Cherenkov detector for triggering the cloud chamber on K- meson decays.

Felix Bloch arrived in Geneva on 11 September 1954. He was still only a candidate highly recommended by the Council of the provisional CERN. On September 29, 1954 when a prescribed point in the ratification procedure was reached, the provisional Organization came suddenly to an end while the permanent Organization was not yet in existence. Therefore all the assets of the provisional CERN became suddenly masterless. For eight days I had the honour - as Secretary General - of sole responsability of all properties and liabilities on behalf of a new born organization. Then the first meeting of the permanent Council assembled in Geneva on the 7 October 1954, the Secretary General presented his Final Report , Bloch was nominated Director General and CERN entered its final permanent form.

The Synchro-cyclotron was operated for the first time in 1958 and a proton beam at full energy circulated in the PS on 24 November 1959, i.e. almost one year before time schedule.

In concluding my speech, I will not try to summarize the many important results and the few discoveries made at CERN during the thirty years or so that passed since its creation. During this period CERN is grown enormously and, in my opinion, in a very satisfactory way, with scientific and technological contributions from all Member States, in particular from United Kingdom. Such a development has been possible only because an adequate funding was provided over the years by the Governments of the Member States. If in recent years Europe has regained a good position on the line where fundamental research is rapidly advancing, this is certainly in great part due to CERN. Criticisms also have been raised and they have received a hearing by a few authorities, in particular by those of United Kingdom, where even the interest and importance of this branch of science seems to be doubted. But if we look back to the history of the investigation of the last constituents of matter, we are struck by the number of first class people orginating from this country, or at least, grown up inside the cultural structures of this country, and have contributed over the centuries to it. Even limiting this historical outline to our century, everybody recognizes that the names of J.J. Thomson, Lord Rutherford, C.T.R. Wilson, P.A.M. Dirac, J. Chadwick, G.P. Thomson, Lord Blackett, C.F. Powell and J.D.Cockcroft, are all connected with fundamental steps forward in this branch of science, which is still flourishing and evolving not less than any other science.

It appears unbelievable that a long and glorious tradition like the one summarized by the names listed above may be voluntarily stopped. The younger generations should continue to have the challenge of participating in this kind of research at a continental level. This means that the senior scientists have the moral duty of trying to do anything possible for avoiding that such an opportunity may go lost.

With this message and wish, for the younger as well for the more senior generations of high energy physicists, from Great Britain and from all other Western European countries, I like to express to all of you my thanks for your patience and attention.

The discovery of the V-particles

G D Rochester

Department of Physics, Durham University, Durham, DH1 3LE

ABSTRACT: A personal account of the early history of the discovery of the strange particles.

1. INTRODUCTION

Important discoveries are often the result of good fortune and good science and in the discovery of the V-particles both elements were present. It was my personal good fortune to be closely associated with three outstanding physicists, Patrick Blackett, Lajos Janossy and Clifford Butler each of whom played a highly significant part in our discoveries. I have, of course, many reasons for being especially grateful to Blackett not least because he persuaded me in 1938 to change my line from band spectroscopy to cosmic rays but also because he taught me the science and the art of the counter controlled cloud chamber. I pay tribute to Lajos Jánossy for initiating me into the study of high-energy interactions in cosmic rays and to my friend and close collaborator Clifford Butler, with whom I jointly built and operated the magnet cloud chamber in which the new particles first appeared.

2. PRE-WAR RESEARCH IN MANCHESTER

Blackett, a sketch of whom is shown in Fig. 1, went to Manchester in 1937 and by the time I joined his group he had established a dynamic school of cosmic-ray physics from staff and students brought from London or recruited in Manchester. Prominent among the staff members were Rossi, Jánossy, Wilson and Lovell and the main lines of research were the mesotron spectrum, extensive air showers, the non-ionising component, penetrating showers and the proton spectrum. Blackett's brilliance and total commitment made the laboratory a stimulating place in which to work as did the stream of distinguished visitors.

Fig. 1 P M S Blackett, (from a design by John Ward, R.A.).

The shadow of war was slowly spreading over Europe, however, and in the

© 1988 IOP Publishing Ltd

summer of 1939 most of the U.K. members of the Blackett School were sent to the N.E. coast to man a secret Radar Station. The work was fascinating and new to those of us who knew nothing of radar but after six weeks the Scientific Manpower Commission decided that university-trained physicists could be better employed than as scanners so the group was assigned to other duties. I was sent back to Manchester because the Commission decided that a few universities should continue to train Honours Physicists. So I went back to teach but in addition ran a small Fire Brigade and did the usual Civil Defence duties.

3. WARTIME RESEARCH IN MANCHESTER

My Fire Brigade duties entailed my being at the University for long periods and because of the infrequency of air raids there was much spare time. Jánossy was also often at the University for his Civil Defence duties and so, instead of spending our time playing bridge and billiards, we spent much time together discussing some of the major problems of cosmic rays. A photograph of Jánossy is shown in Fig. 2 from which it will be seen that he was a man of slight build with a pale complexion usually partly obscured by a shock of jet black hair. Jánossy was already a very experienced cosmic ray physicist having worked not only with Blackett in London but also with Kölhorster. Moreover he was equally at home as an experimentalist or a theorist.

Fig. 2 L. Jánossy

Our problem was to find something we could do on the research side which would not entail the use of electric power or the purchase of equipment from outside the University. In two respects we were fortunate, firstly, because the department was well equipped and, secondly, because there was a friendly glass-blower quite close to the University who was prepared to make us a few G.M. tubes when we needed them. On the personnel side we were also fortunate because we could use the honours students for small investigations and get help from some former physics students, who did fire-watching on large city buildings in lieu of military service.

The result of our discussions was to conclude that the most important unsolved problem was the origin of the main component of cosmic rays at sea level, the so-called penetrating component, which was known to consist of mesotrons. The other component at sea level, the electron-photon component was already well accounted for by the theories of Carlson and Oppenheimer and Bhabha and Heitler. Examples of these two components are shown in Figs. 3 and 4.

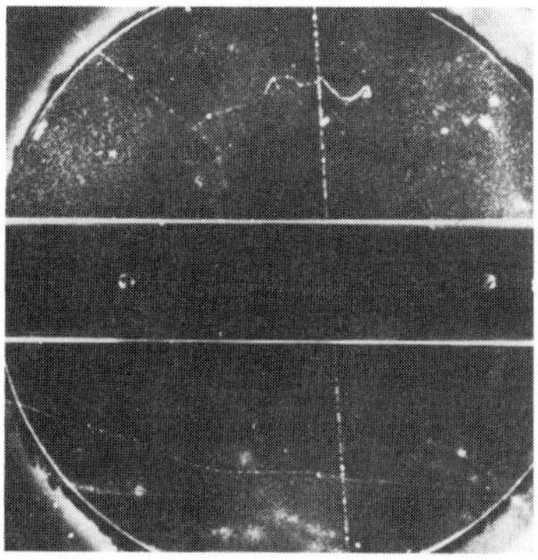

Fig. 3 A cloud chamber photograph of a high energy mesotron penetrating a 3cm lead plate.

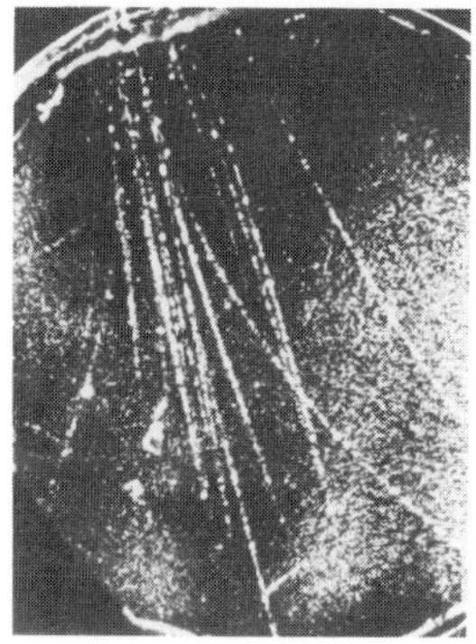

Fig. 4 A cloud chamber photograph of an electron shower, Blackett and Occhialini (1933).

It was known that the mesotron was a particle carrying the elementary charge, had a mass of the order of 200m where m is the mass of the electron, and was unstable with a lifetime of a few microseconds. Its origin was a mystery for it had properties which resembled the particle predicted by Yukawa in 1935 and yet it was different. The confusion lasted for some ten years - called by Oppenheimer "the ten year joke" - and, as is well known was not cleared up until the discovery of the pion in 1947.

To Jánossy and myself in 1940 it seemed that the most promising place in which to look for the origin of the mesotron was in high-energy interactions - high energy because of the relatively large rest mass of the mesotron. Moreover, we knew that Jánossy and Ingleby had already found a way of isolating such interactions from the many spurious events which could simulate them by the use of a complex shower selector. The original set-up, Jánossy and Ingleby, (1940), and Jánossy (1942) is shown in Fig. 5. A somewhat similar array had been developed at almost the same time by Wataghin et al. (1940) in Brazil. The basic selection in the Jánossy set up was five-fold but there was a tray of hodoscoped counters (H) in the middle of the lead to help find the multiplicity of charged particles there. Thus there began a long series of experiments by Jánossy and myself to investigate these showers, by then called "penetrating showers" (ps), and to seek answers to the following questions:-

1. Did the ps sets select showers of penetrating particles (termed pp)?

2. Could the ps showers be produced locally?

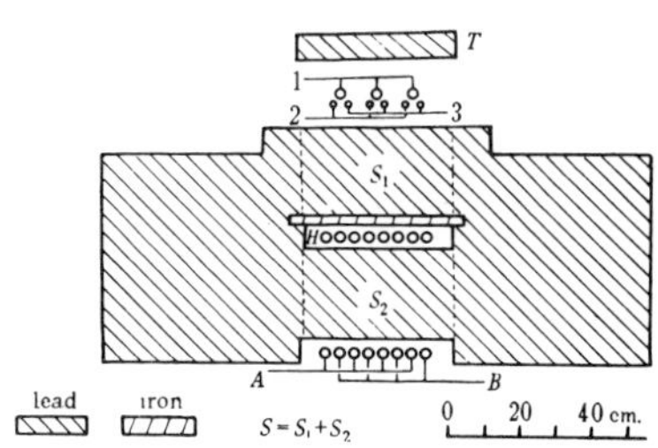

Fig. 5 The original ps set.

3. What were the primaries?

The experiments involved counter-sets usually sevenfold, in a large mass of lead, (an arrangement which effectively eliminated electron-photon showers) and cloud chambers. The cloud chamber work was carried out largely by Broadbent, McCusker and the writer, and by later standards the chamber was somewhat primitive in that the lighting was by incandescent lamps and there was no magnetic field. Thus the momenta and the signs of the charged particles could not be determined.

The results were published in a series of papers between 1941 and 1946 and were briefly as follows:

1. The showers were complex being a mixture of electron showers and what we took to be pp, that is, charged particles penetrating the lead. Pictures of the showers are to be found in Jánossy et al. (1941), and Rochester (1946). An example is shown in Fig. 6.

2. There was a marked transition effect when the lead T, (see Fig. 5) was increased in thickness. For values of T greater than 5cm the rate increased by 50%. Thus the showers could be produced locally.

3. The primaries were non-photonic and indeed probably nucleonic, Jánossy and Rochester (1943). This result was in line with the balloon experiments of Schein and his co-workers (1943).

Fig. 6 Cloud chamber photograph of ps. (The lead plate across the chamber was 2.3cm in thickness).

It should be stated that our cloud chamber pictures of cosmic ray interactions or 'stars' were not unique for fine examples were published by Bostick, Hazen, Herzog, Wilson, Powell and Wollan. However, as far as we can judge most of these interactions were of relatively low energies, perhaps as low as tens of MeV, much too low for the production of the strange particles. Later work showed that our ps had energies in the GeV region.

On several cloud chamber photographs, (Daudin 1945, Jánossy et al., 1945) charged particles, ionizing at minimum, were seen to change direction through quite large angles in the gas of the chamber without any visible recoil. Their nature could not be established because of the absence of a magnetic field but it is thought now that they were probably $\mu \rightarrow e$ or $\pi \rightarrow \mu$ decays.

4. MAGNET CLOUD CHAMBER STUDIES OF PENETRATING SHOWERS

As soon as the War was over Blackett and other members of the pre-war Cosmic Ray School returned to Manchester and initiated a period of intense activity and great fruitfulness. New members of staff were appointed, there was a large influx of outstanding postgraduate students and the undergraduate intake was increased by a factor of three. The result was a large and stimulating department in which the high level of criticism gave every encouragement to originality and excellence. Blackett himself, though still taking a very keen interest in cosmic rays, switched his research interests to geomagnetism and paleomagnetism. The work on ps much impressed him, especially the 'peculiar' events seen in some of the cloud chamber photographs, and he readily agreed that the next urgent step was to place a cloud chamber controlled by a ps set in a magnetic field so that the momenta and changes of the particles could be determined. Jánossy continued his counter work but soon left Manchester to take up a chair in Dublin. I was therefore given an assistant, a bright young physicist with considerable skill in technical matters, Clifford Butler. Moreover, we were allowed to take over the one piece of large equipment in the Cosmic Ray Laboratory, the Blackett Magnet which had been used for beautiful work on the mesotron spectrum. A picture of the original magnet cloud chamber is shown in Fig. 7.

Fig. 7 The original Blackett magnet and cloud chamber.

The old cloud chamber was quite unsuitable for ps work and so Butler and I completely re-designed it introducing many new technical devices developed during the War.

The new chamber was 9cm in depth and 30cm in diameter and was placed in close contact with one pole of the magnet. Across the chamber was a lead plate 3cm in thickness to help identify mesotrons. Illumination was provided by Siemens Flash Tubes and stereoscopic photographs were taken through one pole of the magnet using a fast fine-grained film. The clearing field was always switched off before photography and many precautions were taken to keep the chamber clean and to avoid temperature gradients in the gas. A picture of Clifford Butler working on the chamber is shown in Fig. 8. Photographs were taken with two types of fivefold ps triggers, one of which is shown in Fig. 9, in magnetic fields of strengths 3500 and 7200 gauss. As usual, the ps sets were set in lead. Typical photographs are to be found in Rochester and Butler (1948 a and b), and Rochester and Wilson (1952). The basic philosophy behind the experiment was the same, firstly to find out if mesotrons were created in high energy interactions and, secondly, to investigate 'peculiar' events. The parameters which could be determined for a single particle were the momentum, the sign of the charge, the ionisation (which had to be determined visually) and finally the scattering in the lead plate. Effort was concentrated on the slow particles especially a comparison of the observed quantities against the values expected for protons, pions and muons. That there were many slow particles of meson-like mass was clear from the outset (Rochester et al. 1947a), but it was not known what kinds of mesons they were. For the calculated ionisation - momentum curves the theoretical values for protons (mass 1837m), pions (mass 320m) and muons (mass 200m) were used. The value of 320m for the pion mass came from the value in current use at the time at Bristol. With this value most of our slow mesons fall closer to the 200m line than the 320m line - leading us to the erroneous conclusion that 200m particles were in the showers and were therefore being created. The matter was put right by Barker and Butler in later work when the mass of the pion was known more accurately. Nonetheless I have often reflected on our slowness at appreciating the significance of the Italian work which had already appeared in print and our lack of contact with good elementary particle physicists!

Fig. 8 Clifford Butler adjusting the cloud chamber.

5. THE FIRST V-PARTICLES

The most significant outcome of the first work with the magnet cloud chamber was, however, the discovery of the V-particles, the V^o-particle on the 15th October 1946 and the $V^{\pm}$particle on the 23rd May 1947. The original photographs are reproduced in Figs. 10 and 11 and it is seen that in both cases there is a V in the gas of the chamber and the arms of each V are at minimum ionisation showing that the particles (of the elementary charge) were travelling with velocity of the order of c. Stereoscopic photographs showed that for each V the arms of the V were strictly copunctal and lay well within the gas of the chamber. This proved that the V's were not due to the accidental crossing of two unrelated tracks. Moreover detailed analysis showed that all collision processes were excluded and that the events did not depend upon the amount of matter traversed but on the distance travelled, that is, there were decays.

Attention was then directed to possible decay schemes and it was thought that the most likely for the V^o was decay to two muons or two pions and for the $V^{\pm}$-decay to a muon and a neutral particle. For decay products of masses 200 and 400m the masses of the V^o came to (850 ± 200)m and (1100 ± 150)m. It was most unlikely that the positive particle in the V^o decay could have had a mass as high as the

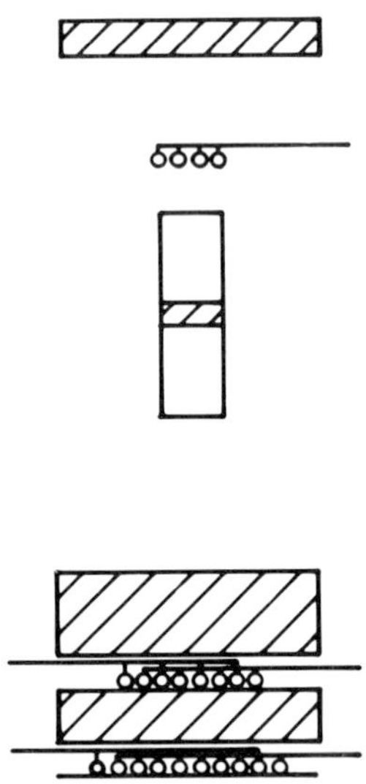

Fig. 9 Disposition of the counters, the chamber and the lead absorber (shown shaded).

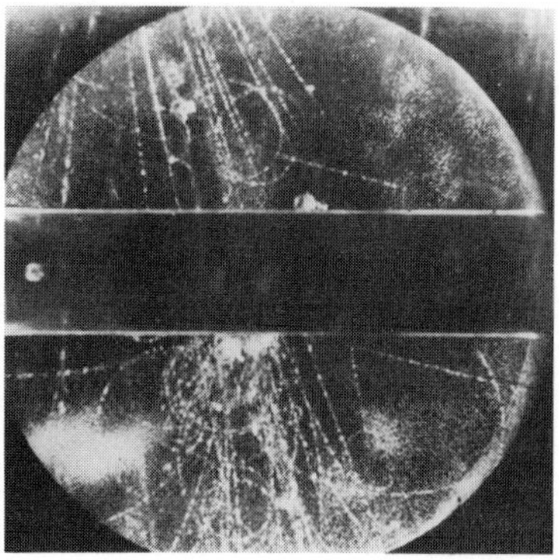

Fig. 10 The V^{o}-Decay

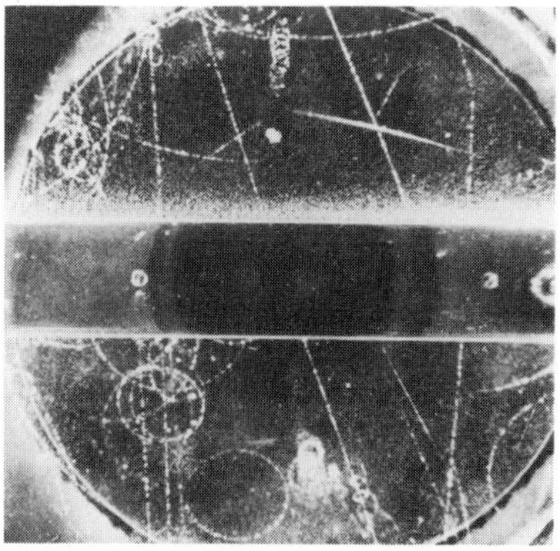

Fig. 11 The V^{+}-Decay

proton for then it would have had an ionisation density 5 times that observed. For decay of the charged V, to particles of mass 200m the primary mass came to (1080 ± 100)m the experimental data and the detailed analyes were published in 1947, Rochester and Butler, (1947).

The V^{o}-particle was the first example of an unstable neutral particle of mass about 1000m but evidence was presented by Leprince-Ringuet and Lheritier (1944) for a charged particle of mass about 1000m which did not show any decay and whose identification depended upon the occurrence of an exceptionally rare knock-on electron.

Copies of our Nature paper were sent out to many distinguished physicists, including, Anderson, Auger, Bethe, Brode, Fermi, Heisenberg, Heitler, Jánossy, Leprince-Ringuet, Marshak, Oppenheimer, Pauli, Powell, Rossi, Schein and Weisskopf and interesting replies came from Fermi, Heitler, Jánossy, Powell, Rossi and Wheeler all of whom were at least sympathetic to the idea of 1000 mass unstable particles though it was clear that there was an air of caution also. Wheeler was the only critic, raising the question of some form of neutron reaction on nitrogen or oxygen. Clearly he had not realised the importance of the non-appearance of recoils of any kind. Powell accepted the evidence and commented that Bristol seemed to require a neutral particle of mass about 130m. The most interesting letter was, however, from my old collaborator, Jánossy, who pointed out, what I had already noticed, that a minimum mass for the V^{+} could be found from the invariant transverse momentum which only involved the deflection and the momentum of the secondary both of which could be found accurately. The result was a mass of (970 ± 150)m, well above the pion or proton masses.

6. THE YEARS OF FRUSTRATION

After the early discoveries which promised so much there followed several frustrating years, a period of strain for Clifford Butler and myself, when no further examples of the V-particles were found. Indeed, the next V-particles to be found by any Manchester group were in the summer of 1950. How the deadlock was broken by Manchester will be described in due course, but in retrospect even in the year 1948, which at the time, seemed bleak indeed, two events occurred which were to inaugurate the enormous developments of later years. The first was the Symposium on cosmic rays held in honour of Millikan's 80th birthday at the California Institute of Technology in June, 1948, and the second the production of the first electron-sensitive emulsion in November by the research staff of English Kodak Ltd., which led to the discovery of the τ-meson. Blackett sent me to the Millikan Conference and among the participants were almost all of the leading experimental and theoretical physicists in cosmic rays. A copy of the Symposium photograph is reproduced in Fig. 12. References were made in several of the papers to the possible existence of particles of mass between the pion and the proton but a Leprince Ringuet particle found in nuclear emulsion seemed to attract most attention. I talked in detail about ps (Rochester 1949) and showed many cloud chamber photographs including the $V^{\pm}$decay. Abram Pais recently pointed out that I did not emphasise the V-decays and this may have been because I had nothing new to report. Nonetheless, later events and some press cuttings of reports in local and national newspaper, which I have seen only recently, showed that the work did attract the attention not only of the press but of some important cosmic ray experts, notably, Anderson, Rossi and Brode.

Fig. 12 The 1948 Millikan Symposium photograph, (after E P Ney).

Back in Manchester there was constant heart-searching and discussion as to why we had seen only two events. Three possible causes were identified, namely,

1. The producing layers of lead were too far from the cloud chamber.
2. The cloud chamber was too small.
3. The rate of high energy nucleons was too low for efficient detection.

Even in 1947 Blackett had anticipated these problems for he formed a group under the leadership fo Anthony Newth to design and build a large magnet cloud chamber suitable for operation at the Jungfraujoch where the rate of nucleons was expected to be at least 15 times greater than at sea level. Unfortunately the Newth project progressed only slowly and indeed was not operational until 1951. This worried Blackett as did the possible competition from the new emulsions developed by Messrs. Ilford and Kodak, which, it seemed, might make the cloud chamber work of minor importance. Indeed, he asked me to get some experience of emulsion work and Clifford Butler to form a new team to modify the original set-up and consider moving to a momentum site in France. As Clifford will describe various sites were considered by him and Braddick as a result of which the Pic du Midi was chosen. Again, after much weighing of the pros and cons the Manchester experiment was stopped in the summer of 1949 and by October the magnet was on the top of the mountain. The magnet was first energised in November, 1949, and then used all winter for emulsion work. The cloud chamber and its associated equipment were installed in the Spring of 1950.

7. THE ANDERSON V-PARTICLES

Nothing was known of the practical outcome of my talk in Pasadena in 1948 until a very exciting letter arrived in Manchester on 28th November 1949, stating that the Anderson group had found 30 cases of forked tracks similar to those of Rochester and Butler and that "so far as we can see their interpretation of the events as caused by new unstable particles seems to be borne out by experiment". Detailed data were not available until one year later when the Anderson paper, (Seriff et al. (1950)), was ready for publication. This paper listed the observation of 34 V-particles from two ps experiments, one at sea level and the other on White Mountain (3200m). Of the 34 particles 30 were of the neutral kind and 4 charged.

The Anderson paper was of decisive importance not only for the welcome confirmation of the existence of V-particles but also for the following reasons:

1. A comparison of the rates of production at sea level and at mountain altitudes confirmed that the V-particles were produced by the nucleonic component.
2. Some of the decay products were shown to be of approximately pionic and protonic mass.
3. The distribution of decay points for the main group of V^o-particles suggested a lifetime of the order of 10^{-10}sec.

Anderson never published his photographs because of severe gas distortion in the cloud chamber but he did allow Wilson and myself to publish examples in our Atlas, Rochester and Wilson (1952). An example of a V^o decay with secondaries that appear to be a proton and a pion is shown in Fig. 13.

8. SIGNIFICANT FACTORS IN THE CLOUD CHAMBER WORK

Two factors were of unique significance in the V-particle discoveries, namely penetrating shower selection and counter-control. Penetrating shower selection ensured that rare high-energy interactions in the energy range, 5-15 GeV were selected and spurious events excluded. This may seem obvious now but with hindsight, even Anderson was to reflect at the first Fermi-Lab Historical Conference in 1980, Anderson (1982), that in 1946 his group should not have rushed into cloud chamber work in a B29 at 30-40,000 ft. but "have spent a week in the High Sierras arguing only about cosmic rays. All the clues were present ... one of the most important being the experiments by Lagos Jánossy".

The second factor was the use of the counter-controlled cloud chamber, invented by Blackett and Occhialini (1933), which also ensured not only that rare events could be selected, and unwanted background excluded, because the time delay between the counter set pulse and the expansion was very short ($\sim 10^{-2}$ sec) but also that a V^o seen the chamber was associated with the penetrating shower. This was not possible with nuclear emulsions for there was no way of associating an event which simulated a V^o-decay with the interaction from which it came. Moreover in 1947 nuclear emulsion was not able to record charged particles ionizing at minimum. Again, by good fortune the physical dimensions of the counter set and the cloud chamber (i.e. $\sim$ 1m) were of the right order for there to be a reasonable chance of picking up decaying particles with lifetimes in the range 10^{-8} - 10^{-10}sec.

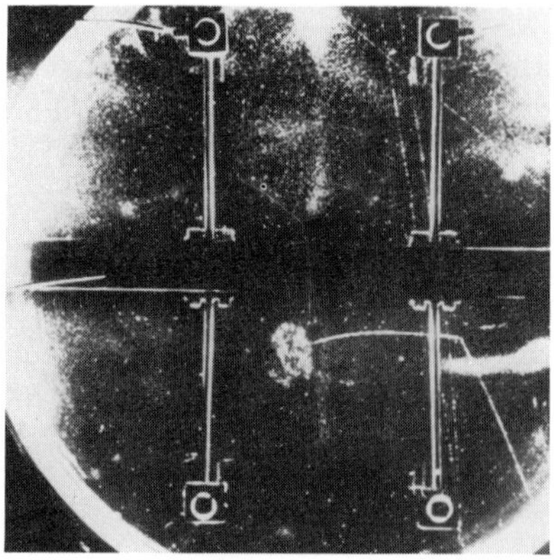

Fig. 13 An early Anderson photograph of a V^o-decay, possibly an early example of a Λ^o-hyperon.

9. POSTSCRIPT

Kinematical tables for all the principal two-body decay modes over a wide range of energies have been published by CERN, Salmeron (1963), and it is therefore possible to look again at the original V^o- and $V^{\pm}$ decays. It is found that V^o fits $K^o_{\pi 2}$ closely with Λ^o definitely included, and V^+ fits the $K^+_{\mu 2}$-decay closely but $K^+_{\pi 2}$ is not completely excluded. A $K^+_{\mu 2}$-decay is supported by the fact that the charged particle from the decay traverses the 3cm lead plate without deflection.

10. ACKNOWLEDGEMENTS

I wish to thank the Head of the Physics Department for the use of departmental facilities. Most diagrams and photographs have been acknowledged elsewhere. The photograph of the participants at the Millikan Symposium in 1948 was kindly supplied by Professor E.P. Ney of the University of Minnesota. Readers will recognise many well known physicists. On the front row are R. Oppenheimer, W. Heitler, Pierre Auger, Robert Millikan, W.F.G. Swan, H.P. Robinson, J. Clay and Bruno Rossi. Carl Anderson and G.D.R. are at the top right hand corner. The Royal Society is thanked for the use of the Ward design in Fig. 1.

11. REFERENCES

Anderson C D 1983 The Birth of Particle Physics, ed. L M Brown and L Hoddeson (Cambridge) 151
Blackett P M S and Occhialini G P S 1933 Proc. Roy. Soc. A 139 699
Daudin J 1944 Ann. de Phys. 11e Ser 19 (Planche IV, Cliche 16)
Jánossy L 1942 Proc. Roy. Soc. A 179 361
Jánossy L, Broadbent D and Rochester G D 1945 Nature 155 142
Jánossy L and Ingleby P 1939 Nature 145 511
Jánossy L, McCusker C B and Rochester G D 1941 Nature 148 660
Jánossy L and Rochester G D 1943 Proc. Roy. Soc. A 182 180
Leprince-Ringuet L and Lheritier M 1944 Compt. Rend. 219 618
Rochester G D 1946 Proc. Roy. Soc. A 187 464
Rochester G D 1949 Rev. Mod. Phys. 21 20
Rochester G D and Butler C C 1947 Nature 160 855
Rochester G D and Butler C C 1948a and b Proc. Phys. Soc. A 161 307 and 535
Rochester G D, Butler C C and Runcorn S K 1947 Nature 159 227
Rochester G D and Wilson J G 1952 Cloud Chamber Photographs of the Cosmic Radiation (London : Pergamon)
Salmeron R A 1963 Kinematical Tables for Two-body Decays (CERN : Geneva)
Schein M, Iona K and Tabin H 1943 Phys. Rev. 63 253
Seriff A J, Leighton R B, Hsiao C, Cowan E W and Anderson C D 1950 Phys. Rev. 78 290
Wataghin G, De Souza M D and Pompeia P A 1940 Phys. Rev. 57 61

Cloud chamber physics on the mountain tops

Clifford Butler

Formerly of Manchester University and Imperial College, London

1. INTRODUCTION

The advantages accruing from exposing cloud chambers to cosmic rays at high altitudes were widely recognised before the discovery of V-particles. For example, Carl Anderson's group built a chamber 30cm in diameter and 11cm deep with a magnetic field of up to 7500 gauss which could readily be transported from sea-level to mountain tops in California (up to 3000m). The nucleonic component of the cosmic ray flux increases substantially with altitude but it was generally accepted that a height of about 3000m was needed to obtain a worthwhile benefit. In the U.S.A. this was a readily achievable target (3200m at White Mountain, California and at Echo Lake, Colorado). In Europe, however, the only possibility in 1947 appeared to be at the observatory on the Jungfraujoch in Switzerland (3500m). Unfortunately, access from the railway terminal to the observatory was difficult for heavy loads, which prevented the Manchester magnet and cloud chamber moving there after the discovery of the first two V-particles. Consequently in 1947 Patrick Blackett decided to build a new magnet and cloud chamber specifically for use at the Jungfraujoch.

The new chamber was planned to be 55cm square but, initially only 9cm deep, much larger than the 30cm diameter chamber used by George Rochester and myself. The new 15 tonne electro-magnet could be broken down into small units which could readily be moved to the Jungfraujoch. Unfortunately the new project took a considerable time to complete. The magnet was installed during the summer of 1950, but the chamber was not completed until early in 1951, which history shows was too late for the new apparatus to confirm the V-particle discovery. The chamber was installed and operated successfully for some years by a Manchester group led by the late J.A. Newth. In particular they made an extensive study of the lifetimes of V-particles.

Early in 1949 Occhialini and Cosyns set about persuading Blackett that the existing Manchester magnet and cloud chamber could be moved quickly to the Pic-du-Midi in the French Pyrenees at a height of 2867m, high enough to increase the flux of penetrating showers 10 to 15 times the sea-level rate. I described the events which led to Blackett's favourable decision on the 9th August 1949 at some length during the July 1982 conference (Butler 1982); consequently I will not repeat all the details here. Figure 1 shows Blackett at "Les laquets", on the way up the Pic-du-Midi with Occhialini, Cosyns and Roesch (the Director of the Observatory) on the 18 September 1949. The magnet reached the mountain top by the end of October 1949, despite the discovery that the base plate was much heavier

© 1988 IOP Publishing Ltd

Figure 1 Blackett near the top of the Pic-du-Midi, with Occhialini, Cosyns and Rosch.

than the French engineers expected! Figure 2 shows some of the equipment on the mountain road on the 6th October 1949 and Figure 3 shows the base plate of the magnet on the final stage of the ascent to the Pic on the 15th October 1949. The magnet was energised by the 24th November and throughout the winter of 1949/50 it was used for emulsion experiments by groups from Manchester and Brussels. The cloud chamber and the electronics were not sent up to the Pic until after the snow melted in the spring of 1950. Figure 4 shows the cloud chamber arrangement installed at the Pic and Figure 5 is a general view of the Observatory in 1949.

It is important to note that the magnet was already at the Pic-du-Midi before Blackett received the remarkable letter from Carl Anderson at the end of November 1949 telling him of the 30 V-shaped tracks observed both at Cal-Tech and on the mountain top. Although we failed to confirm our own discovery, it was clear that the move to the Pic-du-Midi was fully justified and that an exciting future was about to unfold.

In the late spring of 1950 Blackett raised the question of nomenclature for the new particles with Anderson who replied asking him to decide on suitable names, but said that he was strongly against the use of Greek symbols. After consulting with Niels Bohr, Blackett told Anderson that he proposed continuing with the V-particle notation already adopted at Manchester. Anderson replied at once agreeing to this proposal.

At the beginning of 1950 the scene was set for several years of cloud chamber work world wide, mostly on mountain tops, although a few groups obtained excellent results at sea-level. This period of intense and competitive activity lasted until about the end of 1954, thereafter the focus of interest gradually shifted away from cosmic rays to the new generation of large accelerators which, although it could hardly have been foreseen, readily produced the new particles. The Cosmotron started work early in 1953 and the Bevatron operated towards the end of 1954. Before summarising the physics achievements concerning neutral V-particles by the end of 1954, I will describe briefly the programme at the Pic carried out by my own group as well as noting the important installation there of cloud chamber apparatus from the Ecole Polytechnique in Paris.

2. THE PIC-DU-MIDI EXPERIMENTS

The Manchester cloud chamber started regular operation at the Pic in July 1950 and by March 1951 10000 photographs were obtained which included 1000 showers of penetrating particles and 51 neutral and 12 charged V-particles (Armenteros et al 1951a, 1951b and Butler 1952). Remarkably, one of the early pictures of a particle decay, although of poor technical quality, revealed an entirely new phenomenon. A negatively charged particle decayed producing a neutral V-particle which also decayed in the chamber; it is reproduced as Figure 6. This cascade event turned out to be a rare phenomenon and once again the discovery was confirmed later by another group.

In this talk I concentrate on the properties of neutral heavy particles, leaving Professor Rochester to reflect on the cloud chamber contribution to our understanding of new charged particles in his following talk.

The Manchester group's first and therefore preliminary publication of Pic-du-Midi data on neutral V-particles (Armenteros et al 1951a) showed

Figure 2 Parts of the magnet on their way up to the Pic-du-Midi.

Figure 3 The magnet base approaching the top of the mountain.

Figure 4 The cloud chamber arrangement at the Pic-du-Midi.

Figure 5 The Pic-du-Midi Observatory 1949.

that both protons and positively charged mesons were included among the secondary particles. Assuming only two secondary particles were produced in each decay, there was strong evidence that two decay schemes were needed to explain the observations. The following decay schemes were proposed tentatively:

$$V_1^o \rightarrow p^+ + \pi^-$$

$$V_2^o \rightarrow \pi^+ + \pi^-$$

Figure 7 shows a V_1^o -decay with an identifiable proton secondary while Figure 8 contains a neutral V-decay with two high-energy secondary tracks which cannot be directly identified. A detailed consideration of the dynamics of the decay process leads to the conclusion that the decay was that of a V_2^o -particle. Later in 1951 the Manchester group obtained further evidence which was consistent with this initial suggestion and other groups also found supportive evidence. For example, the Indiana group, led by R.W. Thompson, obtained a small number of good quality decays at sea-level which included convincing examples of both types of decay process (Thompson et al 1951).

By the summer of 1951 my group at the Pic-du-Midi reported that each event could not be classified as a V_1^o or a V_2^o straightaway, because both secondary tracks were frequently at minimum ionisation. To effect a separation, it was feasible to use kinematic methods based on the general properties of two-body decays, using the early theoretical work of Blaton and with the help of Podolanski at Manchester. Using this method the initial 26 well-measured neutral V-decays were classified as probably 15V_1^o and 11 V_2^o.

Although the V_1^o and V_2^o decay schemes were originally put forward as a plausible explanation of the early data, developments during later years at the Pic-du-Midi and elsewhere supported this simple model. The possibility, however, that a small proportion of the neutral V-decays involved three secondary particles was a lively issue for a considerable time.

It soon became clear that the Manchester cloud chamber at the Pic-du-Midi was too small, but the pre-war design of the magnet severely limited the possibilities for further development. Although efforts were made to improve the apparatus, its usefulness declined as time passed. We therefore welcomed the arrival of a new large chamber array from the Ecole Polytechnique in Paris in 1953. It consisted of two cloud chambers, one mounted above the other. The top chamber (68cm high by 64 wide and 30cm deep) was provided with a modest magnetic field of 2600 gauss. The lower chamber had the same height and width, but it was 40cm deep and contained 15 plates of copper each 10mm thick. In practice this splendid piece of equipment proved to be most valuable for studying the decays of charged V-particles but, for example, important results on the production of neutral V-particles were obtained. My group worked at the Pic-du-Midi for nine years but, towards the end of this period, the small chamber was closed down and we collaborated with the Ecole Polytechnique group using the large chamber array.

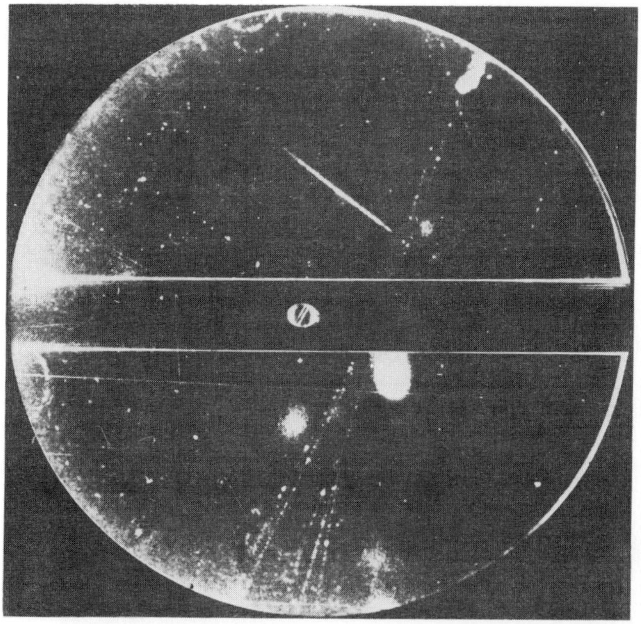

Figure 6 The first Ξ^--decay. (obtained 4 July 1950)

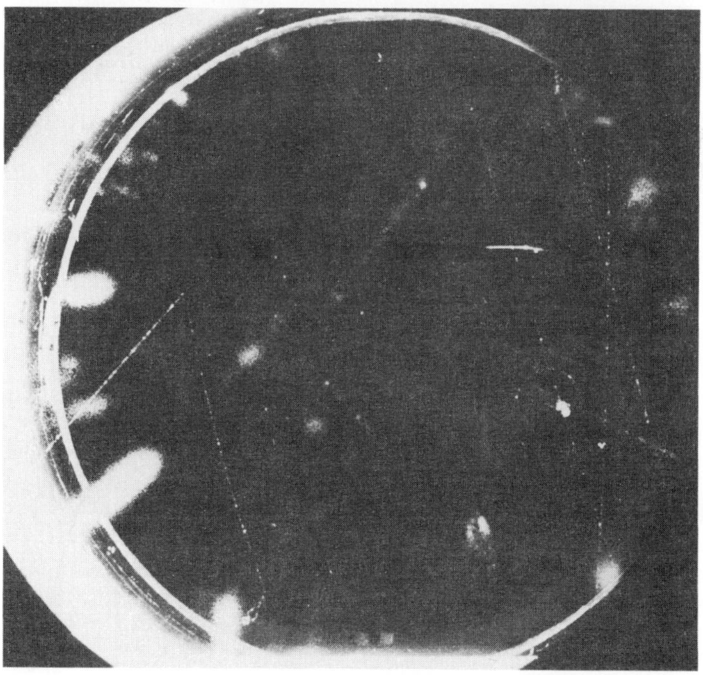

Figure 7 The decay of a V_1^o -particle (obtained 13 January 1951)

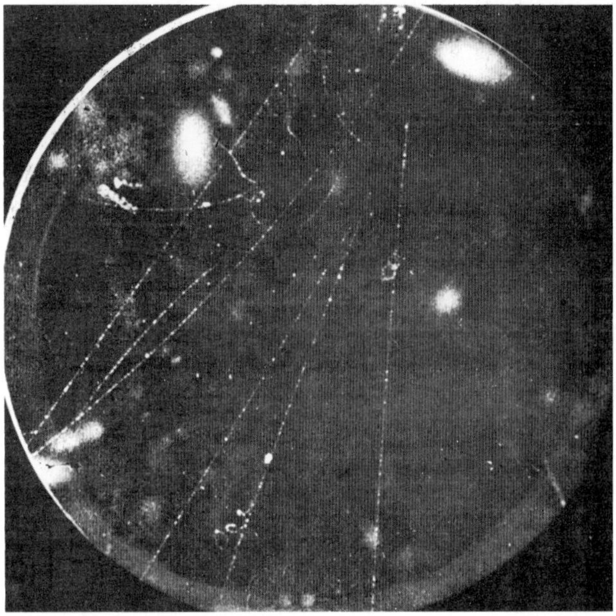

Figure 8 The decay of a V_1^o -particle (obtained 24 January 1951)

3. THE PROPERTIES OF NEUTRAL V-PARTICLES AS UNDERSTOOD AT THE END OF 1954

During the years 1951 to 1954 several large magnet cloud chambers were constructed specially for work on V-particles. One of the most successful was built by Thompson; it was operated at Indiana, at sea-level. The illuminated volume of this chamber was 55cm high, 27cm wide and 12.5cm deep. A uniform magnetic field of 7000 gauss was provided by an iron-cored solenoid. Particularly careful attention was paid to the temperature stability of the chamber to minimise track distortion. Excellent momentum measurements could be made with this chamber which more than compensated for the fairly low rate of obtaining useful events. Using this apparatus enabled the Indiana group to make the most accurate momentum measurements on the secondary tracks of V-particle decays. They further developed the kinematic analysis method first used by my Manchester group and undoubtedly produced the most reliable data on the two kinds of neutral V-decay.

Multiplate chambers were mainly used for studying charged V-particles, but numerous examples of neutral V-particles produced in a plate inside the chambers were obtained. In almost all cases the plane of the V-particle secondaries intersected the origin of the unstable particle, thus indicating that only the two charged secondary particles were produced. Sometimes multiplate chambers were used alone, but several groups placed a large one underneath the magnet chamber, a typical example being the Ecole Polytechnique arrangement at the Pic-du-Midi.

R.W. Thompson collected the available data on neutral V-particles at the end of 1954 and published an excellent review article in 1956, possibly the last to be devoted exclusively to cosmi-ray data. I now describe briefly the main properties of the particles using Thompson's findings.

By the end of 1954 the V nomenclature adopted by Blackett was largely discarded and, despite Anderson's early reservations, was replaced by Greek letters. The V_1^o became the Λ^o hyperon. The decay scheme for the Λ^o was generally believed to be:

$$\Lambda^o \rightarrow p^+ + \pi^- + 37 \text{ MeV}$$

The uncertainty in the Q-value was within 1%. The mass of Λ^o was (2181 ($\pm$) 1)m_e and the best value of the mean life from cloud chamber work was:

$$(3.7 \begin{array}{l} +0.6 \\ -0.5 \end{array}) \times 10^{-10} \text{sec}$$

Attempts have been made to identify Λ^o -decays in photographic emulsions, but unfortunately they are concealed by a large background of 2-prong stars. The Q(p , π)-values of likely candidates, however, were calculated and convincing evidence of clustering around 37 MeV was obtained.

The V_2^o -particles were named θ^o in 1954 and it was generally accepted that they decay according to the scheme:

$$\theta^o \rightarrow \pi^+ + \pi^- + (214 \pm 5) \text{ MeV}$$

This conclusion was derived from cloud chamber data. The process was believed to be a two-body one largely from the unique Q-value. One of the secondaries was known to be a pion and the other was thought likely to be a pion. The mass of the θ^o was (966 ($\pm$) 10)m_e and the mean life

$$(1.7\ ^{+0.6}_{-0.35})\times 10^{-10}\,\text{sec}$$

At the end of 1954 a small number of non-Λ^o-decays have measured Q(π^+,π^-)-values which differ considerably from 214 MeV. These anomolous cases have Q-values from 40 to 125 MeV. Some of these events were probably due to decays into three pions, one of which was neutral. Further work was needed to elucidate this residual problem.

The production processes for the Λ^o and θ^o were not well understood from cosmic-ray data at the end of 1954. By then the Shutt diffusion chamber had been operated in a high energy pion beam at the Cosmotron and the associated production of Λ^o and θ^o convincingly demonstrated. Cosmic-ray evidence supporting the Shutt discovery was obtained in 1954 by the Ecole Polytechnique group at the Pic-du-Midi and by the Indiana group.

The cascade decay process was first found in the Manchester cloud chamber at the Pic-du-Midi. Only a very few events were seen, all negatively charged. The interpretation adopted by the end of 1954 was the two-body decay of a new particle, the Ξ^-, into a Λ^o and a π^-, according to the decay scheme:

$$\Xi^- \rightarrow \Lambda^o + \pi^- + (66 \pm 6)\ \text{MeV}$$

At the Rochester Conference in 1956 it was reported that the word sample of Ξ^--decays was then about ten. The best Q-value was 65 ($\pm$) 4 MeV.

4. SOME PERSONAL REFLECTIONS ON CLOUD CHAMBER STUDIES BETWEEN 1947 AND 1954

The seven years from 1947 were a remarkably successful period for cloud chamber physicists, for the study of showers of penetrating particles and particularly for the discovery and study of neutral V-particles. Although a small number of neutral V-decays were seen in emulsion experiments, undoubtedly the main thrust of the work on these particles was made using cloud chambers in various ways. It was essential to combine data from different experiments quickly and a remarkable international co-operation developed very rapidly.

I leave Professor Rochester to describe the contribution of cloud chamber studies to our understanding of charged V-particles since he has worked with both cloud chambers and photographic emulsions.

With hindsight perhaps we should ponder on the value of the mountaintop experiments. The rate of acquisition of data was increased substantially, particularly for rare phenomena like the cascade decay, but on the debit side work was difficult at high altitude, particularly in Europe, where the groups were working a long way from base and frequently in foreign countries. It now seems to me that the remarkable achievements of the Indiana group at sea-level show that high precision was more readily achieved and maintained at a base laboratory.

The cloud chambers in all their various forms were temperamental instruments, but nevertheless very suitable for momentum measurements, coplanarity tests and for lifetime measurements on the neutral V-particles. They were not particularly suitable for the study of production processes for the new particles, although the high pressure diffusion chamber played an invaluable role in the early experiments with the Cosmotron, particularly for studying production processes in the hydrogen gas filling the chamber. Sadly the cloud chambers all passed out of use once the discovery of the bubble chamber by Glaser in 1952 was understood and exploited on a large scale, which all took place remarkably quickly.

5. ACKNOWLEDGEMENTS

I am particularly grateful to Professor J. Roesch, Director of the Observatory on the Pic-du-Midi, for making the Manchester group welcome in his laboratory and for overcoming so many difficulties, particularly at the beginning of the enterprise. I also wish to thank Professor George Rochester for many valuable discussions. Finally I wish to thank all my colleagues - initially Armenteros et al - who formed the Manchester and later the Imperial College (from 1953) group at the Pic-du-Midi. The names changed with time but they were always a splendid team. It is a cause for sadness that Keith Barker, who contributed so much over the years, died so young.

A SELECTION OF RELEVANT REFERENCES

Armenteros et al (1951a) Nature 167, 501
Armenteros et al (1951b) Phil Mag 42, 1113
Armenteros et al (1952) Phil Mag 43, 597
Armenteros et al (1953) Phil Mag 44, 861
Butler (1982) Journal de Physique, Colloque C8, Supplement 12, vol 43.
Thompson et al (1951) Phys Rev 83, 175

REVIEW ARTICLES

Butler (1952) Heavy Unstable Cosmic Ray Particles
Progress in Cosmic Ray Physics
North Holland Publishing Company

Rochester and Butler (1953)
The New Unstable Cosmic-Ray Particles
Reports on Progress in Physics, 16, 364

Thompson (1956)
Decay Processes of Heavy Unstable Neutral Particles
Progress in Cosmic Ray Physics, Vol III
North Holland Publishing Company

CONFERENCE REPORTS

(1) Congres International sur le Rayonnement Cosmique
Bagneres de Bigorre, July 1953

(2) Pisa Conference 1956
Nuovo Cimento, Supplement to Vol IV, no 2

(3) Proceedings of the Sixth Annual Rochester Conference, April 1956

The role of the cloud chamber in the strange particle saga

G.D. Rochester, Department of Physics, Durham University, Durham DH1 3LE

ABSTRACT: A summary of the contributions of the cosmic-ray triggered cloud chamber to the study of the strange particles.

1. NEUTRAL STRANGE PARTICLES

In the papers by Sir Clifford Butler and myself some of the contributions of the conventional cloud chamber to the study of the strange particles have been described. In this short paper the unique contributions made by this technique will be outlined and compared with those of the emulsion technique. First of all, however, it must be stressed that the two techniques complemented each other to an astonishingly high degree, the cloud chamber being particularly effective in the discovery and study of neutral particles and the emulsion of charged particles.

Historically, the ps-triggered cloud chamber was unique in the discovery of the two main types of V°-particles, the $K^{\circ}_{\pi^+\pi^-}$ -meson and the Λ°-hyperon and in the precise determination of their modes of decay, their masses and lifetimes. In contrast, few observations of emulsion K°-mesons were made for several years, indeed, in a well-known review by Thompson (1956) only four cases were recorded, the best being a well-identified example by Yasin (1954). Because of the nature of the secondary pions these events gave little new information. The situation was very different for the Λ° -hyperon for eventually many fine examples were found in emulsion and from the decays of the highest quality a Q-value was obtained of much higher accuracy than the cloud chamber value, Friedlander et al (1954).

2. CHARGED STRANGE PARTICLES

The cloud chamber also made unique historical contributions to some of the charged strange particles notably the Ξ^--hyperon, the $K^{\pm}_{\pi 2}$ -meson and the $K^{\pm}_{\mu 2}$ -meson.

The Λ° hyperon was discovered by the Manchester Pic du Midi Group in 1952 and was the first example of a particle of strangeness -2. The complete decay mode, $\Xi^- \rightarrow \Lambda^{\circ} + \pi^-$ was established by Cowan (1954).

The resolution of the charged kaon group proved arduous because of the many decay modes of the same particle. After the important discovery of the τ-meson, or $K^{\pm}_{\pi 3}$, and the precise measurement of its mass, the next significant emulsion discoveries were made by O'Ceallaigh (1951) at Bristol who found examples of two and three-body decays to muons and neutral particles from primaries whose masses were shown by Menon and Rochat to be in the region of the kaon mass. Examples have been shown at this Conference by Lock (q.v.) In 1952 and at the Bagnères Congrès Menon and O'Ceallaigh suggested a third form of two-body decay involving

© 1988 IOP Publishing Ltd

a π^+-meson and a neutral particle, a mode which was established by these workers and others in 1954. Details are to be found in Powell, Fowler and Perkins (1959). However, although the positive pion was completely determined, the nature of the neutral particle was not completely known. In spite of this the form of the decay was confidentially expected to be $K^+_{2\pi} \rightarrow \pi^+ + \pi^\circ$. The proof of this decay came in a remarkable cloud chamber photograph taken by Hodson at the Princeton array at Echo Lake where the whole decay was seen and the π°-meson was completely identified by its decay to two Dalitz election pairs, Hodson et al (1954).

Although muons had been shown to be present in some forms of kaon decay definitive answers did not come from emulsion work at first because the emulsion blocks were too small to arrest the charged products of the decays. Bonetti (1956) in his discussion of the G-stack made this point by stating that the G-stack was the first emulsion block in which the long range kaon secondaries could be stopped.

The resolution of the charged kaon group was achieved by the combined work of the M.I.T. and the École Polytechnique cloud chamber groups. Diagrams of their arrays are shown in Fig. 1. from which it will be seen that the important feature of each array was a multiple cloud chamber. The M.I.T. chamber was run at Echo Lake and the École chamber at the Pic du Midi. Details are to be found in the excellent surveys of Bridge (1956) and Peyrou (1982).

The M.I.T. work, published by various authors from a long list of workers - Annis, Bridge, Courant,Dayton, Gregory, Olbert, Peyrou, Rossi, De Staebler, Stafford,Sreekantan and Willard - was part of a long series of excellent studies of high-energy cosmic-ray events in multiplate cloud chambers, in the course of which stopping kaons were found. An example of what appears to be a $K^+_{\pi 2}$-meson decaying to a π^+-meson and a π^o-meson is shown in Fig. 2.

The École Polytechnique set-up was much more elaborate and versatile and consisted of two cloud chambers one above the other, an arrangement used by many cosmic ray workers starting with the early work of Street and Stevenson on the mesotron, Street (1939).

The upper chamber was in a magnetic field and the multiplate chamber contained enough matter, (15 - 1cm copper plates) to stop the kaon and the longest range charged secondary it could produce. As described in Gregory et al (1954) and Armenterous et al (1955) clear evidence was obtained for a kaon decaying to a monoenergetic muon via the mode $K_{\mu 2} \rightarrow \mu + \nu$. An example is shown in Fig. 3.

The evidence for the various modes of decay is shown in Figs. 4 and 5 in which are displayed for the M.I.T. (Fig. 4) and the École Polytechnique (Fig. 5) the number of events against the ranges of the secondaries. Other information is also given. It is seen three main groups occur in which the ranges of the charged secondaries are 15, 45 and 75 g.cm^{-2} of brass. These were identified as $K^+_{\pi 3}$, $K^+_{\pi^+\pi^o}$ and the $K^+_{\mu 2}$-mesons. A very careful analysis of the $K^+_{\mu 2}$ group by the École group proved that the form of the decay is $K^+_{\mu 2} \rightarrow \mu^+ + \nu$.

Later work by many groups showed that the charged kaons occur in the following proportions: 63% $K_{\mu 2}$, 21% $K_{\pi 2}$ and 5% $K_{\pi 3}$.

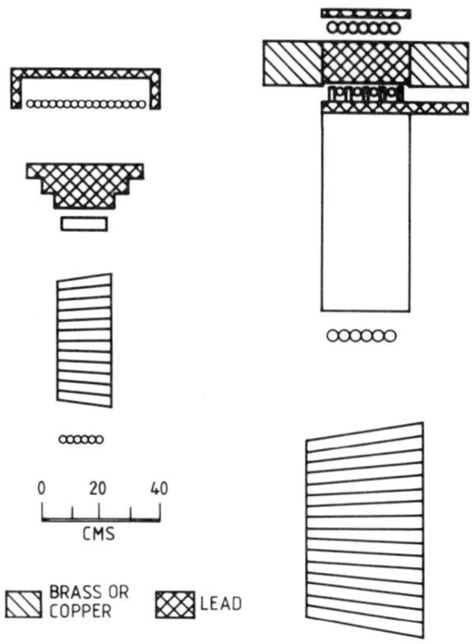

Fig. 1. Diagrams of the M.I.T. (at Left) and the École (at Right) arrays. After Bridge (1956).

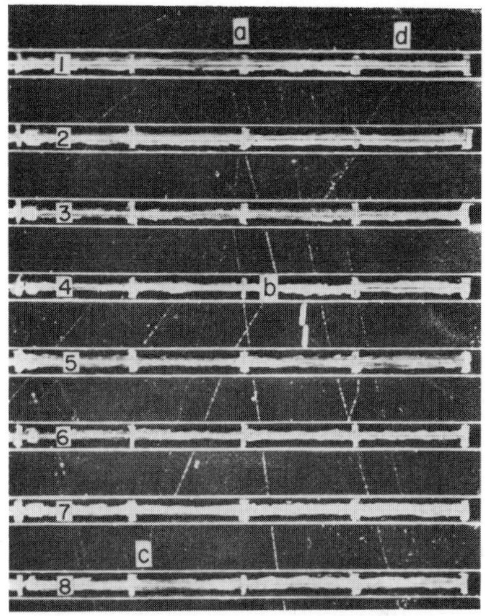

Fig.2. A particle, presumably a kaon, stops at b and produces a charged meson, probably π^+, which stops at c, and a neutral meson, π°, which produces a small electron shower in the third plate. After Bridge (1956).

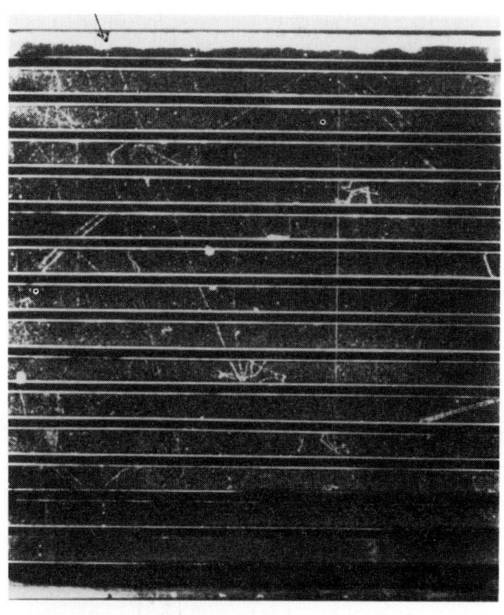

Fig. 3. An example of an École $K_{\mu 2}$ decay. The primary comes into the chamber at the arrow and stops in the 10th plate. The secondary is emitted upwards and stops in the third from the top. Gregory et al (1954).

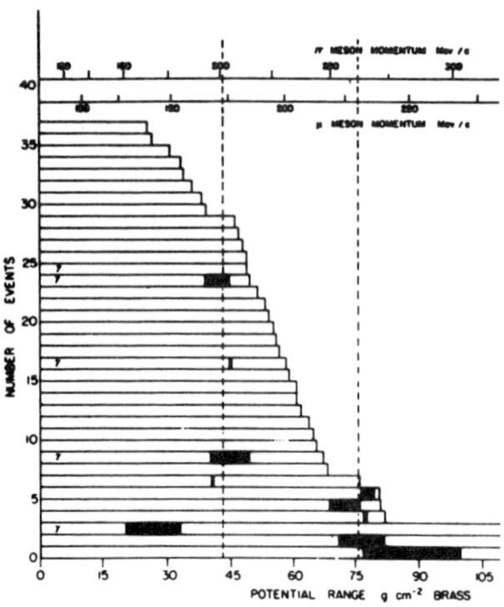

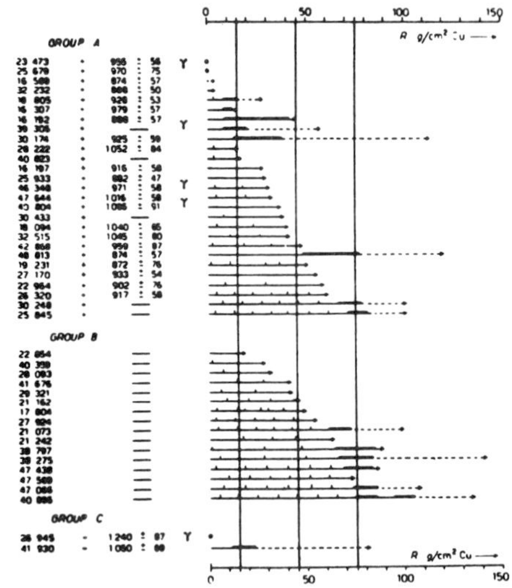

Figs. 4 and 5. In these diagrams the number of events is plotted against range for the most prominent secondaries of charged kaons. After Bridge (1956) and Peyrou (1982).

3. ACKNOWLEDGEMENTS

The Head of the Physics Department is thanked again for the use of departmental facilities. The sources of the diagrams and photographs are indicated in the text; acknowledgement has already been made to the authors. Professors Fowler and Friedlander are thanked for kindly bringing the work of Yasin to my attention.

4. REFERENCES

Armenterous, R., Gregory, B., Hendel, A., Lagarique, A., Leprince-Ringuet, L., Müller, F., and Peyrou, C., 1955, Nuovo Cim. Ser. 10, 1, 915
Bonetti, A., 1956, Nuovo Cim. Ser. 10, 4, (Supp.), 419
Bridge, H.S., 1956, Prog. Cosmic Ray Phys. 3, 182
Cowan, E.W., 1954, Phys. Rev. 94, 161
Friedlander, M.W., Keefe, D., Menon, M.G.K., and Merlin, M., 1954, Phil. Mag. 45, 433
Gregory, B., Lagarique, A., Leprince-Ringuet, L., Müller, F., and Peyrou, C., 1954, Nuovo Cim. 1, 292
Hodson, A.L., Ballam, J., Arnold, W.H., Harris, D.R., Rau, R.R., Reynolds, G.T., and Trieman, J.B., 1954, Phys. Rev. 96, 1089
O'Ceallaigh, C., 1951, Phil. Mag., 42, 1032
Peyrou, C., 1982, J. de Phys. 43, (Supp. 12), C8-7
Powell, C.F., Fowler, P.H. and Perkins, D.H., 1959, The Study of Elementary Particles by the Photographic Method (London : Pergamon)
Street, J.C., 1939, J. Frank. Inst. 227, 765
Thompson, R.W., 1956, Prog. Cosmic Ray Phys. 3, 255
Yasin, M., 1954, Phil. Mag. 45, 413

Discussion

Perkins

I'd just like to ask Clifford about his mentioning the decision to go to the Pic du Midi rather than the Jungfraujoch because of the difficulty of getting large components up the lift at the Jungfrau. I remember Newth, when we went to CERN, went with a CERN-based cloud-chamber to the Jungfraujoch, the Sphinx.

Butler

That was the Manchester one.

Perkins

So how did the Manchester chamber get up the Jungfraujoch?

Butler

The magnet was designed to break into smaller pieces to fit into the lift.

Perkins

So it wasn't that Occhialini preferred the Pic?

Butler

No. I'm sure Occhialini did prefer it, but Blackett tried the idea out on the Swiss engineers and they would not take the risk. They wouldn't take any risks with the lifting equipment in the tunnel and you couldn't drag it up the outside of the Jungfraujoch. The Swiss wouldn't have done the things the French did!

Perkins

You didn't have enough Swiss Francs, Clifford!

Peyrou

I want to make a comment on Clifford's speech. He sort of said that the Cosmic Rays didn't do very much with the K^o_s, which is true, there were a few examples but nothing systematic. It was done on a machine, but still with a cloud chamber, by Lederman. The whole K^o_s in 3π was done by Lederman and Co. at Brookhaven.

Friedlander

There was another major contribution to physics which came from the cloud chambers. Rossi's group had also looked at the S particles, the heavy meson which stopped and gave a single charged secondary. Then later Sreekantan came back and re-examined these photographs and found that in the opposite direction there was a shower which identified the particle as the π^o.

© 1988 IOP Publishing Ltd

The identity of the neutral particle was turned up by Sreekantan re-examining Rossi's pictures. That was also a very important contribution.

Rochester

The photograph which absolutely established the $K^+ \to \pi^+ \pi^o$ was of course the Hodson picture which was taken at Echo Lake with the Princeton array. This is one of the most remarkable cloud chamber photographs ever taken and shows the complete decay $K^+ \to \pi^+ \pi^o$; $\pi^o \to \gamma\gamma$ which form two Dalitz pairs. So there are four electrons and the π^+. I wouldn't say that that was the main contribution because the emulsion people had already established the decay of $K^+ \to \pi^+ \pi^o$ but the Hodson photograph somehow completely confirmed and convinced everyone.

Fowler

It is a very small point: you said we never actually saw a θ^o in emulsion, in a way that could be recognised. We did in fact have one very close to a star, and we could only pick it out because it decayed 20 μm away, which is about $10^{-3} \to 10^{-4}$ of the expectation, and it decayed very asymmetrically, so that the π^- had almost no energy and stopped and the rest had a long track. So we did see one.

Rochester

Good.

Hodson

I'd like to make one comment about Professor Rochester's kind comment about our event. The important point was that the π^o decayed directly into four electrons rather than through two γ-rays. I think it was the first time this decay scheme was seen.

Sard

I'd like to mention another important contribution of the cloud chamber, the picture of LePrince-Ringuet and L'Heritier in 1944 of the first charged K. The mass was clearly determined by an electron knock-on. I believe that event survived much criticism after the war.

Rochester

I always mention this as a possible charged K, but it was very severely examined and criticised by Bethe who thought that there was a few % probability of this being a proton with a knock-on.

Sard

Well, Bethe did criticise it, but I must say I didn't find his criticism convincing.

Peyrou

I am supposed to be an expert on what happened in LePrince-Ringuet's laboratory. I must say that I never believed the criticism of Bethe. It was before the first $K \to 3\pi$ discovery so he wanted to doubt about mesons of mass 1000 so he made a critique of all evidence. His article ends about LePrince-L'Héritier's event that with quite a lot of stretching of errors you can make it a proton. It says no more than that. Since there are K-particles, you cannot say that this event is a phoney. On the other hand I examined the conditions of that picture, and there cannot possibly be a K-particle in the chamber. There is no producer, there is a single particle. So I always write on that picture that it is an unlikely event, in the sense that if it is a proton it is an unlikely knock-on and if it is a K-particle it shouldn't have been there!

G. Kalmus

You mentioned that after about 1956 the cloud chamber more or less died but I remember that there was a brief revival in around the late '60's when, if I remember, McCusker looked for quarks in the cores of showers. That led to a small cottage industry which I happened to be slightly involved with in Berkeley when Wilson Powell activated some chambers and I guess others did too. That was a comment. I also wanted to ask a question: are there any cloud chambers being used in anger now?

Hodson

The Leeds 3 m^2 cloud chamber is being used at this moment for a search for free quarks right in the cores of air showers where it has not been possible to make a search before. We find the position of the shower core at roof level using current imaging spark chambers, trace the direction down to the cloud chamber level 6 m below so that in each event which is scanned we know precisely where we have looked. That search is currently in progress.

Menon

Since a large part of what we have been discussing this morning has essentially been on heavy mesons and hyperons and the cloud chamber work, I'd just like to clarify some of the points that George Rochester raised about the emulsion work. The first of the heavy mesons found in emulsion was in 1949 in the τ mode which was described yesterday in detail. Thereafter the normal practice when the plates came in and the scanners looked at them was to classify the various events seen as heavy primaries, showers, stars etc. There were two categories, one called the ρ mesons, which are those classified earlier in the "Nature" papers, and Peter Fowler showed the breakdown yesterday. The other category was the σ's. These were the ones which interacted and had stars at the end. The ρ's were the ones which decayed. That was the normal classification.

Perkins

The ρ's did nothing.

Menon

In electron sensitive emulsions you saw the decay. I'm talking about the post-τ meson period, not the first C2's. What really happened was that we were looking in various groups. Peter mentioned Cormac O'Ceallaigh who was looking at the spectrum of electrons from muons, which is a continuation of what Lock described, the Davies-Lock-Muirhead experiment. One of the scanners found a so-called ρ, with an electron, and a minimum ionising track going off. It was in the bunch sent off to O'Ceallaigh. Unfortunately Cormac is not here so I thought I'd just mention the way things developed. Of course he immediately looked at the momentum by scattering and found a very high value of pβ, about 250. Of course this was way above anything one could have from μ decay. He had to leave for Dublin, so he asked his wife, Millie, who unfortunately passed away a few months ago, to show it to me and ask me to remeasure it. When I remeasured it I found the same value: there was nothing wrong in the measurements and quite clearly therefore it couldn't be a muon decay. Just then Rochat and I were working on the measurement of primary mass based on scattering range. So we quickly measured and fortunately, that K1, as it was then called, had enough range and we found a mass of around 1200. That's when we took it to Cecil; he said that we would wait until Cormac came back. A few days later the second event was seen, which Peter has on display under the microscope, the so-called K2 in which you had a decay into a muon which stopped and had a long range. In these two events, K1 and K2, one had a high pβ but its nature could not be identified and the other had a secondary which was quite clearly a stopping muon. The first 10 events were soon found, all looked for in the ρ category, i.e. a stopping muon with something at the end of it. The measurement was very simple in a certain sense. What we were trying to do in those days was just look at the secondaries, measure the pβ, identify them on the same curve which Peter Fowler and Owen Lock showed and they were clearly not coming from muons or pions in many cases. The pions all had the same value of energy. The primary masses were all around the same, around 1000 with a small spread. With the first ten events one clearly had, without going into any observations anywhere else, just in this group, evidence for monoenergetic pions and a wide range of energies for the muons. This was clear at the Bristol conference on V particles and thereafter at a succession of meetings. Sometime later we found the electron which suffers bremsstrahlung along the trajectory. I want to emphasise that this entire category emerged on a self-contained basis from just a simple set of measurements at that time.

George also mentioned a point about the range. The Ecole Polytechnique group had reported on the beautiful work with the double cloud chamber system at the Pic du Midi after it was clear that if one wanted to stop the μ from the K-μ-2 then of course one would have to look for the range which Peter mentioned. But on the other hand one could stop the K-π-2, the π secondary, which is the origin of the G-stack, because we would have to go for something of about 10 litres. So for the first time we took a giant step in terms of the volume of emulsion handled. Nobody had ever handled 15 litres before and that was the basic idea which came about at the time that Merlin from Padua was also at Bristol. That's how the G-stack was really born. As Mike Friedlander mentioned yesterday, the first paper was probably the first in which the entire first page consisted of authors' names!

Peyrou

Just to complete what Menon was saying. They discovered the K-π-2, even before M.I.T., it was a very high quality job. What we did was to find a monoenergetic μ among the candidate decaying K-mesons. That was partly shrewdness, partly luck. The shrewdness was to have a chamber with high stopping power, to stop the secondaries. It turns out that momenta for scattering the π from the K-π-2 and the μ from the K-μ-3 are not that different and therefore it is hopeless to disentangle them, whereas the μ, being lighter than the π has a much larger range for a slightly higher energy, so that you were sure that if something passed through 70 g of copper coming from a particle with mass less than 1000, it could not be a π. Our conclusion was extremely easy. That was not planned!

Fowler

One comment on that. The grain density in the K-π-2 is distinctly, about 10%, above that of the μ, and we had that as a help.

Peyrou

I know that's how you did it. But it's tough!

Rochester

One further point to add about what Peyrou said. Many people miss things in the French paper. There are many subtle things there which confirm the K-μ-2. One of the most subtle, which I think is a very telling argument, is that the range of the μ which they found meant that it could not be a pion because the mass of the primary was less than 1000 so that one could not possibly produce a pion of that range. That is a very convincing argument to me that this was a genuine, mono-range, muon.

Menon

Around the time we were doing this work Cecil Powell was not quite well and didn't want to go through the bother of correcting the paper written by a mad group of Irishmen, Indians, Israelis and so on, so he decided that the physics was all right, he had looked at that, but that the English must be looked at. Since he didn't have the time he decided that a friend of his, a poet from here would correct our papers. I would like to say that he made a total hash of all the papers, which had to be recorrected.

Peyrou

I have a small confession. It's well known by the old experts that our mass was a little on the low side, by two standard deviations or so, but we never tried to push it up because the argument was much better if it was lower!

The pion in particle physics today—an overview

R.H. Dalitz

Department of Theoretical Physics, Oxford University

ABSTRACT: Pions play an outstanding role in nuclear and particle physics, whether as fundamental field quanta, as composites made of quarks and gluons, as Goldstone bosons generated by chiral-symmetry breakdown, or otherwise. They contribute dominantly to the deuteron binding energy and quadrupole moment, generate the one-pion-exchange pole apparent in many cross-sections at high energies, and have both charge and parton structure. These and other facets of the pions, in both active and passive roles, are traced to their special features. They have been repeatedly illuminating as a testing-ground for our beliefs about underlying symmetries for the physical interactions.

PREFACE

Let me remark first that I worked in the Bristol Physics Department for the year 1948-49 as research assistant to Professor Mott. At that time, having recognized the fundamental significance of the discoveries being made on the fourth floor of his department, he had the intention in mind to move back into research on nuclear physics, from which he had been diverted in the thirties and the war years, and I was to help him to make this change of direction. Well, he didn't make it back into nuclear physics but that's another story; at least I need feel no guilt at my failure to bring him back into our field, for he subsequently did very well indeed in the field from which he failed to break away.

However, I want to say here that 1948-49 was one of the best years of my life. The personal contacts and friendships I made at Bristol, and the research work I saw going on, determined the later course of my life, and especially of my research interests. I am happy to be back here again today.

1. THE PION AND ITS DECAY MODES

The decay of the charged pion is due to the Weak Interactions. These are carried by the Weak Currents, of which there are two; the vector current V_μ, which has spin-parity (1-), and the axial-vector current A_μ, which has spin-parity (1+). V_μ is a conserved current, but A_μ is not, as we shall now see. The decay $\pi^+ \rightarrow \mu^+\nu_\mu$ is its dominant mode; more generally we may denote this by $\pi^+ \rightarrow \ell^+ \nu_\ell$, where ℓ^+ is the generic charged lepton and ν_ℓ is its associated neutrino.

© 1988 IOP Publishing Ltd

1. $\pi^+ \to \ell^+ \nu_\ell$.

This is a transition '$\pi^+ \to$ vacuum', as far as strongly interacting particles are concerned, which can be induced only by the A_μ current. We may write this in the following form,

$$\langle \pi(k) | A_\mu(k) | vac \rangle = f_\pi k_\mu, \tag{1.1}$$

where $(-k_\mu)$ denotes the energy-momentum four-vector for the outgoing $\ell\nu_\ell$ pair. The right-hand of (1.1) is the only form possible for this transition amplitude, since it must obviously be a four-vector and k_μ is the only four-vector available. If A_μ were conserved, we would have

$$\partial A_\mu / \partial x_\mu = 0. \tag{1.2}$$

In momentum-space, $\partial/\partial x_\mu$ is simply ik_μ and the matrix-element of the left-hand side of (1.2) then reads

$$0 = i \langle \pi(k) | k_\mu A_\mu(k) | vac \rangle, \tag{1.3}$$

which, following eq. (1.1), becomes

$$0 = i f_\pi (k_\mu)^2. \tag{1.4}$$

However, $k_\mu{}^2$ is simply $(\text{mass})^2$ for the pion $\pi(k)$. Since the pion mass is non-zero, we would conclude that f_π must then be zero. Since the π^+ <u>does</u> decay, we can only conclude that the axial-vector current $A_\mu(k)$ is not conserved.

So the decay $\pi^+ \to \ell^+\nu_\ell$ is an allowed process (Taylor 1964). However, the axial-vector current A_μ has the form $(\bar{\ell}\gamma_\mu(1+\gamma_5)\nu_\ell)$, where ℓ and ν_ℓ here denote the Dirac spinor wavefunctions for these particles. The operator $\gamma_\mu(1+\gamma_5)$ has the property of helicity conservation, if the lepton mass m_ℓ is zero. This would mean that the outgoing leptons, which momentum conservation requires to have opposite momenta, must have opposite helicity. The net angular momentum component about the lepton momentum would then be ±1, which is forbidden since the ($\pi^+ \to$ vacuum) transition involves angular momentum zero both initially and finally. This means that the amplitude for this transition must be zero for $m_\ell = 0$. In the real world, $m_\ell \neq 0$, so that the transition amplitude must be proportional to m_ℓ for small m_ℓ, as can readily be verified by direct calculation. This leads at once to the following branching ratio

$$\Gamma(\pi^+ \to e^+\nu_e)/\Gamma(\pi^+ \to \mu^+\nu_\mu) = (m_e/m_\mu)^2 \approx 10^{-5}. \tag{1.4}$$

a striking result, which was first demonstrated empirically by Fazzani et al.(1958), and which provided an important check on the underlying theoretical ideas.

The well-known lifetime for π^+ decay leads to a value of about 90 MeV for the coefficient f_π in (1.1).

1.2 $\pi^0 \to \gamma\gamma$.

The neutral pion decay occurs through the electromagnetic interactions, which are another part of the unified Electroweak Interaction. The graph corresponding to this decay is shown schematically on Fig. 1. The photons interact through V_μ currents, while the pion necessarily interacts through the A_μ current. This A → VV transition involves closed loops of intermediate fermion fields; it was first calculated by Steinberger (1949). This graph has considerable importance since it was the first example of an <u>Anomaly</u>. In terms of Anomaly Theory today, it can be expressed in a general form

$$(2/\pi)(e^2/\hbar c)S/f_\pi \tag{1.5}$$

where S is given by

$$S = \Sigma_i\, G_A(i)Q_i^2, \tag{1.6}$$

$G_A(i)$ being the axial-vector coupling amplitude $\pi^0 \to f_i\bar{f}_i$ for the hadronic fermion of type i, Q_i being its charge value, the sum to be taken over all types i of fermion. The $\pi^0 \to \gamma\gamma$ decay rate is then proportional to S^2, the remaining factors being of kinematic origin.

In 1949, Steinberger knew only the hadronic fermions (p,n) for which S takes the value 1/2. When $\tau(\pi^0)$ was finally measured, almost 20 years later, it was found that the value S = 1/2 actually gave good agreement with its value. Today, however, it is appropriate to take these hadronic fermions to be the quarks, in which case

$$S = \frac{1}{6}\cdot N_C \tag{1.7}$$

where N_C denotes the number of different quarks with the same charge and the same strangeness, in other words the number of "colours", the dimension of "colour space". Agreement of the calculated decay rate with the experimental value now requires $N_C = 3$, which we know to be the case on other grounds.

The π^0 meson has a number of lesser decay modes related with its established decay mode:

(a) $\pi^0 \to \gamma e^+e^-$, resulting from internal pair conversion of one photon. The branching ratio can be rather reliably estimated, with the result $(\gamma e^+e^-)/(\gamma\gamma) \approx 1/84$, in excellent agreement with the empirical value of (1.198±0.030)%.

(b) $\pi^0 \to e^+e^-e^+e^-$, resulting from internal pair conversion for both of the photons. The branching ratio is then $(2e^+2e^-)/(\gamma\gamma) = (1/168)^2$, in good agreement with the empirical value of $(3.18\pm0.30) \times 10^{-5}$. In general, each electron pair emerges with a small angle of divergence, so that the two pairs appear back-to-back, as depicted in Fig. 2; the two momenta of a pair define a pair conversion plane, and the angular correlation between the azimuths of the two planes can be measured, as suggested by Kroll and Wada (1955). The form of this correlation is related with the π^0-parity. In terms of the electric ($\underline{E}$) and magnetic ($\underline{H}$) fields associated with each of the two final photons, the interaction takes the forms

(0+): $(\underline{E}_1\cdot\underline{E}_2 - \underline{H}_1\cdot\underline{H}_2)$, which is zero for perpendicular electric fields.
(0−): $(\underline{E}_1\cdot\underline{H}_2 + \underline{E}_2\cdot\underline{H}_1)$, which is zero for parallel electric fields.

Since the plane of an e^+e^- pair is correlated with the electric field $\underline{E}$ of the virtual photon which gives rise to it, the correlation in azimuth between these planes is a direct indicator of the parity, and this was the most direct means by which the π^0 meson has been shown to have negative parity.

(c) $\pi^0 \to e^+e^-$. This process consists of two steps, $\pi^0 \to \gamma\gamma$ and then $\gamma\gamma \to e^+e^-$, involving the same number of electromagnetic interactions as there are for $\pi^0 \to 2e^+2e^-$. However, the electromagnetic interaction conserves helicity insofar as the electron mass can be neglected. Since the π^0 meson has zero total angular momentum and the e^+ and e^- have opposite momenta in this decay, the argument used above for $\pi^+ \to e^+ e$ holds here for $\pi^0 \to e^+e^-$ decay and we must conclude that the amplitude for $\pi^0 \to e^+e^-$ vanishes in the limit $m_e = 0$. In other words, the amplitude for $\pi^0 \to e^+e^-$ has its leading term proportional to (m_e/M), where M is some mass characteristic of the details of this mechanism, not less than m_π. Calculation shows that its leading term also has the factor $\ln(m_\pi/m_e)$. Hence the branching ratio $(e^+e^-)/(\gamma\gamma)$ is expected to be no more than about $2(e^2/\hbar c)^2(m_e/m_\pi)^2(\ln(m_\pi/m_e)^2 \approx 5\times10^{-8}$, which explains why it has been difficult to detect (the present upper limit is 2×10^{-7}). The essential validity of this argument is exemplified by the case of $K_L{}^0 \to \mu^+\mu^-$ decay. The $K_L{}^0 \to \gamma\gamma$ decay has a known branching ratio $(4.9\pm0.4) \times 10^{-4}$ and the two-step process $K_L{}^0 \to \gamma\gamma \to \mu^+\mu^-$ has a rate suppressed only by $(m_\mu/m_K)^2$ from helicity conservation. The measured branching ratio for $K_L{}^0 \to \mu^+\mu^-$ is $(9\pm2)\times10^{-9}$, in accord with this estimate.

2. THE SPECIAL ROLE OF THE PION

The leading role which the pion is seen to play in particle physics arises from one or more of the following facts, depending on the circumstances:

(a) the pion mass m_π is rather small relative to the mass $M_N \approx 1$ GeV characteristic of hadronic interactions, m_π/M_N being about 1/7.

(b) the pion is rather well separated in mass from other non-strange mesons with isospin I=1, the next lightest being the ρ-meson. Boson masses tend to occur as $(\text{mass})^2$ expressions and we have $(m_\pi{}^2/m_\rho{}^2) \approx 0.03$. This is to be contrasted with the situation for strangeness ±1 mesons with I=1/2, where the corresponding ratio is $(m_K/m_{K^*})^2 = (494/891)^2 \approx 0.3$.

(c) the great strength of the pion-nucleon coupling. This is usually measured by its pseudoscalar coupling constant $G^2/4\pi \approx 13.5$, corresponding to the interaction form

$$G\phi_\pi(\bar{N}\gamma_5 N), \qquad (2.1)$$

where N denotes the nucleon field. However this is an interaction which vanishes for zero momentum transfer, so that this large value for $G^2/4\pi$ rather overstates the case. The interaction is therefore often specified in its non-relativistic pseudovector form

$$f(\bar{N}\underline{\sigma}N)\cdot\underline{\nabla}\phi_\pi/m_\pi, \qquad (2.2)$$

the coupling coefficient f then being $(m_\pi/2M_N)G$, so that the coupling constant is $f^2/4\pi \approx 0.08$. This rather overstates the case in the opposite direction. It invites comparison with the electromagnetic coupling constant, their ratio being only $f^2/e^2 \approx 12$, but this assumes

Fig. 1. Anomaly graph for $\pi^0 \to \gamma\gamma$ decay

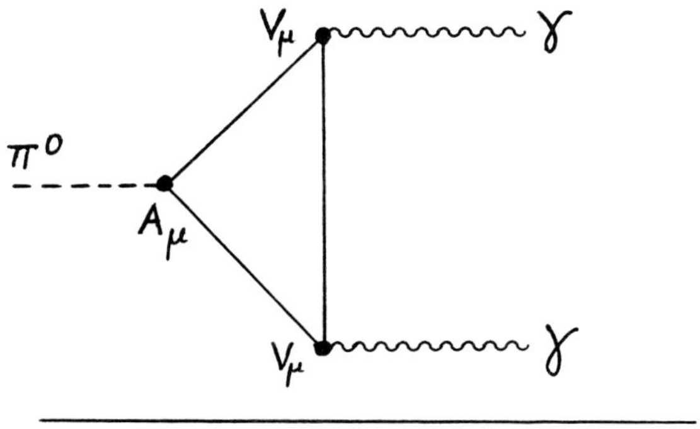

Fig. 2.

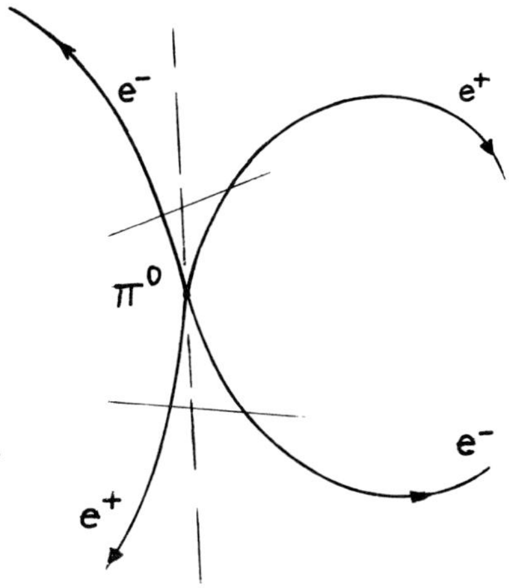

Fig. 2. A typical configuration for $\pi^0 \to 2e^+2e^-$ decay. The light dashed line shows the net momentum for each pair. The light lines across each pair are intended to indicate the two planes referred to in the text.

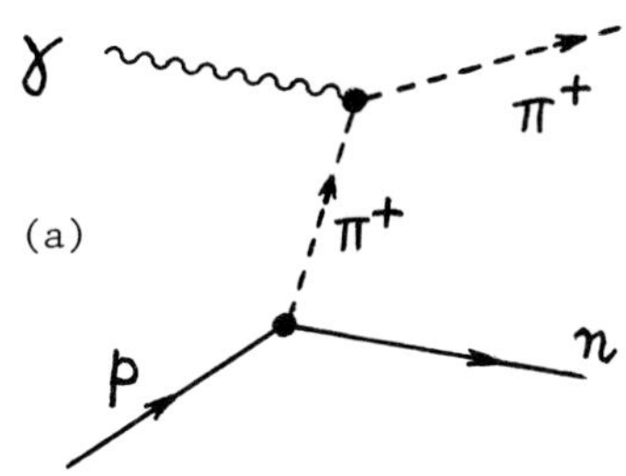

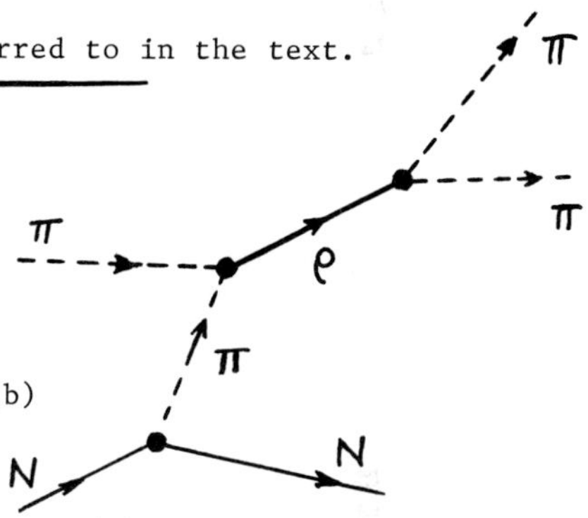

Fig. 3. OPE graphs for (a) $\gamma p \to n\pi^+$, and (b) $\pi N \to \rho N \to \pi\pi N$.

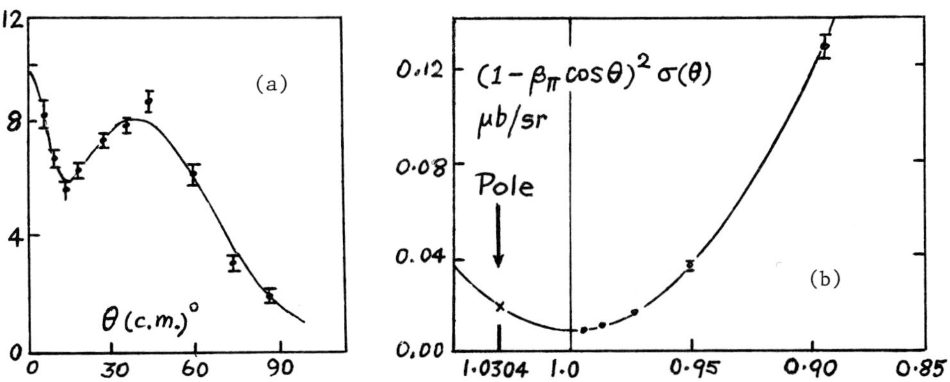

Fig. 4. For the process $\gamma p \to n\pi^+$, we show (a) the c.m. differential cross section measured by Walker (1960) for lab. photon energy about 1050 MeV, showing the OPE peak at 0°, and (b) the extrapolation of $(1-\beta_\pi \cos\theta)^2\sigma(\theta)$ to the pion pole at $\cos\theta = 1.0304$.

that the typical momentum transfer is of order m_π. In fact, the interaction (2.2) is momentum dependent, being proportional to $\underline{\sigma}.\underline{k}$, where $\underline{k}$ denotes the pion momentum, and the intermediate (virtual) momentum in any process can readily rise arbitrarily high; since the latter situation is unacceptable, the interaction (2.1) is usually modified by a damping factor $\rho(k)$ to limit the magnitude of the intermediate momenta to $k < K$, where the cut-off momentum K is to be determined empirically. This was the basis of the Chew-Low model for calculating pion-nucleon interaction processes, such as pion photoproduction $\gamma+N \rightarrow N+\pi$. Of course, the expression (2.2) is a static approximation to the interaction, in that the nucleons are treated as fixed sources or sinks for the meson field; the inclusion of nucleon recoil suppresses the intermediate momenta to some degree but a cut-off momentum K is still necessary. The use of the relativistic interaction (2.1) removes the need for a cut-off momentum, since the theory is then renormalisable and finite. However, firstly, it was never possible to find an adequate procedure for calculating the consequences of (2.1) as a quantum field theoretic interaction. Secondly, as we shall mention below, the rise of gauge theories for the elementary interactions, which have rather attractive motivations, has shifted our attention to quarks and gluons for the hadronic interactions, and especially to the gauge theory, quantum chromodynamics, and the pion has been displaced from the central role it appeared to have as a fundamental field in the early 1960s.

2.1 The pion as the nuclear force meson.

The fact that m_π is so small means that its exchange between nucleons provides the term of longest range in the nucleon-nucleon interaction. This one-pion-exchange (OPE) term also dominates in $V_{NN}(r)$ because of the great strength of the πNN interaction, whether (2.1) or (2.2). Further, because of the spin-dependence explicit in the form (2.2), the OPE-potential includes a very strong tensor component, its form being non-central,

$$V_T(x) = \frac{f^2}{4\pi}\left(\sigma_1\cdot\hat{r}_{12}\,\sigma_2\cdot\hat{r}_{12} - \frac{1}{3}\sigma_1\cdot\sigma_2\right)\left(\frac{1}{x}+\frac{3}{x^2}+\frac{3}{x^3}\right)\exp(-x), \qquad (2.3)$$

where $x = m_\pi r_{12}$ and $\hat{\underline{r}}$ denotes a unit vector in the direction $\underline{r}$. It was recognised quite early that such a strong tensor interaction was necessary in the neutron-proton potential to account for the electric quadrupole moment observed for the deuteron. This tensor interaction is strongly singular, which is a consequence of the strong momentum dependence of the interaction (2.2), the large momentum regime corresponding roughly to the regime of small distances r_{12}, and it is a major contributor, if not the dominant contributor, to the binding of the deuteron.

During the forties, the pseudoscalar meson theory, corresponding to interaction (2.1) or its approximation (2.2), was the only meson theory to predict a strong tensor NN potential. However, there was great reluctance towards accepting pseudoscalar meson theory in those days, partly because of the relativistic character of its interaction (2.1) and partly because of its great coupling strength, even after the quantum field theory of the pseudoscalar pion and its interaction (2.1) was shown to be renormalizable.

A further pointer towards the pseudoscalar meson theory in that period was the large anomalous magnetic moments measured for the neutron and the proton, namely −1.91 n.m and +1.79 n.m., respectively. It was pointed out by Wick (1935) that the constraints of angular momentum and parity conservation implied that the pion emission process $N \rightarrow N\pi$ led dominantly to a p-wave pion, and that such a meson with charge Qe would contribute $Qe\hbar/2m_\pi c = Q(M_N/m_\pi)$ n.m. to the nucleon. For the neutron, $n \rightarrow p\pi^-$ would generate a negative magnetic moment; for the proton, $p \rightarrow n\pi^+$ would generate an opposite magnetic moment, thus positive, to be added to the Dirac moment of +1 n.m. arising from its charge +e. This mechanism therefore accounted both for the sign of the neutron magnetic moment, and for both sign and magnitude of the anomalous magnetic moment of the proton, given the observed neutron magnetic moment . Further, since this pion emission process does not conserve energy, the emitted pion is necessarily virtual and can be present "in the air" for only a fraction of the time. The probability that there is a p-wave pion in the nucleon is related with the pion-nucleon coupling constant $f^2/4\pi$, and must vanish with $f^2/4\pi$. Since $(M_N/m_\pi) \approx 7$ and $f^2/4\pi \approx 0.1$, we see that Wick's remark leads to an estimate with the right order of magnitude.

2.2 One-pion exchange (OPE) processes

There are many physical processes at high energies which can be mediated by one-pion-exchange. We show two of them on Fig. 3, namely the processes

$$\gamma + p \rightarrow n + \pi^+, \tag{2.4a}$$

$$\pi^+ + p \rightarrow \rho^+ + n. \tag{2.4b}$$

In both cases, the next lightest meson whose exchange could contribute to the process is much heavier, being the ρ^+ meson (mass 776 MeV) for (2.4a) and the ω meson (mass 782 MeV) for (2.4b). The amplitude for each exchange process has a factor $1/(k^2-m^2)$, where m denotes the mass of the exchanged meson and k^2 is (four-vector momentum-transfer)^2. The pole at $k^2 = m^2$ cannot be reached for any configuration physically possible for the initial and final particles, even if only because the meson emission process $N \rightarrow (N + \text{meson})$ cannot satisfy the constraints of energy and momentum conservation. The precise location of the pole depends on the mass m of the exchanged meson but the smaller m^2 is, the closer it lies to the physical domain permitted for the energy-momentum variables. The effect of the closest pole is often quite apparent from inspection, as is the case for the angular distribution for reaction (2.4a), shown in Fig. 4(a). It is then possible, at least in principle, to extrapolate the physical data into the unphysical region to reach the pole location and so to determine the residue of this pole. For the reaction (2.4a), we have

$$(k_\mu{}^2-m_\pi{}^2) = (k_\pi-k_\gamma)_\mu{}^2-m_\pi{}^2 = -2E_\pi k_\gamma(1-\beta_\pi\cos\theta_\pi), \tag{2.5}$$

where $k_{\pi\mu}$ and $k_{\gamma\mu}$ are the energy-momentum vectors for the pion and the photon, β_π is the velocity of the created pion in the over-all c.m. frame and θ_π is the angle between the outgoing pion and incident photon in this frame, so that the differential cross section necessarily has a factor $(1-\beta_\pi\cos\theta_\pi)^{-2}$ corresponding to a pole of the amplitude at the unphysical angle $\theta_\pi = \cos^{-1}(\beta_\pi{}^{-1})$. The extrapolation of $(1-\beta_\pi\cos\theta_\pi)^2\sigma(\gamma p \rightarrow n\pi^+, \theta_\pi)$ into the unphysical region for $\cos\theta_\pi$ is shown on Fig. 4(b). Its value at the pole location is then proportional to the vertex amplitudes $(pn\pi^+)$ and $(\gamma\pi^+\pi^+)$; the first is essentially the coupling amplitude $G(k_\pi{}^2 = m_\pi{}^2)$, while the second is the charge amplitude $eF_\pi(k_\gamma^2)$ for the pion, where $F_\pi(k^2)$ denotes the charge form factor of the pion. In the present case, $k_\gamma^2 = 0$ and $F_\pi(0) = 1$ by definition, so that a determination of the pole factor constitutes a measurement of $G^2/4\pi$. The value obtained in this way is consistent with other determinations (see below) but its accuracy is not high.

This "polology" can be taken further, as shown on Fig. 5 for the case of π^+ production by inelastic electron scattering on a proton, where there occurs a product of two poles, one due to photon exchange and the other to pion exchange. However, in this case, we are interested primarily in the pion-exchange pole. The scattered electron is selected in direction and energy so as to give a definite (four-momentum-transfer)^2 = k^2 to the photon, while the final pion direction and energy are selected and the cross section then measured, with a view to extrapolating to the pion pole. The residue at this pole will have factors proportional to the electron charge form factor, which is essentially unity (as is already well known), the $\pi\pi$ amplitude $eF_\pi(k^2)$, and the πNN amplitude G, already deducible from the reaction (2.4a) or even from this reaction $ep \rightarrow e'n\pi^+$ itself. Hence it is possible to evaluate the electric charge form factor for the pion empirically, for the region $k^2 < 0$. These experiments have been done, for example with NINA at the Daresbury Nuclear Physics Laboratory, with the result that the simplest possible form factor,

$$F_\pi(k^2) = 1/(1 - k^2/m_\rho^2) \tag{2.6}$$

has been found to fit the data rather well. It corresponds to an r.m.s. charge radius of $R_\pi = 0.7$ f for the pion.

This OPE mechanism makes a strong contribution to many important processes and allows the determination of particle-particle cross sections not accessible in other ways. For example, with a proton target, the various processes

$$\pi + p \rightarrow \pi + \pi + N, \tag{2.7}$$

illustrated by Fig. 3b, are mediated in part by the pole numerator $A(\text{"}\pi\text{"}+\pi \rightarrow \pi+\pi).G_{\pi NN}$, where A denotes the amplitude for scattering of the incident pion by a virtual pion "π" from the proton to give a final π-π system. Extrapolation of $(k^2-m_\pi^2)^2$x(observed cross section) as function of k^2 for the virtual pion to the value $k^2 = m_\pi^2$, as was first proposed by Chew and Low (1959), leads, at least in principle, to a determination of the cross section $\sigma(\pi+\pi \rightarrow \pi+\pi, \cos\theta)$, where the scattering angle θ is determined from the final configuration in the final π-π rest frame. The dominant term in the π-π cross section is that due to the excitation and decay of the (776) meson

$$\pi^- + p \rightarrow n + \rho^0 \tag{2.8a}$$

$$\rho^0 \rightarrow \pi^+\pi^-, \tag{2.8b}$$

which occurs in the p-wave of the π-π system, and this has been a prominent and very informative reaction. It has however proved more difficult to determine reliably in this way the lesser partial waves contributing to $\sigma(\pi+\pi \rightarrow \pi+\pi; \cos\theta)$ with the limited statistics available in practice. The same OPE mechanism is effective in the process of $K^*(891)$

excitation by K collisions with protons.

The OPE term is also identifiable in high-energy neutron-proton scattering for angles close to the forward and backward directions so this process yields a furrher estimate for $G_{\pi NN}$. This term parallels the strong role of OPE at low energies, where it generates the strong tensor potential which contributes so much to the deuteron binding and structure, as we noted in Sec. 2.1.

2.3 The structure of the pion

In Sec. 2.2, we have described how the pion charge form factor $F_\pi(k^2)$ can be determined by polology. However, it can be determined more precisely by direct methods using a very high energy pion beam. The collision of such a pion with an electron in hydrogen target atoms through their Coulomb interaction will project the electron at a very high energy in a forward direction if the c.m. π-e scattering angle is near to 180°. The amplitude for this scattering includes the factor $F_\pi(k^2)$ whose magnitude can thus be determined as function of k^2 by observing the recoil electron intensity as function of θ_{Lab}. With 300 GeV pions, Amendolia et al. (1984, 1986) have measured $F_\pi(k^2)$ from $\surd(-k^2)$ = 170 to 450 GeV/c quite recently, obtaining an r.m.s. radius $R_\pi = 0.66 \pm 0.01$ f of high accuracy.

The pion form factor $F_\pi(k^2)$ has also been obtained for the time-like domain $k^2 > (2m_\pi)^2$ from cross section measurements on the electron-positron annihilation process

$$e^+e^- \rightarrow "\gamma" \rightarrow \pi^+\pi^- \tag{2.9}$$

The virtual photon "γ" has spin (1-) and the $\pi^+\pi^-$ system is therefore limited to its c.m. p-wave. This is dominated at low k^2 by the ρ-meson pole. The amplitude for reaction (2.9) includes the same pion form factor $F_\pi(k^2)$, which is now complex and dominated by a p-wave Breit-Wigner resonance amplitude up to $k^2 \lessapprox 1.5$ (GeV/c)^2. The function $F_\pi(k^2)$ is an analytical function of k^2 in the k^2-plane with a branch cut from $k^2 = (2m_\pi)^2$ at the threshold for the process (2.9) to $k^2 = +\infty$. Its magnitude is measurable from $k^2 = 0$ to $k^2 = -\infty$, but not within the interval $k^2 = 0$ to $(2m_\pi)^2$. It has been the subject of much theoretical calculation, which takes into account the constraints of analyticity in the k^2-plane, the measurements of $|F_\pi(k^2)|$ along the negative k^2-axis and along the right-hand cut, together with a unitarity relation on the latter. These calculations find the R_π value just quoted to be quite acceptable.

The charge structure just discussed is concerned with the coherent elastic scattering of the pion. However, at high energy and high momentum transfer, we have to do with the granular structure of the hadrons. This is the domain of deep inelastic scattering, where the struck parton (= quark) acts independently and incoherently in the collision, as illustrated in Fig 6a. The description of the nucleons by parton structure in this regime is well known, of course, but this picture is also applicable to the pion, experimentally by measurements of the Drell-Yan process (cf. Fig. 6b)

$$\pi^- + p \rightarrow (\mu^+\mu^-) + X, \tag{2.10}$$

in which the muon pair is created by a virtual photon resulting from a

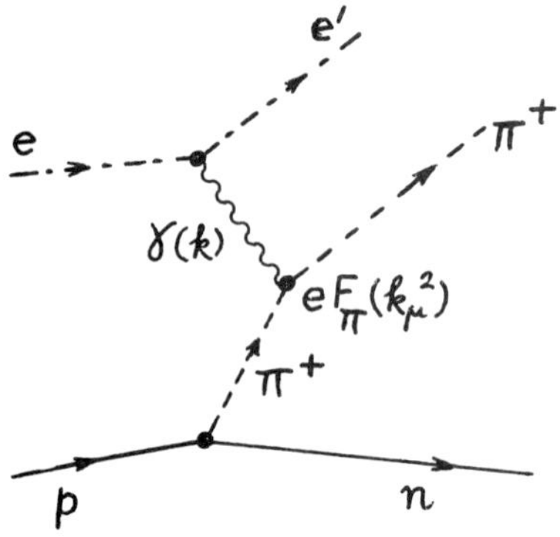

Fig. 5. Depicts inelastic electron scattering leading to π^+ emission from the target proton

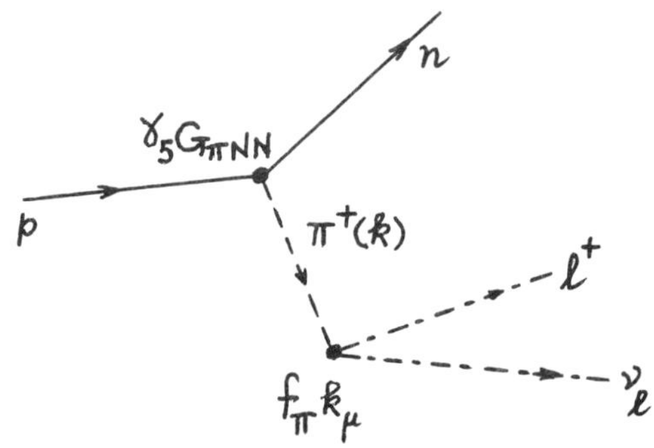

Fig. 7. Depicts Nambu's interpretation of the second term of eq. (2.17)

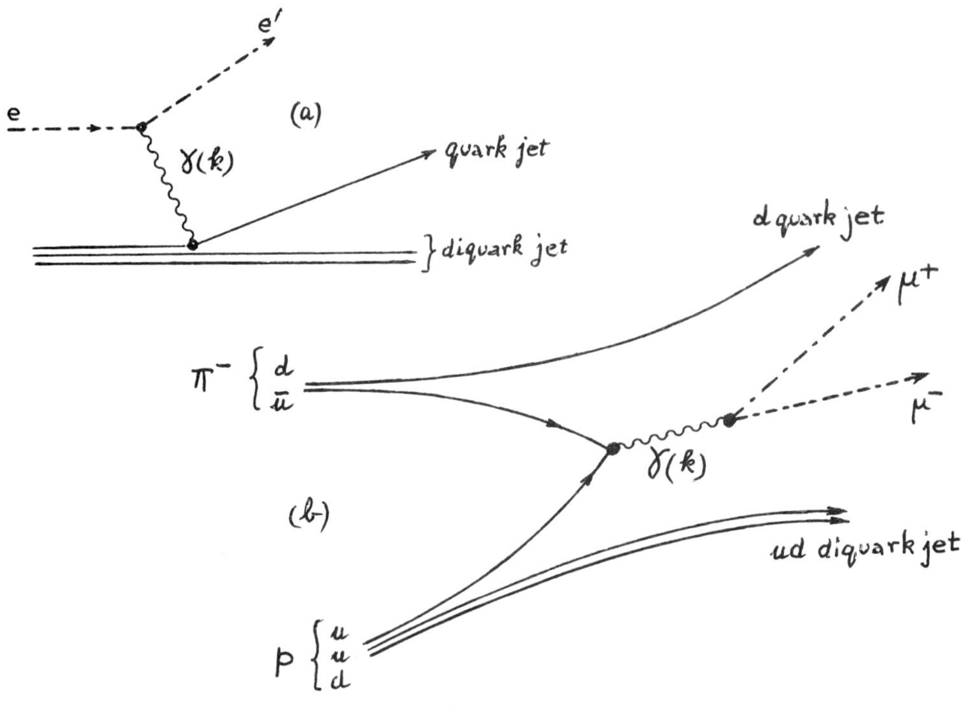

Fig. 6. Deep inelastic reaction processes at high energy for (a) inelastic electron scattering off a proton, and (b) initial π^- meson ($\bar{u}d$) incident on a proton (uud), generating a virtual photon leading to muon-pair creation, a typical Drell-Yan process.

quark-antiquark annihilation process. The dominant interaction here is $u\bar{u} \rightarrow "\gamma" \rightarrow \mu^+\mu^-$, where $\bar{u}$ comes from the π^- and u from the proton, so that the cross section is proportional to the structure functions $U_\pi(x_\pi)$ and $U_p(x_N)$, where x denotes the fraction of the hadron momentum carried by the quark in the relevant initial hadron. The function $U_p(x_N)$ is well known from analyses of the deep-inelastic scattering of electrons by nucleons, and $U_\pi(x_\pi)$ is to be determined by comparison with the data. This structure function will depend on $M_{\mu\mu}{}^2$ and $p_{\mu\mu T}$, but the data allow only a mean to be determined at present, and even that only for $x_\pi > 0.4$, so that only the valence quark in the pion contributes significantly. The present data are adequately fitted by $U_\pi(x_\pi) = (1-x_\pi)$, but this does not exclude theoretical forms of the type $x_\pi(\gamma+(1-x_\pi)^2)$ suggested by QCD calculations.

No precise relationship between this pion structure function and the pion charge form factor is yet known. These two representations of the pion reflect different features of the pion and are therefore both significant for a complete understanding of it. However, it is clear that this understanding will need the determination of U_π for smaller x_π values and therefore of the distribution of the sea quarks in the pion.

2.4 Forward Dispersion Relation

As a consequence of causality, the forward pion-nucleon elastic scattering amplitude f(E) is known to be an analytic function of the c.m. energy E everywhere in the complex E plane apart from the real axis. There is a branch cut on the positive E axis from $(M+m_\pi)$ to positive infinity, the amplitude $f_+(E)$ for π^+p scattering being the value of f(E) reached by approaching this branch cut from the upper half-plane. There is also a branch cut on the negative E axis from $-(M+m_\pi)$ to negative infinity, the amplitude $f_-(E)$ for π^-p scattering being given by the value of f(-E) reached by approaching the negative branch cut at -E from the upper half E-plane. There is one further singularity, a pole on the real axis, representing the contribution from the sequence of real but unphysical processes

$$\pi^- + p \rightarrow n \rightarrow \pi^- + p, \tag{2.11}$$

which lies at $E_N^* = -m_\pi^2/2M_N$. Its residue is given by $f^2/4\pi$, the pion-nucleon coupling constant, apart from known factors.

When these remarks are put together, the Cauchy theorem for an analytic function f(E) leads to the well-known subtracted Forward Dispersion Relation

$$\mathrm{Re}(f_+(E) - f_-(E)) - \int_{M_N+m_\pi}^{\infty} \frac{\sigma_+(E) - \sigma_-(\epsilon)}{(E-\epsilon)\epsilon^2} d\epsilon = A + BE + \frac{Cf^2/4\pi}{(E-E_N^*)} \tag{2.12}$$

where C is a known constant, while A and B are two constants which have to be determined by fitting the data. The left-hand side of (2.8) can be determined as a function of E from experimental data. Baillon et al. (1976,1977) have provided a number of very accurate measurements of $\mathrm{Re}(f_\pm(E))$ which are invaluable for this purpose, and we have many good total cross section measurements up to energies as high as are needed, even if $\sigma_\pm(E)$ behaves asymptotically like $(\ell nE)^2$. These values are to be fitted to a very specific functional form on the right-hand side, having only three free parameters A, B and the coupling constant $f^2/4\pi$. It is from this dispersion relation, or from adapatations of it, that we have our most precise evaluation of the pion-nucleon coupling constant, $f^2/4\pi = 0.078\pm0.001$ (Kroll, 1981).

2.5 The partially conserved axial-vector current

We have already noted above that the axial-vector current A_μ is not conserved, from the evidence on weak decay processes. However, let us reconsider this question for the case of the nucleon. For momentum transfer k, the most general form that the matrix-element of A_μ can take, on the basis of angular momentum and parity conservation, is

$$\langle N'|A_\mu(k)|N\rangle = \langle N'| -g_A\gamma_\mu\gamma_5 + g_P\gamma_5|N\rangle. \tag{2.13}$$

where the coefficients g_A and g_P are functions of k_μ^2. Henceforth, for convenience, we shall drop the bra and ket in eq. (2.9) but they are to be understood as being always present.

Now, if A_μ were to be conserved, we would have from (2.13),

$$\begin{aligned} k_\mu A_\mu(k) &= 0 \\ &= -g_A \not{k}\gamma_5 + g_P k_\mu{}^2\gamma_5 \end{aligned} \tag{2.14}$$

Since $k_\mu = p_{N\mu} - p_{N'\mu}$, the matrix-element $\langle N'|\not{k}\gamma_5|N\rangle$ equals $2M\langle N'|\gamma_5|\rangle$, and we see that eq. (2.14) can hold only if

$$2Mg_A = g_P k_\mu{}^2 \tag{2.15}$$

or, in other words, that g_P must have a singularity at $k_\mu{}^2 = 0$, having the form

$$g_P = 2Mg_A/k_\mu{}^2. \tag{2.16}$$

The axial-vector current would then have the form

$$A_\mu(k) = -g_A\gamma_\mu\gamma_5 + \frac{(2Mg_A)}{k_\mu{}^2}\cdot\gamma_5 k_\mu \tag{2.17}$$

It was then the genius of Nambu (1960) to interpret the pion as the Goldstone Boson associated with the breakdown of chiral invariance, since it is the chiral invariance of a theory which would imply that its axial-vector current must be conserved. However, the pion does not have zero mass, as should be the case for a Goldstone boson, arising from tbe breakdown of an exact symmetry, but it does have a small mass, which is a further consequence of the chiral-symmetry breakdown. Nambu therefore proposed that the pion mass should be put into (2.17) by hand, interpreting the factor $(k_\mu{}^2)^{-1}$ as the pion propagator $(k_\mu{}^2 - m_\pi{}^2)^{-1}$, leading to the partially-conserved axial-vector current form

$$A_\mu(k) = g_A\gamma_\mu\gamma_5 + \frac{(2Mg_A)}{k^2_\mu - m_\pi{}^2}\cdot\gamma_5 k_\mu \tag{2.18}$$

The second term now has a direct interpretation as the additional contribution of the pion intermediate state to the beta decay interaction, as depicted in Fig. 7. On this we have noted that the two interactions occurring for the pion are already known to us, namely as $\gamma_5 G_{\pi NN}$ and $f_\pi k_\mu$, so that by equating this product with the numerator of (2.18), we have

$$(2Mg_A)/(f_\pi G_{\pi NN}) = 1, \tag{2.19}$$

the Goldberger-Treiman relationship.

When the values known empirically are inserted into the left-hand side of (2.19), the value obtained is actually 0.92±0.02, rather than 1. This reflects the fact that the insertion of a finite m_π^2 in expression (2.18) will not be the only manifestation of the breakdown of chiral invariance but will be rather its most striking manifestation. We attribute the difference between the right and left sides of (2.19) in the physical world as being due to other chiral-symmetry breaking effects not included here.

In the Sections 2.1-2 and 2.4, we saw the pions as the quanta of a fundamental quantum field theory, whose manifestations were notably the result of its small mass value and its strong coupling. In this section we have seen the pions as the quanta of a very particular aspect of this field theory, that of its Goldstone Boson, the result of spontaneous breakdown to it from some deeper theory of the fundamental hadronic interactions.

2.6 Pionic and other mass splittings

Today, the pions and other hadrons are viewed as states arising from quantum chromodynamics, the theory of quarks interacting with the gluonic colour field, the gluons themselves also interacting with the gluonic field in a markedly non-linear way. There are various types of quark, u, d, s, c, b and probably more. Each has its characteristic mass value, most probably arising from the breakdown of symmetries of the Electroweak Interaction in a way not yet understood. The two lightest quarks are u with charge +2e/3 and d with charge −e/3, their mass values being only a few MeV. The next lightest is the strange quark s, with charge −e/3 and mass about 200 MeV. The others are much heavier.

The u and d quarks do not have the same mass. Let us consider the mass differences of some pairs of hadronic particles in terms of the u and d masses. For example, we have

Particle structures	Model Mass differences	Experimental value
$p = (uud)$, $n = (udd)$	$M_n - M_p = (m_d - m_u)$	1.3 MeV
$\Sigma^+ = (uus)$, $\Sigma^- = (dds)$	$M_{\Sigma^-} - M_{\Sigma^+} = 2(m_d - m_u)$	8 MeV
$K^+ = (us)$, $K^0 = (ds)$	$M_{K^0} - M_{K^+} = (m_d - m_u)$	4 MeV

No doubt there will be some additional contributions arising from the electromagnetic energy associated with the charge distribution within each hadron. The latter alone are not able to account for all of the mass differences observed for particles considered to be in the same isospin multiplet. It has been known for a long time that the empirical mass difference (M_n-M_p) cannot be accounted for by electromagnetic interactions since the calculations of their effect have always predicted the proton to be heavier than the neutron. Today we have learned to understand the n-p mass difference to be due primarily to the mass difference (m_d-m_u). We see from the above list that the mass differences observed are consistent with (m_d-m_u) being positive and of magnitude several MeV.

These hadronic mass differences have always been interpreted in the past as being due to an electromagnetic violation of charge independence, but we see now that this violation of charge independence extends down to the quark level. Indeed, the $SU(2)_T$ isospin symmetry is no longer considered to be of fundamental origin, but to be accidental and approximate, reflecting the fact that the u and d quarks have about the same mass, indeed the fact that their mass values are both close to zero. In other words, isospin symmetry is no longer of interest from a fundamental standpoint, but is of interest only to nuclear and intermediate-energy physicists, who find it a convenient formalism.

For the pions, the u-d mass difference does not contribute to the observed mass difference $(m(\pi^{\pm})-m(\pi^0))$. For the $\pi^{\pm}$ states, which have structures $(u\bar{d})$ and $(d\bar{u})$ respectively, the quark masses contribute the amount $(m_u+m_d)/2$; the same is true for the π^0 state, whose structure is $(uu-dd)/\sqrt{2}$. However, since the $\pi^{\pm}$ mesons have a higher mass than the π^0 meson, their difference might still be due to electromagnetic effects; at least the sign is right.

3. STRANGENESS AND THE QUARK MODEL

The strange particles were discovered very early in cosmic ray observations, using cloud chambers and nuclear emulsion. The earliest observations were those by Rochester and Butler at Manchester, and we know now that the neutral particle they observed was a K_S^0 particle with its characteristic V-tracks due to $K_S^0 \rightarrow \pi^+\pi^-$; Λ particles were first observed and identified a few years later. The first charged strange particle was observed by Brown et al. (1949) in nuclear emulsion, an example of the decay $K^+ \rightarrow \pi^+\pi^+\pi^-$, not long after their discovery of the π^+ meson, the subject of our celebration here.

3.1 The strange particles

These are now interpreted as due to a third quark, the strange quark s. The patterns of states observed are exemplified by the octet of baryons and the nonet of pseudoscalar mesons depicted in Fig. 8. The Λ and Σ particles have one s-quark replacing a u or d quark in the neutron or proton; the Ξ particles have two s quarks. From this, we can conclude that the s quark has mass about 200 MeV.

The octet and other such patterns were considered to be evidence of an SU(3) symmetry in the space of (u,d,s) quarks. In quantitative terms, this SU(3) symmetry is badly broken by the large difference between m_s and the masses m_u and m_d, but the qualitative picture is good. The particle masses satisfy the Gell-Mann-Okubo mass formula rather well, even when it is extended to take into account the electromagnetic contributions to the particle masses (Glashow 1964). Definite relationships are predicted connecting the coupling coefficients $G_{\pi NN}$, $G_{\pi\Lambda\Sigma}$, $G_{KN\Lambda}$ and others, which are not far from the physical facts, when these are reliably available. Similarly, there are relationships predicted between $G_{\pi\pi\rho}$ and $G_{\pi KK^*}$, the parameters which determine the hadronic decay widths for the vector mesons $\rho(776)$ and $K^*(891)$, which are not grossly in error.

With the quark model, we can also compare states from multiplets with different spin-parity values. For example, we can compare corresponding states from the pseudoscalar octet (or nonet, if we include $\eta'(958)$ with $q\bar{q}$ structure $^1S_0$, with those for the vector nonet, with $q\bar{q}$ structure $^3S_1$.

$K(494)^+ = {}^1S_0(u\bar{s})$, $K^*(891)^+ = {}^3S_1(u\bar{s})$, $m(K^*)-m(K) = 400$ MeV,
$\pi(140)^+ = {}^1S_0(u\bar{d})$, $\rho(776)^+ = {}^3S_1(u\bar{d})$, $m(\rho)-m(\pi) = 636$ MeV,

but there is serious question whether we should not be comparing $(\text{mass})^2$ values rather than mass, in which case the two differences would be much closer in value. The spin-dependence of the mass values for these states is believed due to a Fermi-Breit interaction due to one gluon exchange, a $\underline{\sigma}.\underline{\bar{\sigma}}$ contribution to the $q\bar{q}$ interaction. Similarly, the $q\bar{q}$ Fermi-Breit interaction separates the baryonic spin states; for example, we have

$N(938) = {}^2S_{1/2}(uud)$, $\Delta(1232) = {}^4S_{3/2}(uuu)$, $m(\Delta)-m(N) = 294$ MeV,

a mass difference comparable with that found between the mesonic states just above.
In these remarks, we see the Quark Model in operation. We see that this model describes and relates the low-lying hadronic states, without specific SU(3) hypotheses. The patterns of states seen follow from the existence of these quarks (u,d,s), their mass values (of which we have no fundamental understanding) and the fact that the gluon field couples equally with each kind of quark. The model has no call for SU(3) flavour symmetry ($SU(3)_F$) and also goes far beyond $SU(3)_F$ symmetry besides.

In this Section, we have seen the pion as a composite object, consisting of quark-antiquark pairs and gluons. In this QCD view the π and K particles are closely related. However, the K-meson does not stand out in the way that the pion does. Its $(\text{mass})^2$ is much greater, $(m_K/m_\pi)^2$ 12.5, and the next massive state $K^*(891)$ is much closer to it than is $\rho(776)$ to the pion, with $(m_K/m_{K^*})^2 \approx 1/3$ in comparison with $(m_\pi/m_\rho)^2 \approx 1/30$. It may be that PCAC relationships hold also for the strange particles but circumstances make it difficult to establish whether this is the case or not.

3.2 Excited hadrons with opposite parity

For both mesons and baryons, many excited multiplets are now known which have parity opposite to those for the low-lying multiplets just discussed. For the mesons, the best known is $a_2(1320)$, which decays particularly to $\rho\pi$ and $\eta\pi$, and which has spin-parity (2+), but there are also $a_1(1270)$ and $b_1(1235)$ with (1+) and $a_0(980)$ with spin-parity (0+), to mention a few others. For the baryons, we will mention only a few:

$(5/2^-)$ $N(1675) \rightarrow \pi N$ and $\pi\Delta$; $\Sigma(1765) \rightarrow KN, \pi\Lambda, \pi\Sigma$; $\Lambda(1830) \rightarrow \pi\Sigma$,
$(3/2^-)$ $\Lambda(1520) \rightarrow KN$ and $\pi\Sigma$; $N(1520) \rightarrow \pi N$ and $\pi\Delta$; $\Lambda(1670) \rightarrow KN$ and $\pi\Sigma$,
$(1/2^-)$ $\Lambda(1405) \rightarrow \pi\Sigma$; $N(1535) \rightarrow \pi N$ and ηN

At least two kinds of excited state can be imagined:

(i) the excitation of an additional s-wave $q\bar{q}$ pair. This pair necessarily has negative parity, and it can contribute additional angular momentum S=0 or 1 from the quark spins. This is the kind of excitation appearing most natural in a field-theoretic approach.

(ii) the excitation of orbital motions for the quarks and antiquarks already present in the ground state. Thus, excited baryonic states would then be visualised as being like the states of a triatomic molecule (qqq). In particular, the lowest excitation would be that with internal orbital angular momentum LP = 1-, which would imply that these lowest excitations

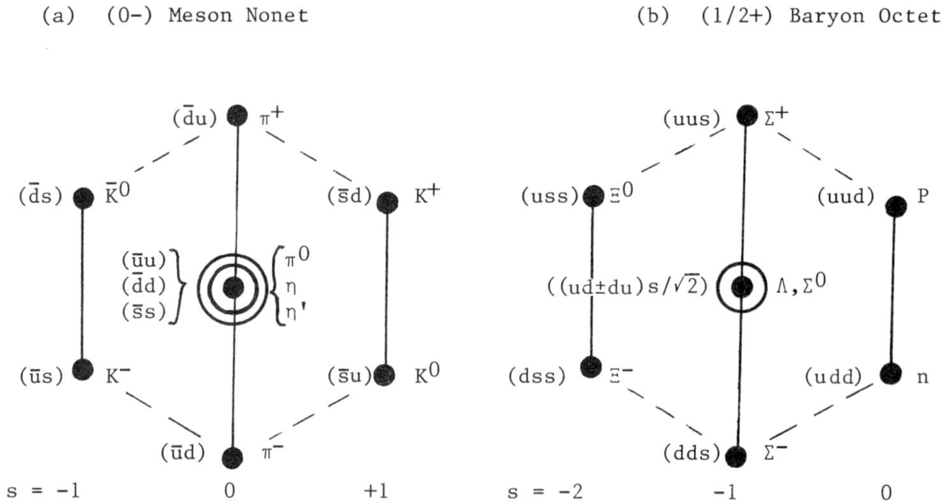

Fig. 8. Depicts the lowest meson and baryon multiplets formed from (u,d,s) quarks, for (a) the nonet of pseudoscalar mesons, and (b) the octet of spin-1/2 baryons. The lines show isospin multiplets.

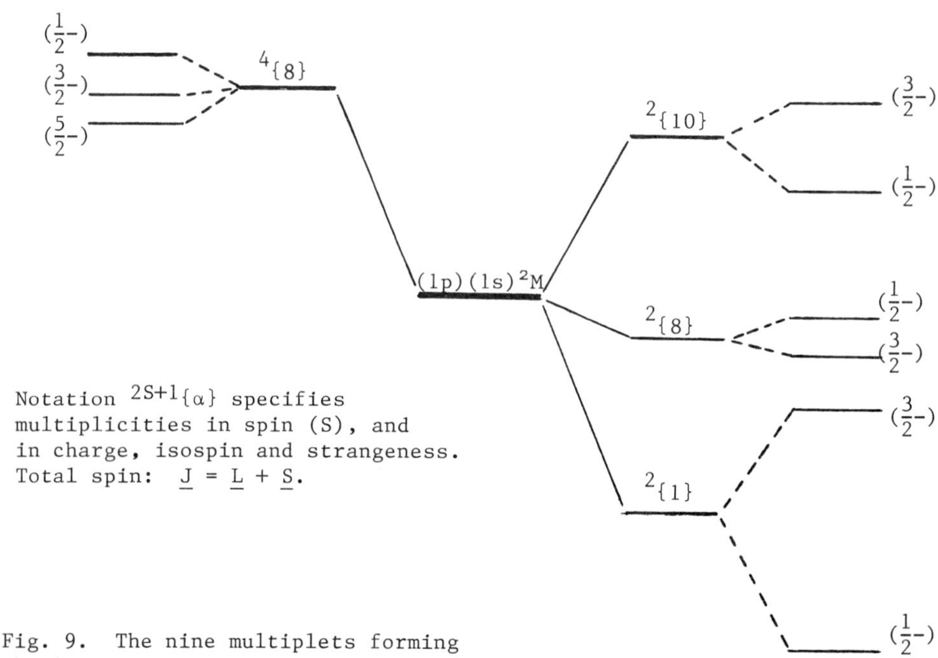

Fig. 9. The nine multiplets forming the lowest L = 1- supermultiplet of three-quark baryons having mixed (M) permutation symmetry for the colour factor in their wavefunction. The first step in their mass separation is due to the Fermi-Breit interaction; the second step is due to spin-orbit and tensor forces.

also should have negative parity and allow a range of total spin J values from 1/2 to 5/2, as observed. Similarly, mesonic states would be visualised as like those of a diatomic molecule (qq), where the first rotational excitation would have LP = 1-.

Fortunately, the LP = 1- excitation for the baryons leads to a unique supermultiplet of SU(3) multiplets, which are laid out on Fig. 9. It contains two unitary singlets, five octets and two decuplets with prescribed spin values. The two singlet states are included in the list of well-established states above. The other states mentioned there all have a definite place in the pattern of Fig. 9, as do all of the other established states with negative parity in this mass region.

I remember being present at the Pion-Nucleon Scattering Conference held at the University of California at Irvine in December 1967. Lovelace was there, bringing the phase-shifts which he and Donnachie had just deduced from the new pion-nucleon scattering data which had been obtained by experimental groups at the Rutherford Laboratory, at the Lawrence Berkeley Laboratory, at CERN, at Saclay and elsewhere. As he noted down the various resonances they had established from these data, giving their spin-parity and mass values, I realised that they fitted the pattern of Fig. 9, with only the (3/2-)Δ(1690) state missing, and that most of the N^* and Δ^* states reported were matched correctly by Λ^* and Σ^* states corresponding to the SU(3) multiplet predicted by the quark model, and even by a known Ξ^* state in one case.

I had to take part in a Round Table Discussion about the Pion-Nucleon Resonances the next day. We were each asked to give some short introductory remarks about our views on this topic, as a means of getting the discussion going. I spoke about the quark model and its predictions for excited baryon states, giving Fig. 9 as the basis for this discussion. I chided the audience for their myopia in confining their attention to <u>nucleonic</u> resonances and for not ever mentioning the words "Unitary Symmetry" at the conference. I explained that the new data put the Quark Model for baryonic resonances right in the centre of the picture, the excited states observed indicating clearly that the second type of excitation, involving a molecular picture of the three-quark system, was that required by these data. It was a rather long introductory talk, but it was clear that the Quark Model was vindicated (Dalitz 1969).

After I sat down, there was silence. Nobody spoke for a few minutes. Then the chairman called for the next introductory remarks. In fact, nobody else mentioned unitary symmetry during the remainder of the Discussion, nor during the remainder of the conference, for that matter. It seemed that strangeness was out of bounds! At the very least, it should have been realized that the established Λ^* and Σ^* resonances gave significant support to the deductions being made about N^* and Δ^* states from the πN scattering analyses. Sometimes, for example, strange members of some unitary multiplet are easier to excite and study than are their corresponding nucleonic members.

Of course, both types of excitation are still possible. There is no sign of this yet for the baryonic states, but it seems quite probable that the (0+) mesons may have a large admixture of the $(q\bar{q})^2$ configuration corresponding to $(q\bar{q})$ pair excitation additional to the primary $(q\bar{q})$ structure for this state. This conclusion has been argued by Jaffe (1977) in an attempt to account for the mass relationships between the strange and non-strange (0+) mesons and to account for the strength of the decay $f_0(975) \rightarrow K\bar{K}$, and also by Barnes (1985) to account for the strong excitation of $f_0(975)$ by $\gamma\gamma$ interactions, since it is natural for each photon to excite a $(q\bar{q})$ pair from the vacuum, leading naturally to states whose $(q\bar{q})(q\bar{q})$ content is relatively large. There is still much to be learned, both experimentally and theoretically, about the spectroscopy of hadrons in this regime of excitation energies of 1-2 GeV.

4. SYMMETRIES IN SPACE-TIME

In the Cosmic Ray Era, by 1953, two meson-decay processes had become well established, namely

$$\text{(a)} \quad \tau^+ \rightarrow \pi^+\pi^+\pi^-, \qquad \text{(b)} \quad \theta^0 \rightarrow \pi^+\pi^- \tag{4.1}$$

It was known that the two parent particles had comparable mass values, although the θ^0 parent mass was much the less precisely known, but that their lifetimes were very considerably different, the θ^0 parent having lifetime about two orders of magnitude less than that for the τ^+ parent. Many other decay processes were known for strange mesons, but these states were much less well known in mass and lifetime. The older physicists urged us to recognize that we were dealing with phenomena which could well have great complexity, and that we must not identify different decay modes as being due to the same parent particle unless we had overwhelming evidence of the identity of the two parents. Indeed, this was good advice (although rather late) in the case of the V^0 particles, where it took some years before it was realized (and proved definitely by R. Thompson <u>et al.</u> (1953)) that there were at least two different parent particles, one mesonic and the other baryonic. So we had a long series of names for the mesonic particles, defined by their decay modes, in order to guard us against any inclination to identify any two of them: such as $K_{\pi 2}$, $K_{\mu 2}$, $K_{3\pi}$, K_{e3}, etc., with charge values 0 or ±1. The climate of opinion was against hypotheses involving the assumption of identity for any two of these phenomenologically defined mesons. However the $K_{\pi 3}{}^+$ (previously known as τ^+) was very well-established by 1953, and there was a question whether or not there was a $K_{\pi 2}{}^+$ with decay to $\pi^+\pi^0$, a charged counterpart to the $K_{\pi 2}{}^0$ (<u>alias</u> the θ^0 meson). The evidence from counter experiments at M.I.T. certainly suggested the existence of an S^+ meson with a highly energetic π^+ secondary, but the nature of the neutral secondary accompanying it was difficult to establish, especially as a prominent decay process $K_{\mu 2}{}^+ \rightarrow \mu^+\nu_\mu$ was seen in work at Paris. In 1954, an event was observed which really settled this matter in the minds of particle physicists, a decay event $K^+ \rightarrow \pi^+\pi^0$, where the π^0 was identified clearly by its decay to $(e^+e^-)(e^+e^-)$, discussed in Sec. 1.2 above.

Historians who write about this period in elementary particle physics all over-simplify the issues of that time, puzzled to understand how physicists could have been so stupid as not to realize at once that there was simply one K^+ meson and one K^0 meson, both spinless, each having a long list of competing decay modes, and that the existence of the two decay modes $K^+ \rightarrow \pi^+\pi^0$ and $\pi^+\pi^+\pi^-$ was therefore telling us quite clearly that parity conservation must be violated in the decays of the K^+ meson. It wasn't at all like that. Things were by no means so straightforward. It is easy to see now what questions should have been asked by 1954 and what conclusions could have been reached, but this is so only with hindsight. There was in fact much confusion over the mass values and the lifetimes to be associated with each of the decay modes observed. Different experiments, with or without different techniques, did not reach consistent mass values, even for the same mode, and there was a strong psychological barrier in the air, militating against any premature hypotheses. The existence of two neutral K^0 mesons, $K_S{}^0$ and $K_L{}^0$, the latter with many competing decay modes, was not understood until long after this period and was also a source of confusion. It was difficult to measure the lifetimes for the various K^+ decay modes, except for the most prominent modes, since there were so few events available from cosmic ray studies; the equality of the lifetimes associated with the $K_{\pi 2}{}^+$ and $K_{\pi 3}{}^+$ decay modes was only established much later, from experiments using the K^+ beams generated using the Bevatron at Berkeley. The possibilities inherent in the early data were very many and it would have been a brave man indeed who would have made any firm assertions by 1954.

By the end of 1955, it was possible to draw some strong conclusions, which could be based on the cosmic ray data alone:

(i) The $\theta^+ (= K_{\pi 2}{}^+) \longrightarrow \pi^+\pi^0$ mode definitely existed,

(ii) the final state reached in $\tau^+ (= K_{\pi 3}{}^+) \longrightarrow \pi^+\pi^+\pi^-$ decay definitely had opposite parity to that of the final state reached in $\theta^+ \longrightarrow \pi^+\pi^0$ decay.

(iii) the $\pi^+\pi^+\pi^-$ state resulting from $K_{\pi 3}{}^+$ decay most probably had spin-parity (0-).

(iv) The masses for the $K_{\pi 2}{}^+$ and $K_{\pi 3}{}^+$ modes were quite close, separated by not more than 10 MeV and quite consistent with equality.

(v) The lifetimes for all K^+ decay modes were roughly consistent with equality, with a value about 10^{-8}s, although the uncertainties in their measurement were rather large, especially for the lesser modes. The lifetime associated with the θ^0 mode was about two orders of magnitude shorter, quite different from $\tau(K^+)$.

By this date, K^+ data from Bevatron experiments were beginning to come in, although not yet at all overwhelmingly. These conclusions did not depend on this new source of data, although this data did strengthen them. Of course, the situation changed rapidly in favour of the accelerator experiments, after that time.

Why was the conclusion that parity was violated in K^+ decay not reached by the end of 1955? I believe that we were all rather confused about the relationship between parity conservation and the properties of space-time. For example, we all knew that angular momentum conservation was a consequence of the isotropy of space, i.e. the property of rotational-invariance of space-time, and it was almost universally believed that parity conservation was similarly a consequence of the reflective invariance of space-time. However, if this were so, it would follow that if parity conservation were observed to fail in one class of experiment, there would then be no reason why it should hold in any other class of experiment - it should therefore be observed to fail in all interactions. In this period 1954 to 1956, this was the unanswerable objection which I heard over and over again.

Even Lee and Yang, by their own accounts (Lee 1971, Yang 1983), took a long time before they realized that this question was irrelevant. Initially they favoured the idea that all strange particles might be parity doublets, occurring in pairs with closely comparable mass values, so that there would be parallel channels for strong interactions, e.g.

$$\pi N \begin{array}{c} \nearrow \Lambda_+ + K_+ \searrow \\ \searrow \Lambda_- + K_- \nearrow \end{array} \text{final states,} \tag{4.2}$$

the suffix denoting the parity for the hadron specified. Parity conservation would still hold for all interactions, and reaction processes, and the production cross sections for K_+ and K_- would be the same, since the $(N\Lambda_+K_+)$ and $(N\Lambda_-K_-)$ interactions would have the same form; however, the $K \rightarrow 3\pi$ decay observed would stem from $K_- \rightarrow 3\pi$, while the $K \rightarrow 2\pi$ decay would stem from $K_+ \rightarrow 2\pi$. The stumbling block for this proposal was there was no reason why the K_+ and K_- lifetimes should then be so closely similar. Although the two lifetimes had not been so accurately measured, a difference in lifetime would have revealed itself by a dependence of the apparent branching ratio $(3\pi)/(2\pi)$ with the circumstances of the experiment. Finally, Lee and Yang came to consider a very sophisticated hypothesis, namely that different classes of interaction might not obey the same discrete conservation laws: thus we might have the following table, where P and C denote parity and charge conjugation invariance.

Interaction	Strong (hadronic)	Electromagnetic	Weak	Gravitational
Isospin I	✓	I_3 only	X	X
P and C	✓	✓	not P, nor C	?

The curious fact is - and it is a sign of their genius - that Lee and Yang chose to invoke parity nonconservation first of all for the weak, leptonic decay modes of non-strange particles, rather than for the nonleptonic decay modes for particles with strangeness, the area of phenomena which had given rise to these questions. This strongly-sideways jump led them to consider particularly

$$\text{(a)} \quad n \rightarrow pe^-\bar{\nu}_e, \qquad \text{(b)} \quad \pi^+ \rightarrow \mu^+\nu_\mu, \ \mu^+ \rightarrow e^+\nu_e\bar{\nu}_\mu. \tag{4.3}$$

It was only after considerable calculation concerning the first of these processes that they realized the essential point, that no experiments involving weak interaction processes had ever been carried out to measure a pseudoscalar quantity; parity nonconservation could only be demonstrated explicitly by observing a non-zero value for some pseudoscalar quantity. It was for these two processes (4.3a,b) that experimenters subsequently sought, and found, substantive and explicit evidence for parity non-conservation. It was realized later by Lee, Yang & Oehme (1957) that the non-zero result of these experiments also required the violation of charge-conjugation parity but that this could happen in such a way that CP, the product of the two operations was still conserved.

Of course, it would have been more logical to look first at $\Lambda \rightarrow p\pi^-$ decay, since this is more directly linked with the original τ-θ puzzle. If we had assumed that there was only one K-meson, which allowed both $K \rightarrow 2\pi$ and $K \rightarrow 3\pi$ decay modes, we might have calculated the amplitudes associated with the two graphs of Fig. 10, which lead to final $p\pi^-$ states with opposite parity. We would have then found that the interference between these two amplitudes would have given rise to pseudoscalar terms in the decay rate, terms like $\underline{\sigma}.\underline{k}_\pi$ in the decay rate for polarized particles, k_π being the final c.m. π^- momentum, giving the decay angular

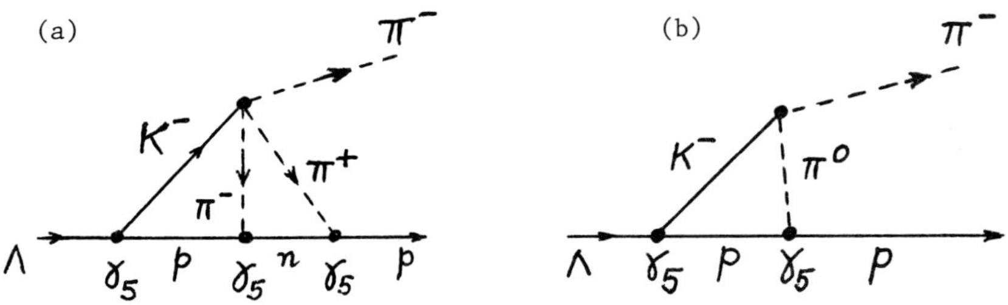

Fig. 10. Two graphs describing $\Lambda \to p\pi^-$ decay on the basis of (a) the τ-mode interaction $K^- \to \pi^-\pi^+\pi^-$ and (b) the θ-mode $K^- \to \pi^0\pi^-$, and the strong πNN†interactions. Graph (a) leads to a pseudoscalar coupling (i.e. p-wave decay) and graph (b) leads to scalar coupling (i.e. s-wave decay). Their interference leads to pseudoscalar spin-dependent terms in the decay rate.
†and $\bar{K}\Lambda p$

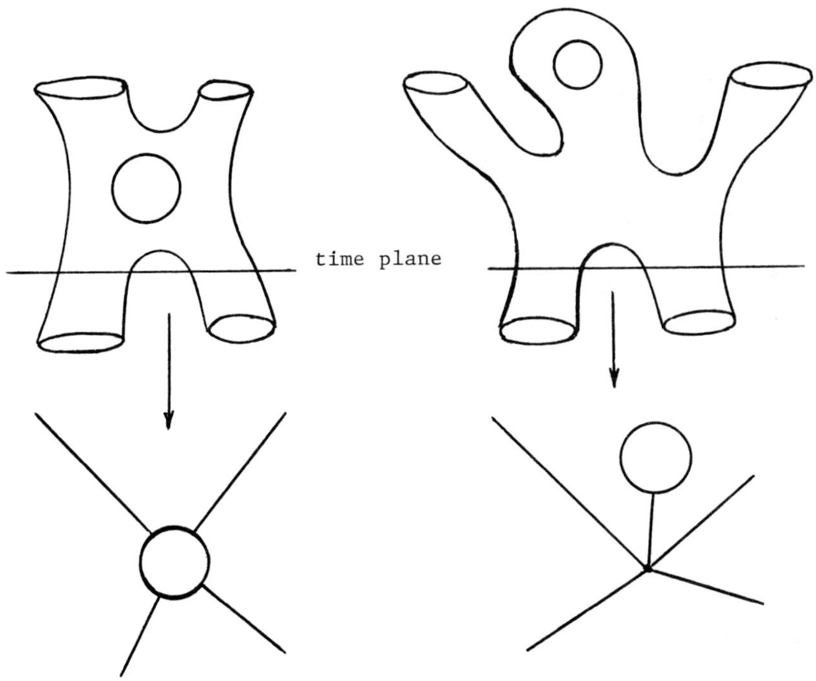

Fig. 11. The upper shapes depict surfaces traced out by two closed strings. They collide, coalesce and divide as time proceeds (up the page). Shape (b) is topologically equivalent to shape (a) but the sequence of intermediate processes is different. The graphs below show the Feynman graphs which result when each string is shrunk to a point.

distribution a forward-backward asymmetry relative to the initial polarization, or terms like $\underline{\sigma}_p.\underline{k}_\pi$ which corresponds to the generation of a longitudinal proton polarization in unpolarized Λ decay. However, although I thought about making this calculation at the time, I didn't actually do it, mostly because I felt that it would not have been regarded as a realistic calculation, being based on perturbation theory and using only two graphs out of hundreds which one could easily make up, neglecting the relationship between Λ and Σ particles and their π,σ,ρ,K,K^*.. couplings, and so on. However, in retrospect, the question whether or not this calculation was firm, or unique, or definitive, was not the point. The essential point of such an explicit calculation would have been the realization that what had to be done experimentally was to measure a pseudoscalar quantity, <u>any pseudoscalar</u>, and to find a non-zero value for it. For this, a definitive, or even a quantitatively convincing calculation was not required; <u>any calculation</u> which generated both scalar and pseudoscalar couplings in the $\Lambda \rightarrow p\pi^-$ decay interaction, giving both s-wave and p-wave states in the final $p\pi^-$ system as seen in its c.m. frame, would have sufficed to make this point clear. We did make such a calculation (Dalitz 1971) about ten years after this crucial time and verified that the final state calculated did have s- and p-wave amplitudes of comparable magnitude and would have made this point explicitly clear.

5. QUANTUM CHROMODYNAMICS $(QCD)_C$ AND QUARK FLAVOURS

Returning to the strong interactions we recall that QCD represents an exact colour symmetry, an $SU(3)_C$ symmetry in a three-dimensional colour space. The interactions are transmitted by the gluons, the quanta of the colour gauge field. The coupling of the gluon is to the current of the colour variable, and this interaction has the same structure and the same strength for every type of quark. Apart from the the quark mass values, there is only one free parameter in this theory, which is usually chosen to be the momentum scale Λ_{QCD}. Since each type of quark, say q_F where F denotes quark 'flavour', has its strong coupling only to the gluon field, the number N_F of quarks q_F minus the number $\bar{N}_F$ if antiquarks $\bar{q}_F$ is constant. Consequently there is an additive quantum number

$$n_F = (N_F - N_F) \tag{5.1}$$

for each quark 'flavour' F, and the total baryon number n_B, given by

$$n_B = (\Sigma_F n_F)/3 \tag{5.2}$$

is necessarily conserved by the strong interactions. Of course, the Electroweak interactions do not conserve flavour, as we have seen already from specific weak interaction processes such as $n \rightarrow pe^-\bar{\nu}_e$ which does not conserve hadronic charge, but they do conserve baryon number.

We have spoken already about isospin conservation, the $SU(2)_\tau$ symmetry which stems ultimately from the near-equality of the u and d quark masses, both being essentially zero. For these quarks, the additive flavour quantum numbers are n_u and n_d, which are more usually specified in terms of charge,

$$\begin{aligned} Q &= (2n_u - n_d)/3 & (5.3a)\\ &= (2N_u - 2\bar{N}_u - N_d + \bar{N}_d)/3 & \mathbf{(5.3b)} \end{aligned}$$

and isospin component

$$I_3 = (n_u - n_d)/2 \qquad (5.4a)$$
$$= (N_u - \bar{N}_u - N_d + \bar{N}_d)/2 \qquad (5.4b)$$

The $SU(2)_\tau$ symmetry goes beyond these conservation laws, for it stems from the interaction of the gluon field A_μ with these quarks, being schematically

$$(\bar{u}C_\mu u + \bar{d}C_\mu d)A_\mu, \qquad (5.5)$$

where C_μ denotes the colour current operator for the quarks, since this is indeed invariant under SU(2) transformations in the (u,d) space, the transformations specified above as $SU(2)_\tau$. In order to be clear, we shall refer to this transformation group after its generalisation to N quark flavours as $SU(N)_F$, the corresponding colour current being schematically

$$\sum_{i=1}^{N} (\bar{q}_i C_\mu q_i) . A_\mu \qquad (5.6)$$

With the inclusion of the strange quark s whose electric charge is −e/3, the symmetry group is $SU(3)_F$ and this symmetry became well known in the sixties as 'unitary symmetry', considered then to have some subtle and deep origin of great significance, as a generalisation of isospin. In fact, we see here that the symmetry is valid only to the extent that the mass-difference $\delta = \{m_s - (m_u + m_d)/2\}$ can be neglected, whereas we know that δ is about 200 MeV. It is therefore somewhat surprising that the unitary symmetry of the model holds as well as it does in the real world, but it is apparent from our discussion above that many of its successes are mainly the reflection of specific aspects of the Quark Model and of the simplicity of structure of the L=0 mesons and baryons.

The next quark to be discovered was the <u>charm</u> quark, with charge +2e/3 and mass about 1500 MeV, whose existence first became apparent in 1974 through the astonishing phenomena of the narrow J/ψ particle. Although one could in principle speak of an $SU(4)_F$ symmetry when the c quark is added to (u,d,s), the mass term

$$\sum_{i=1}^{4} \bar{q}_i m_i q_i \qquad (5.5)$$

is no longer even approximately invariant under $SU(4)_F$, in view of the enormous mass difference $\{m_c - (m_u + m_d)/2\} \approx 1500$ MeV involved. Nevertheless, the pattern of L=0 states observed still retains the underlying SU(4) structure. The fifth quark, the <u>bottom</u> quark b, has charge −e/3 and mass about 4500 MeV, while the <u>top</u> quark t, expected to have charge +2e/3, is still to be discovered and now appears most likely to have mass in the range 50 to 200 GeV. There may well exist further still heavier quarks, although we have no specific reason to expect them. If a further doublet family exists, it could well be that its −e/3 member (b', say) might have mass lying below that of the t quark. We summarize as follows:

Family =	1	2	3	4?
Charge +2e/3	u(5)	c(1500)	t(?)	
Charge −e/3	d(10)	s(200)	b(4500)	b'(?)

At this point it is clear that there is not much to be gained by discussing the $SU(N)_F$ symmetry of the quark colour current, in view of these enormous mass differences which violate it. From a practical standpoint, we might just as well forget the flavour symmetry $SU(N)_F$, since so little trace of it can survive in the real physical world. We have already seen that you might just as well forget even the symmetry $SU(2)_T$, unless you are a nuclear physicist, or work only in intermediate-energy physics, since we see no fundamentasl significance for it any more. What really matters at the deepest levels of the phenomena we see is the existence of quarks, their parameters and especially the uniform coupling of their colour currents with the gauge field of colour, or in a word their quantum chromodynamics. It is our expectation, based on our experience to date, that QCD and the known pattern of quarks as three doublets with charge values (2/3,-1/3)e will be sufficient for the description and understanding of all hadronic phenomena. In the immediate future, our attention must necessarily be focussed on them. It is possible, of course, that the quarks themselves might be composite objects - there are sufficient of them known for us to query their fundamental character - but the present data, which indicates that quarks are point-like down to a distance of about 10^{-3}f, give no hint of this.

In any GUT framework, the pion appears to have the same special role, associated with the (u,d) sector of QCD. In the limit of massless quarks, QCD is chiral invariant. The breaking of the Electroweak symmetry generates mass values for the quarks, in violation of the chiral symmetry, but the masses for the u and d quarks are rather small, although not zero. The pion can then be seen as the Goldstone boson associated with the small chiral symmetry breaking in QCD for the (u,d) sector, its mass being non-zero because of this chiral symmetry breaking but being small because the violation of chiral symmetry is small.

6. RECENT AND CURRENT DEVELOPMENTS

6.1 The Standard Model and Grand Unified Theories

The minimal group structure for the interactions known to us today is $SU(3)_C \times SU(2)_{Wk} \times U(1)$, now known as the minimal or standard model. The U(1) part has to do primarily with the conservation of electric charge, while the SU(2) part involves the weak gauge bosons and their relationship with the photon. The weak bosons $W^{\pm}$ and Z^0 are very massive, about 82 and 93 GeV respectively, while zero mass for the photon is ensured by the requirement of exact charge conservation. The standard model has a large number of parameters. These are:

(i) the coupling parameters, expressed as $e^2/\hbar c$ and Λ_{QCD},

(ii) the electroweak parameters M_W, M_Z and G_F, the latter being alternatively expressed in terms of the Weinberg angle θ_W as follows:

$$G_F = (e^2/\hbar c)/(4\surd 2M_W^2 \sin^2\theta_W) \qquad (6.1)$$

(iii) the masses of the fermions, both quarks and leptons (possibly including non-zero masses for the neutrinos ν_e, ν_μ and ν_τ),

(iv) the elements of the KM matrix (Kobayashi & Maskawa 1973) relating the eigenstates of weak isospin (the $SU(2)_{Wk}$ symmetry of the Electroweak Interaction) for the quarks with their mass eigenstates. There will be a similar matrix for the charged leptons if the neutrinos happen to be massive.

(v) the Higgs particle mass m_H, which is poorly constrained. Its coupling to other particles has the unusual property that this coupling has a factor M^2 for a particle of mass M. This means that a very massive Higgs particle may have rather strong interactions with other heavy particles, which means that it might have a very large width, its decay leading to heavy particles which have complex decay modes and which are difficult to recognize without large statistics. The production cross section for the Higgs particle may also be rather small since the initial particles in this reaction are necessarily our familiar particles with very low masses on the $\approx$100 GeV energy scale appropriate to Weak bosons and the Higgs particle, and with the consequent small coupling to the Higgs particle.

It is clear that a model with as many parameters as the standard model cannot be accepted as a final, fundamental theory. There has therefore been much theoretical interest in Grand Unified Theories involving a much more extensive group structure, whose symmetry-breaking proceeds in one or more stages until $SU(3)_C \times SU(2) \times U(1)$ is reached, as a residual symmetry. These GUTs may have features not present in our low-energy theories today. For example, the minimal SU(5) theory, which was the prime GUT candidate for a considerable time, permitted spontaneous decay of the proton to non-baryonic particles. Since it had no free parameters, it predicted definite rates for the various decay modes of the proton - for the dominant mode, $p \rightarrow e^+\pi^0$, the predicted rate is $2 \times 10^{-30\pm1.7}$ per year (Marciano 1982). There have been a number of large experiments devoted to the search for proton decay; the present situation (Goldhaber 1987) is that the upper limit for $\Gamma(p \rightarrow e^+\pi^0)$ is 3×10^{-33} per year, significantly lower than the prediction from the minimal SU(5) GUT. Unfortunately, other possible GUTs have free parameters and are therefore more difficult to exclude, although this may be possible when a sufficient number of possible decay modes have been searched for and excluded.

In all these GUTs, the pion has no special role beyond that which it already has in QCD and the Electroweak Interaction, as described earlier.

6.2 Superstring Theories

Supersymmetry has received much attention over the past decade. Its attraction lies in the fact that the patterns of particle states in its supermultiplets include both fermions and bosons. A particularly popular model includes fermions with spins 1/2 and 3/2 and bosons with spins 0,1 and 2, and so offers a prospect of relating gravitational phenomena with strong and electroweak phenomena through some unified supersymmetric theory, a possibility going beyond those offered by previous GUTs. However, to date, all of this theoretical work has had little, if anything, to do with experiment, but that is not where its motivation lies.

Earlier by a further decade, there was a great deal of work on field theories of a relativistic string, inspired by the S-matrix amplitude which Veneziano (1968) invented to illustrate in an explicit way the concept of duality, since these string theories could lead directly to S-matrix-elements with that form. Although Veneziano's amplitude was most interesting, there were a number of problems with it. First, it did not have the property of unitarity and no convincing procedure could be devised to make it so. Even more important, since it might still be possible to botch up unitarity in some unsatisfying way, these string field theories generally suffered from a number of unpleasant diseases, such as the prediction of faster-than-light tachyon states and the occurrence of negative probabilities, and these diseases could only be avoided if these theories were developed in a space-time with a rather large number of dimensions, many more than the four we are accustomed to. So those old string theories were laid aside by most people, after a short passage of time.

In recent years, however, Green and Schwarz have investigated the possibility of combining these two theoretical elements to make sypersymmetric string field theories, or more briefly superstring theories. The present activity and excitement concerning these theories arose because Green and Schwarz (1984) discovered that such theories can have some particularly attractive features. They can be made anomaly-free and they can be finite in all orders of calculation. However, they are still free from the diseases of string field theories only in a space-time with high dimensionality, and they can be used in our more limited space-time only if the additional dimensions can be shown to compactify spontaneously, curling up with some radius of curvature comparable with $\hbar/M_{\mathcal{P}}c$, where $M_{\mathcal{P}}$ denotes the Planck mass, given by $(\hbar c/G_N)^{1/2}$ where G_N is the well-known gravitation constant $(6.707\pm0.004)\times10^{-39}\,\hbar c^5/(\mathrm{GeV})^2$, so that $M_{\mathcal{P}} = 1.22\times10^{19}$ GeV/c^2 and $\hbar/M_{\mathcal{P}}c = 1.62\times10^{-20}$f.

Witten (1987) has emphasized the positive aspects of these superstring theories, while emphasizing that we are only at the beginning of the trail and have almost no idea where it will lead us. We note:

(i) that the motion of a closed heterotic string in time generates a smooth surface, and that this is to be contrasted with conventional field theories whose interactions are point-like, generating infinities when two or more of these interaction points coincide. Since superstring theories do not have these point-like interactions, there is a good chance that they may be finite.

(ii) the topological structures of these string-generated surfaces is all important. As closed strings join or separate, the surfaces develop holes and handles, as we illustrate here in Fig. 11, which depicts the scattering of two closed strings (representing two particles). As shown there, one topological structure can lead to many different Feynman graphs in the limit where the area of each closed string tends to zero. This may have a connection with our experience in conventional quantum field theory. Feynman perturbative calculations generate a large number of graphs in n-th order, the number of graphs increasing more or less like n!, a behaviour which has suggested to many that the perturbation expansion may never converge. Dyson (1952) has given physical reasons why this might be the case for quantum electrodynamics, while Cvitanovic (1977) has given intuitive reasons suggesting that these expansions may have a finite radius of convergence. Be that as it may, it has often been noticed that after a number of complicated graphs have been evaluated with great labour, and then put together with other contributions of the same order, the result can often be expressed in quite a simple form. This has been noted recently by Broadhurst (1985), for example, in his high-order calculations for the case of a simple field theory. But which graphs are to be taken together? - this we do not know, although Cvitanovic has put forward a view about it for quantum electrodynamics, which may owe a lot to Kinoshita's experience with his high-order calculations for the magnetic moments of charged leptons. These experiences lead to the response: "there must be a better way to do these calculations!" In some deep sense, we are not viewing the problem correctly. Perhaps superstring experience may help us to find the proper way, and some current work has suggested that this may turn out to be the case

(iii) these supersymmetric theories are the first to incorporate the gravitation on the same footing as the photon and W bosons and the electroweak interactions.

There are a substantial number of steps beyween these supersymmetric string theories and the real world.

Heterotic String → Compactify by → GUT: $E_8 x E_8$ or SO(32)
(n = 10 dim.)　　Calabi-Yau

In these GUTs, the scale of excitation energy is $M_g \approx 10^{19}$ GeV, with respect to which all of the particles known or envisaged have zero mass. Particular schemes of GUT symmetry-breaking start from this point and provide energy scales appropriate to the very many particle states which exist in the lowest supermultiplet, such as our known particles π, K, ρ, p, Λ, W, e, etc. and their supersymmetric partners, like $\tilde{e}$, $\tilde{\gamma}$, $\tilde{W}$, etc. which experiment tells us must have masses of at least 50 GeV or more, in fact often very much more.

The E_8'xE_8 GUT has a special interest, in that the two E_8 spaces are coupled only by gravitation. One space would be like our universe, while the other would be a 'shadow universe' linked with our universe only by gravitational forces, a possibility of a kind once mentioned, but only in a philosophical vein, by Wilkinson (1979). Such a situation could be the origin of a current puzzle, in that the matter in the E_8' space could be the 'dark matter' we are aware of in our universe, affecting the motion of our galaxies by its gravitational effects but invisible to us with respect to all other radiations, although astrophysical opinion (Kolb, Seckel & Turner 1985) appears to be that this possibility cannot account for the present observations.

The SO(32) GUT is of a standard type, and the pion plays the same role in it as it does for any GUT, as discussed above. With specific symmetry-breaking mechanisms, this GUT can be related with physical data in the "low energy regime". There are many possibilities but only few which can meet all the requirements of the data; and there is a good deal of exploration still going on, concerning these questions (Weinberg, 1987, Greene et al. 1986 & 1987).

7. CONCLUSION

To conclude, we may repeat that the pion has a special significance through the coincidence of its small and isolated mass value, its role as a Goldstone Boson, and its very strong coupling to the nucleons. The K-meson is also probably a Goldstone Boson, but its mass is not small, nor is it far isolated from the K^* meson, and it has not been possible to demonstrate this aspect of its nature. The pion has been immensely important for us, in enabling us to focus attention on the outstanding and most significant aspects of the data and on the detailed properties of a large class of the interactions we know today. The development of our knowledge of particle physics and of the fundamental forces in Nature has been remarkable in the 40 years which have passed since the discovery of the pion in 1947 and a large part of this development has been made possible by striking features of its properties.

REFERENCES

Amendolia S R et al. 1984 Phys. Lett. 146B 116
——— 1986 Nucl. Phys. B277 168
Badier J et al. 1983 Z. Phys. C 18 281
Baillon P, Bricman C, Ferro-Luzzi M, Jenni P, Perreau J M, Tripp R D and Ypsilantis T 1976 Nucl. Phys. B105 365
Baillon P, Declais Y, Ferro-Luzzi M, Jenni P, Petiau J M, Seguinot J and Ypsilantis T 1977 Nucl. Phys. B107 189
Barnes T 1985 Phys. Lett. 165B 434
Broadhurst D J 1985 Phys. Lett. 164B 356
Brown R H, Camerini U, Fowler P H, Muirhead H, Powell C F and Ritson D 1949 Nature 163 82
Chew G F and Low F E 1959 Phys. Rev. 113 1640
Cvitanovic P 1977 Nucl. Phys. B127 176
Dalitz R H 1969 Pion-Nucleon Scattering, Proc. πN Scattering Conference 1-2 December 1967 at Univ. Calif. (Irvine), eds. G.L. Shaw and D.Y. Wong (New York: Wiley) p.187
——— 1971 Atti del Convegno Mendeleeviano 'Periodicita e Simmetrie nella Structure Elementare della Materia' ed. M. Verde (Torino: Accad. Sci. Torino & Accad. Naz. Lincei) p.341
Dyson F J 1952 Phys. Rev. 85 631
Fazzani T, Fidecaro G, Merrison A W, Paul H and Tollestrup A V 1958 Phys. Rev. Lett. 1 247
Glashow S 1964 Group Theoretical Concepts and Methods in Elementary Particle Physics (New York: Gordon & Breach) p.303
Goldhaber M 1987 Proc. XXIII Conf. on High Energy Physics, ed. S.C. Loken (Singapore: World Scientific) p.248
Green M B and Schwarz J H 1984 Phys. Lett. 149B 117
Greene B R, Kirklin K H, Miron P J and Ross G G 1986 Nucl. Phys. B278 667
——— 1987 Nucl. Phys. B292 606
Jaffe R L 1977 Phys. Rev. D15 267 and 281
Kobayashi K and Maskawa T 1973 Progr. Theor. Phys. 49 652
Kolb E W, Seckel D and Turner M S 1985 Nature 314 415
Kroll N M and Wada W 1955 Phys. Rev. 98 1355
Kroll P 1981 Physics Data 22-1 (Fachinform-Zentrum Karlsruhe)
Lattes C M G, Occhialini G P S and Powell C F 1947 Nature 160 453
Lautrup B E and Smith J 1971 Phys. Rev. D3 1122
Lee T D 1971 Elementary Processes at High Energy, ed. A. Zichichi (New York: Academic) p.827
Lee T D, Oehme R and Yang C N 1957 Phys. Rev. 106 340
Marciano W 1982 Quoted by M. Goldhaber (1987)
Nambu Y 1960 Phys. Rev. Lett. 4 380
Palestrini S 1985 Phys. Rev. Lett. 55 2649
Steinberger J 1949 Phys. Rev. 76 1180
Taylor J C 1964 Rep. Progr. Phys. 27 407
Thompson R W, Buskirk A V, Etter L R, Karzmark C J and Rediker R H 1953 Phys. Rev. 83 329; ibid 90 1122
Veneziano G 1968 Nuovo Cim. 57A 190
Walker R L 1960 Proc. 1960 Ann. Intl. Conf. on High Energy Physics at Rochester, eds. E.C.G. Sudarshan, J.H. Tinlot and A.C. Melissinos (New York: Interscience) p.17

Weinberg S 1987 Proc. XXIII Intl. Conf. on High Energy Physics, ed. S.C. Loken (Singapore: World Scientific), p.271
Wick G C 1935 Accad. Lincei 21 170
Wilkinson D H 1979 Scientific Models and Man, ed. H. Harris (Oxford: OUP) p.86
Witten E 1987 Proc. Salt Lake City Meeting, eds. C DeTar and J. Ball (Singapore: World Scientific) p.155
Yang C N 1983 Selected Papers 1945-1980 with Commentary (New York: Freeman)

Discussion

McCubbin

You made it clear that in 1966 at the Irvine conference you felt that the quark model was vindicated. Nevertheless at that stage you were talking essentially about the the spectroscopy of particles. I would like to ask you what you thought happened when two protons collided, or what did you think was happening from a scattering point of view at that time?

Dalitz

There was a sense in which you had two quarks interacting directly but the quark-quark cross-section was fairly small. The total proton-proton cross-section was quite large but you could work backwards to see what the individual quark-quark scattering was and it was quite small. You could discuss multiple scattering processes in the forward direction and quite a lot of that was done. The system was broken up, of course; what the fragments' ultimate fate was, was not determined. We still do not have this capability today.

Broadhurst

I was fascinated yesterday by the two rival etymologies we heard from Frank and from Payne. As a newcomer, I don't know the etymology of Kaon. Is there agreement on this, and if not could we hear rival etymologies as we did yesterday?

Dalitz

There was a commission on naming new particles at the 1953 Bagneres conference. By that time the K had been used but I have no idea why.

Peyrou

I had a comment on that. The Bagneres conference settled the name Λ. It called the K_S^0 the θ; that has disappeared. There was a rule at this time set by Rossi that particles which had definite properties should have a Greek name, even if they didn't exist, and particles which certainly existed but had no definite properties should have a Latin name. When Gregory, Lagerrique, LePrince-Ringuet, Muller and I decided we had a particle (it was actually a 'decay mode', but let's forget about that), we wanted to be modest so not to give it a Greek name. Since we maintained it was emitting a μ meson we called it K-μ, which was a pun on the French writer, Camus. Rossi jumped on that and decided that since it was clear we were heading towards one particle with several decay modes, to come back to K. The first thing of O'Ceallaigh was called the κ. He came to K and decided that they would be identified by their decay and the number of particles involved in the decay. So our K-μ became K-μ-2. O'Ceallaigh's became K-μ-3, the χ of Menon and O'Ceallaigh became K-π-2. That settled the K definitely and eventually erased the τ, θ, etc.

© 1988 IOP Publishing Ltd

Friedlander

My recollection is clear that after the Bagneres conference Powell, Amaldi and LePrince-Ringuet signed a letter to "Nature" suggesting a nomenclature. If anybody wants to check it, it should be in Nature in the second half of 1953.

Williamson

The reference is Nature 172 (1954) 123 and the authors are

Amaldi, Anderson, Blackett, Fretter, Peters, Powell, Rochester and Rossi. They decided on a nomenclature.

Concluding remarks

D.H. Perkins

Nuclear Physics Laboratory, Keble Road, Oxford, OX1 3RH

1. INTRODUCTION

We are here to recall some stirring days of UK physics in 1947; the twin achievements of the discovery of the pion and of V-particles, at Bristol and Manchester.

1947 was indeed an amazing year. The table shows some of the relevant publications. The experimental papers were on the positive and negative pions observed in emulsion experiments, on the fate of negative muons in counter experiments, and on the discovery of V-particles in a cloud chamber. We should not forget the theoretical contributions, by Marshak, Bethe and Weisskopf, pointing out the difficulties of reconciling the high rate of production of mesons in the atmosphere with their relatively slow rate of decay, and its resolution in terms of the 2 meson hypothesis.

The other important theoretical paper was that of Charles Frank, seeking to explain the π-μ events in terms of capture of a negative meson by a nucleus and its subsequent re-emission with an energy release of 4MeV. Frank's paper rejected this explanation but did predict μHD fusion (with the "wrong" energy release of 5.5MeV), a process to be independently and experimentally discovered 10 years later, in a hydrogen bubble chamber by Alvarez. In the most recent experiments in DT mixtures, thermonuclear catalysis by a single muon can occur up to 100 times, and just now experiments on this process are starting at ISIS at RAL. Indeed, had the muon mass been half what it is, and the lifetime 50-100 times longer, muon-induced fusion could be a viable new energy source. What a near miss for high energy physics! Muon fusion as a growth activity in published papers is shown in the figure.

2. SCIENTIFIC AND TECHNOLOGICAL ASPECTS

The scientific and technological developments leading to the discovery of pions and V-particles have been described already and I want only to make some brief remarks.

In 1947, the pion was hailed as the long-sought Yukawa quantum of the strong nuclear field; that view was to hold until finally displaced by the quark picture in 1964, although the field theory of interquark forces - quantum chromodynamics - did not arrive until 1973. And pion exchange is still used to describe the long-range part of the nucleon-nucleon interaction.

The 2 V-events (in retrospect one was $K^0 \rightarrow \pi^+\pi^-$, the other was a $K^+ \rightarrow \mu^+ + \nu_\mu$ event) presented a real puzzle; we could not know at that time that from a handful of

© 1988 IOP Publishing Ltd

such V-events, a new quark degree of freedom (strangeness) was to emerge, to give us the u,d,s quarks of Gell-Mann and Ne'eman in 1964; or that we would have to await until 1974-77 to discover two new quark flavours, c and b. (We are still awaiting the "top" quark.)

In retrospect, we can see indeed that the discovery of the V-particles and the muon started the "generation puzzle". As Rabi had said of the muon "Who ordered that?" - and we can ask the same question of the other "heavy electron", the τ, as well as the exotic s,c, b(and t) quarks. We are no nearer answering Rabi's question than we were when he asked it.

The emulsion and cloud chamber techniques which led to the discovery of pion, muon and strange particles were, in a sense, fortuitous by-products of other objectives. The development of the cloud chamber by CTR Wilson between 1894 and 1911 was originally meant for the investigation of cloud formation (and particularly the optical "glory" phenomenon) and it was a happy outcome that, with carefully chosen expansion ratios, fast charged particles - α-particles and β-particles - could also promote, via charged ions, the formation of vapour droplets delineating the particle tracks. A subsequent development, 30 years ago, was the invention of the bubble chamber by Glazer, and although now practically defunct, the bubble chamber played a unique role in the cataloguing of hadron resonances, the essential input for the quark picture to emerge.

The emulsion technique to record tracks of charged particles dates back to Kinoshita in 1911, but the special "nuclear" emulsions were developed in the mid-40's under contract to the then Ministry of Supply, by Ilford & Kodak Ltd., as part of the nuclear energy programme. It was again a fortunate outcome - as a result of the initiative and vision of Powell - that these same emulsions could eventually be made sensitive enough and thick enough to record relativistic charged particles, and that the lifetimes of many of the "elementary particles" of the 40's and early 50's were short enough that interesting decays could be observed on a scale of microns to millimetres, in emulsion.

Remarkably, both the cloud chamber and nuclear emulsion were developed virtually to the "state of the art" within a period of a few years. Cloud chambers and bubble chambers have now all been phased out (except for the 15′ chamber at Fermilab); pulsed cyclic accelerator beams have given way to (almost) CW colliding beams. Emulsions however have retained a small niche in modern particle physics, as described by Don Davis. They are compact, portable and do not need teams of 70 people to operate them (as did the Big European Bubble Chamber). Examples of the use of nuclear emulsions today are in the detection of short-lived hadrons, such as heavy-flavour decays, where they are not yet completely submerged by silicon strips and CCD's.

I should say that even I have used emulsions as recently as 10 years ago. That was to calibrate solid state counters used to monitor the muon flux (and hence the neutrino flux) in neutrino experiments at CERN. Another method in a parallel experiment employed beam current transformers and integrating gas Cerenkov counters to measure the intensity of the parent pions and kaons of the muons/neutrinos. The two methods gave somewhat different results. I am happy to tell you that, as a result of a long-standing discrepancy between CERN and Fermilab cross-sections - only resolved 3 years ago - it was found that the emulsion calibration method had in fact been accurate!

3. COSMIC RAYS - THEN AND NOW

The cosmic-ray results of 1947-53 stimulated a programme of building high-energy accelerators, mostly in the USA, and this put the subject of high energy physics on a quantitative experimental basis. This accelerator programme appeared at just the right time: it benefitted from some apparent confusion in the minds of the funding agencies that basic research in particle physics was somehow connected with nuclear energy and was strategically important - a not completely misguided view, as exemplified by the near miss of muon-induced fusion mentioned above.

From the early 50's, the big strides in high energy physics were exclusively at the large accelerators, but cosmic rays itself is still a very productive field, and it still produces surprises. Examples are the observation of intense sources of gamma rays from binary pulsars and other stellar sources, with energies to 10^4TeV or beyond. How do pulsars produce γ rays of such enormous energies? These "beam dumps in the sky" are something of a mystery and there may very well be new physics in the stars, which we might even be able to unravel from Earth observations.

Another example is the observation of the interactions of neutrinos from the supernova SN1987a, marking the birth of neutrino astronomy and providing an astonishing verification of our ideas about collapse of massive stars - as well as a few questions. Finally of course we have had, for two decades, and closer to home, the solar neutrino problem, which just will not go away, and which might well give us crucial information on neutrino masses and flavour oscillations. Here, the scale of extra-terrestrial phenomena allows us to probe regions of the neutrino mass scale which are just not accessible to terrestrial accelerators or reactors.

So, non-accelerator high energy physics, which was front-line stuff 40 years ago, has staged something of a come-back, with monopole, proton decay and dark matter searches and the study of extra-terrestrial sources of high energy particles. Some of the crux questions in particle physics - the hierarchy problem and the origin of particle masses - may yet find answers, or at least clues, from non-accelerator experiments.

4. SOCIOLOGY AND INTERNATIONAL COLLABORATION

Cosmic ray research with emulsions in the late 40's and early 50's led to the first very large-scale international scientific collaborations, using balloon-borne emulsion stacks and dividing the technical and physicist effort among many countries. For example, the Sardinia expedition in 1953 involved 22 labs. from 12 countries. It is a tribute to the vision of Powell and to the flexibility of the emulsion technique, that these large collaborations could be made to work effectively and allow small groups with meagre resources to contribute at the forefront of research in particle physics.

1947

JAN 25	**PERKINS**	*Nature* **159**, 126	**First "σ-star"** (π^-) $m = 100 - 300m_e$
FEB 1	**CONVERSI** **PANCINI** **PICCIONI**	*Phys. Rev.* **71**, 209	**Negative mesons** **Decay in Carbon** (μ^-)
FEB 8	**OCCHIALINI** **POWELL**	*Nature* **159**, 186	**6 σ-stars** $m = 350 \pm 100m_e$
MAY 24	**LATTES** **MUIRHEAD** **OCCHIALINI** **POWELL**	*Nature* **159**, 694	**2** $\pi - \mu$ **Decays**
SEPT 15	**MARSHAK and** **BETHE** **WEISSKOPF**	*Phys. Rev.* **72**, 506 *Phys. Rev.* **72**, 510	**2 Meson Hypothesis** **(High Rate of Prodn.** **of Mesons, to be** **reconciled with W.I.)**
OCT 4	**LATTES** **OCCHIALINI** **POWELL**	*Nature* **160**, 453	**644 Mesons** – **105 σ-stars** – ($40\pi - \mu$) – **11 Complete** $\pi \rightarrow \mu$ – **499 ρ-mesons** ($\mu^\pm$)
OCT 18	**FRANK**	*Nature* **160**, 525	**Meson-induced** **fusion** (**μHD**)
DEC 20	**ROCHESTER** **BUTLER**	*Nature* **160**, 855	**V-particles**

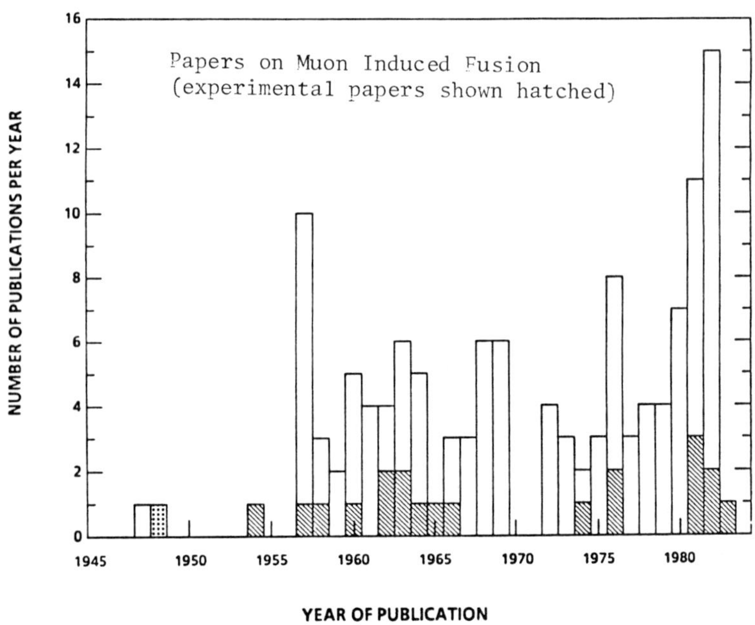

These developments in international collaboration formed an essential background to the setting up of large international laboratories like CERN. Indeed, it is no accident that in the CERN convention (Article II, 2a) work in the field of cosmic rays is specifically mentioned as one of the aims of the organisation. This type of work did in fact continue for some time: the first experiment searching for proton decay using liquid Cerenkov counters was carried out in 1960 by a CERN group.

In a meeting such as this, we are looking back to happy days, long ago when we were all much younger and brighter and full of enthusiasm, and when the life of a physicist was somehow much simpler and less complicated. Just to mention one incident: the first, rather heroic effort, to measure the pion lifetime was carried out with emulsion in cocoa tins mounted on bamboo poles stuck in the Aletsch glacier at the Jungfraujoch. It even got the answer (τ_π=8ns) correct within a factor 3! But the achievements of those days were not just what happened then, at that time, but the implications for the progress and promotion of particle physics in the subsequent decades. The work of 40 years ago set a pattern of endeavour in a field of science which is still producing excitement and surprises, which will continue long after we are all gone.

Lastly, it is my pleasant task to thank all of you for coming here for these two days and for making our meeting such an enjoyable occasion. Finally, on behalf of everyone I would like to thank our hosts at Bristol University for organising the occasion, and the many outside bodies who provided funds that made it all possible.

and with this new reinforcement build high and commanding a world structure of peace which no one dare challenge, the most awful crisis of history will have passed away and the high road of the future will again become open. . . . If during the next five years we can build a world structure of irresistible force and inviolable authority, there are no limits to the blessings which all men may enjoy and share."

Mr. Churchill makes it clear that in our own country this must be the work of both parties, for he tells us that "Mr. Attlee has declared 'Europe must federate or perish' and he does not readily change his opinions".

So since the *Nature* article was published the leadership to action which we needed and for which we have waited so long has been given.

All Souls College, Oxford. LIONEL CURTIS
Jan. 13.

Nuclear disintegration by meson capture

RECENTLY, multiple nuclear disintegration 'stars', produced by the cosmic radiation, have been investigated by the photographic emulsion technique. Plates coated with 50 μ Ilford B.1 emulsions[1] were exposed in aircraft for several hours at 30,000 ft. One of these disintegrations was of particular interest, for whereas all stars previously observed had been initiated by radiation not producing ionizing tracks in the emulsion, the one in question appears to be due to nuclear capture of a charged particle, presumably a slow meson.

The star consists of four tracks *A*, *B*, *C* and *D* (Fig. 1). *A*, *B* and *D* lie almost in the plane of the emulsion, whereas *C* dips steeply (at about 40°) and ends in the glass. *D* is due to a proton of energy 3·7 MeV., and *C* also corresponds to a proton, of more than 3 MeV., and most likely about 5 MeV. Track *B* was most probably produced by a triton of 5·6 MeV. A short track, about 1 μ long, between *A* and *B* is apparently due to the residual recoil nucleus.

Track *A* appears to enter the emulsion surface about 150 μ from the star centre. On account of the relatively large distances between consecutive grains at this range, the track cannot be distinguished at all easily against the spontaneous background of grains, and only the last 100 μ of track (below arrow) can be traced with certainty. Assuming it to be singly charged, the mass of the particle producing track *A* has been roughly evaluated by the following methods.

(1) Both ionization and scattering increase towards the origin of the star, hence the particle was definitely travelling *towards* the disintegration point.

An electron is discounted because the observed ionization is far too high (an electron track of this range would, in fact, not be detected at all), and the scattering too small. On the other hand, a proton is discounted since the observed scattering is too great (Fig. 2). We must, therefore, conclude that the particle had a mass intermediate between that of electron and proton.

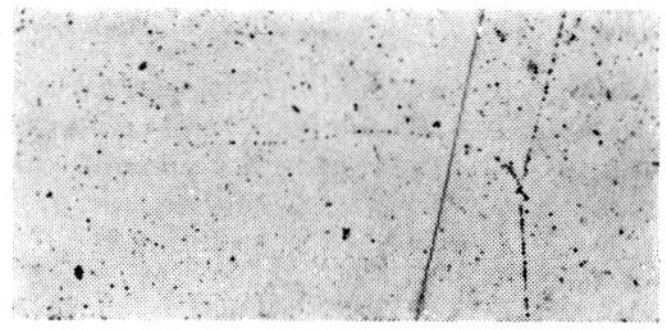

Fig. 1 *a*. PHOTOMICROGRAPH OF CENTRE OF STAR, SHOWING TRACK OF MESON PRODUCING DISINTEGRATION. (LEITZ 2 MM. OIL-IMMERSION OBJECTIVE. × 500)

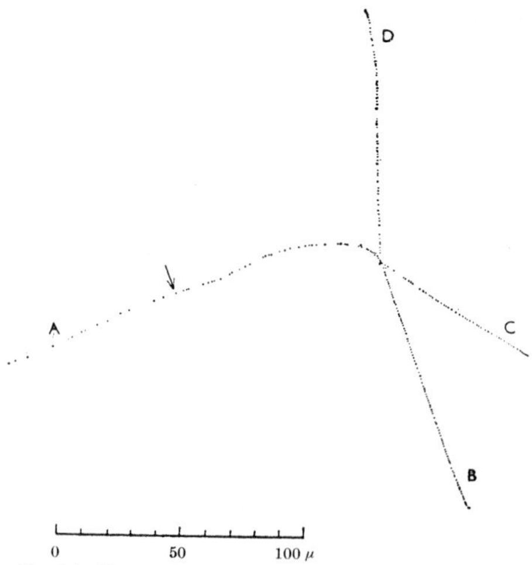

Fig. 1 *b*. TRACE OF COMPLETE STAR ON SCREEN OF PROJECTION MICROSCOPE, SHOWING PROJECTION OF THE TRACKS IN THE PLANE OF THE EMULSION. TRACK *A* CANNOT BE TRACED WITH CERTAINTY BEYOND THE ARROW

The grain density along track *A* does, in fact, agree well with that to be expected of a meson of the observed range of about one tenth of the proton mass. The range-energy curve for mesons in the emulsion has been obtained from that for protons (kindly lent by Dr. C. F. Powell), using the ratio of the masses of the two particles.

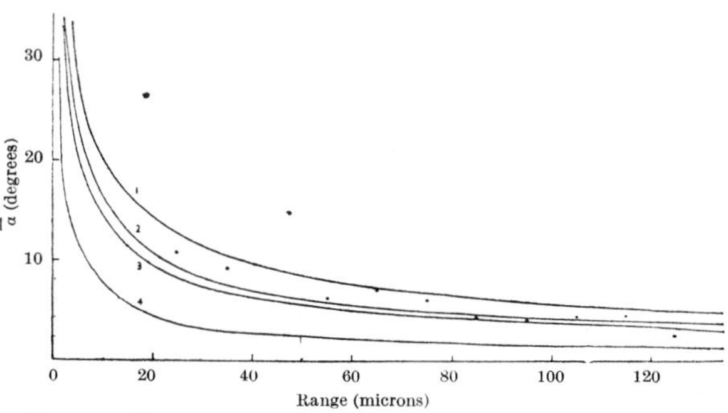

Curve 1 : $M = 100\ m_e$; curve 2 : $M = 200\ m_e$; curve 3 : $M = 300\ m_e$; curve 4 : $M = 1{,}800\ m_e$

Fig. 2. MULTIPLE SCATTERING OF MESONS AND PROTON IN THE EMULSION. $\bar{\alpha}$ DENOTES THE MEAN ANGLE OF SCATTERING FOR A 10 μ LAYER OF THE EMULSION. CURVES CALCULATED FOR PARTICLES OF MASS 100 m_e, 200 m_e, 300 m_e AND 1,800 m_e (PROTON). m_e BEING THE ELECTRON MASS. EXPERIMENTAL POINTS, FROM MEASUREMENTS ON TRACK *A*, INDICATED BY DOTS

Reprinted with permission from *Nature* **159** 126–7 (1947)
© 1983 Macmillan Journals Limited

(2) The elastic scattering of mesons and protons can be calculated from the known emulsion constitution, using Williams's formula[2] for the mean square angle of multiple scattering, and is shown plotted against range in Fig. 2. Bearing in mind the fact that the single scattering will introduce large fluctuations, the experimental points appear to be consistent with a meson mass between 100 m_e and 300 m_e, m_e being the electron mass. Only the horizontal projection of the scattering along the track can actually be measured; on the average, multiplying this by a factor $\sqrt{2}$ will give the true value. The scattering of fast electrons cannot be found from any simple formula, but it is certainly much greater than the observed values.

(3) A value of the meson mass can also be arrived at from the energy required in the nuclear disintegration. In this process, the emitted particles must have sufficient energy to surmount the potential barrier, so that the disintegrating nucleus in this case cannot be that of silver or bromine, and must, therefore, be either carbon, oxygen or nitrogen (in the gelatin).

An approximate calculation of the momentum of the recoil nucleus indicates that at least one neutron must be postulated to conserve momentum; the kinetic energy of this neutron must be around 4 MeV. If we assume that a negative meson is captured by a nucleus nearly at the end of its range and is annihilated, so that the rest energy of the meson becomes nuclear excitation energy, we could have disintegration schemes such as

$$Y^- + O_8^{16} \rightarrow N_7^{16*} \rightarrow H_1^3 + 2H_1^1 + n_0^1 + Be_4^{10*}$$
$$(\text{or } 2n_0^1 + Be_4^9)$$
$$Y^- + N_7^{14} \rightarrow C_6^{14*} \rightarrow H_1^3 + 2H_1^1 + n_0^1 + Li_3^{8*}$$
$$Y^- + C_6^{12} \rightarrow B_5^{12*} \rightarrow H_1^3 + 2H_1^1 + n_0^1 + He_2^{6*}$$

As the recoil nucleus would be expected to have a fairly high excitation energy (5–10 MeV.) above the ground-state, it must be relatively stable against further disintegration into charged particles. With this limitation, there are still a large number of possible reactions (considering all isotopes of carbon, oxygen and nitrogen), but it appears that in general the mass excess of the recoil nucleus $\sim$ 15 MeV., whereas that of the initial one $\sim$ 5 MeV. or less. The negative Q value of the reaction, allowing for excitation energy, is then found to be $\sim$ 60 MeV., and the total kinetic energy of the ejected particles $\sim$ 20 MeV. The total excitation energy of the original nucleus would then be $\sim$ 80 MeV., probably with an error of $\pm$ 20 MeV. (to allow for the various numbers of neutrons emitted, etc.).

On the above hypothesis, the meson should, therefore, have a rest energy of 60–100 MeV., that is, a mass of between 120 m_e and 200 m_e.

Near the end of the meson track, a small number of grains are observed slightly off the main track. If these are due to fast secondary electrons, their ranges appear to be considerably greater than would be expected from the energy of the primary.

I am indebted to the A.O.C., Royal Air Force, Benson, Oxon., for kindly exposing the plates.

D. H. PERKINS

Imperial College of Science and Technology,
London, S.W.7.
Jan. 8.

[1] Powell, Occhialini, Livesey and Chilton, *J. Sci. Instr.*, **23**, 102 (1946).
[2] Williams, *Proc. Roy. Soc.*, A, **169**, 531 (1938).

Angular Distribution of Protons Ejected in the Disintegration of Nitrogen by α-Particles

USING a powerful polonium source and an especially well-defined geometry, we have recently obtained 850 expansion-chamber photographs showing 90 proton tracks produced on the disintegration of nitrogen by α-particle bombardment. Our arrangement, therefore, gave on the average one disintegration for every ten photographs, thus greatly reducing both the expense of the photographic materials used and the labour of the examination of the photographs.

The number of disintegrations decreases rapidly as the energy of the α-particles is reduced, but the photographs show that disintegrations occur quite definitely for energies less than 1·7 MeV., and probably at energies less than 1 MeV.

The angular distribution shows a marked change with energy. For α-particles of energy less than 3 MeV., the distribution is almost uniform in numbers between 30° and 120°, while at 3·5 MeV. there is a marked preponderance at about 90° with proportionately smaller numbers of protons ejected symmetrically at angles greater and less than this value. It is interesting to recall that the well-known disintegrations recorded by Blackett and Lees[1] for 3–4 MeV. α-particles in nitrogen both show the ejection of the proton in the neighbourhood of 90°.

A full report of the experiment, together with selected photographs, will be published elsewhere.

F. C. CHAMPION
R. R. ROY

Wheatstone Laboratory
King's College,
London.
Dec. 10.

[1] Blackett, P. M. S., and Lees, D. S., *Proc. Roy. Soc.*, A, **136**, 325 (1932).

A Revised Estimate of the Age of the Earth

SINCE the publication of my preliminary account of a new method of estimating the age of the earth[1], based on Nier's isotopic analyses of samples of lead from galena and other lead minerals of known geological age, I have revised the calculations, using Glaisher's Exponential Tables[2], and greatly increased the number of solutions. 1,419 solutions for t_0 (the time since the isotopic constitution of the earth's primeval lead began to be modified by additions of lead isotopes generated from uranium I, actinium U and thorium) and for x and y (the respective isotopic abundances of Pb^{206} and Pb^{207} in primeval lead) have been obtained from the following combinations of data (Table 1). The numbers are those of the lead samples as listed in the original article[1], and the italicized figures in brackets are the approximate ages in millions of years (t_m) of the parental lead minerals.

TABLE 1

Number of lead sample	Combined with Nos.	and also with Nos.
25 (Great Bear Lake, *1330*)	1, 2, 3, 5, 6 (*25*)	19 and 21
23 (Broken Hill, N.S.W., *1200*)	7 (*60*)	19 and 21
22 (Broken Hill, N.S.W., *1200*)	9 and 11	19 and 21
21 (Quebec, *800*)	(Joplin, *100*)	19
19 (Ivigtut, *600*)	13–18 (*220*)	—

together with combinations of each of the above five sets with the others.

The data for six samples have not been used: either because the ages of the parent minerals are unknown or because the constitutions of the leads are so abnormal as to give highly discordant and mutually divergent results.

The Division of Building and Housing Research will be concerned mainly with the physical aspects of building and housing research. This may include investigations on the strength of building materials, acoustics of buildings, heat- and sound-insulation of buildings. Housing in the tropics and air-conditioning are expected to form an important part of this Division's work, as research on this aspect of India's housing needs is most urgently required.

The Division of Hydraulics Research is another engineering division in which the National Physical Laboratory's work will be concerned mainly with the physical aspects of the many problems. The importance of this Division will be very great in view of the large river development schemes which the Government of India is to launch very soon.

The Division of Analytical Chemistry will undertake analytical work for all the Divisions of the Laboratory.

Standardization

A very important aspect of work which has not so far received sufficient attention in India is that of industrial standardization. But with the recent inauguration of the Indian Standards Institution, India, it is hoped, will develop its own industrial standards in accordance with its special requirements. The National Physical Laboratory, it is expected, will be frequently called upon to assist the work of standardization by providing trained personnel and facilities for investigation by technical committees of the Standards Institution. The Council of Scientific and Industrial Research, it may be noted here, has been associated actively with the setting up of the Institution and was represented at the British Commonwealth Standards Conference and the International Conference on Standardization held in London in October last.

Location of the Laboratory

The question of the location of the Laboratory for a vast country like India is a very important one. The Planning Committee examined this question with great care, and made out an overwhelming case in favour of locating the Laboratory at the headquarters of the Central Government. It was realized that in any new constitution which India may have, the importance of contact between science and State will be even greater. Delhi has, in addition, a number of scientific departments of the Government of India, and close contact between them and the National Physical Laboratory will be mutually beneficial. In New Delhi, the Laboratory will be situated on a spacious site of more than sixty-six acres in close proximity to one of India's largest centres for agricultural research.

Very careful thought has been given to the design and lay-out of buildings for the Laboratory. Air-conditioning will be carried out practically throughout the entire Laboratory, and all precautions will be taken to eliminate dust. Since the great problem in India is to exclude heat from buildings as much as possible, the main wings of the Laboratory will run east and west, which will permit the use of natural light through north-facing windows without direct sunlight coming in. Provision is being made for a large lecture theatre and several seminar rooms. In addition to the central workshops, which will be equipped with the best and latest precision machinery, the several Divisions of the Laboratory will be provided with individual workshops for their day-to-day work. Staff amenities like cafeteria, social club and recreation grounds are being provided, as the necessity of human contacts between members of the staff is being constantly kept in view. On the question of internal lay-out and choice of apparatus, we had the benefit of the advice and experience of the National Physical Laboratory, Teddington. In the Division of Metrology, where accurate control of laboratory temperature is vitally necessary, the plans have been discussed at length with the Teddington staff.

Staff and Finance

The Laboratory will be under a director, who will be assisted with nine assistant directors—one in charge of each Division of the Laboratory. Each Division will have its complement of senior and junior scientific officers and scientific assistants, and for a start a scientific staff of ninety-eight has been recommended. The director will be assisted by an administrative officer to look after the routine non-technical work, and by an assistant director to look after routine technical work.

The capital cost of the Laboratory is estimated at four million rupees (£300,000), and a recurring grant of one million rupees (£75,000) annually is recommended for the first five years.

Money for the capital costs of the Laboratory has already been made available by the Government of India, and construction of the Laboratory is commencing very shortly.

Nuclear disintegration produced by slow charged particles of small mass

By Dr. G. P. S. OCCHIALINI

AND

Dr. C. F. POWELL

H. H. Wills Physical Laboratory, University of Bristol

In studying photographic plates exposed to the cosmic rays, we have found a number of multiple disintegrations each of which appears to have been produced by the entry of a slow charged particle into a nucleus. Mosaics of photomicrographs of three of these events are given in Figs. 1, 2 and 3. The edges of the individual photographs have not been trimmed so that the components of the mosaics can be distinguished. Three grains of a track in Fig. 1, indicated by three arrows, which were out of focus in the original negatives, have been blackened with ink, but the photographs are otherwise completely unretouched.

It will be seen from Fig. 1 that, associated with the 'star', there is one track, marked *m*, which shows frequent changes in direction. The points of scattering are most frequent near the centre of the 'star', and become progressively fewer in moving away from it along the trajectory. This behaviour suggests that the particle approached the disintegrating nucleus. The conclusion receives additional support from the observation that the number of grains per unit length of the track, which can be taken as a measure of the ionization produced by the particle, is greatest

Reprinted with permission from *Nature* **159** 186–90 (1947)
© 1983 Macmillan Journals Limited

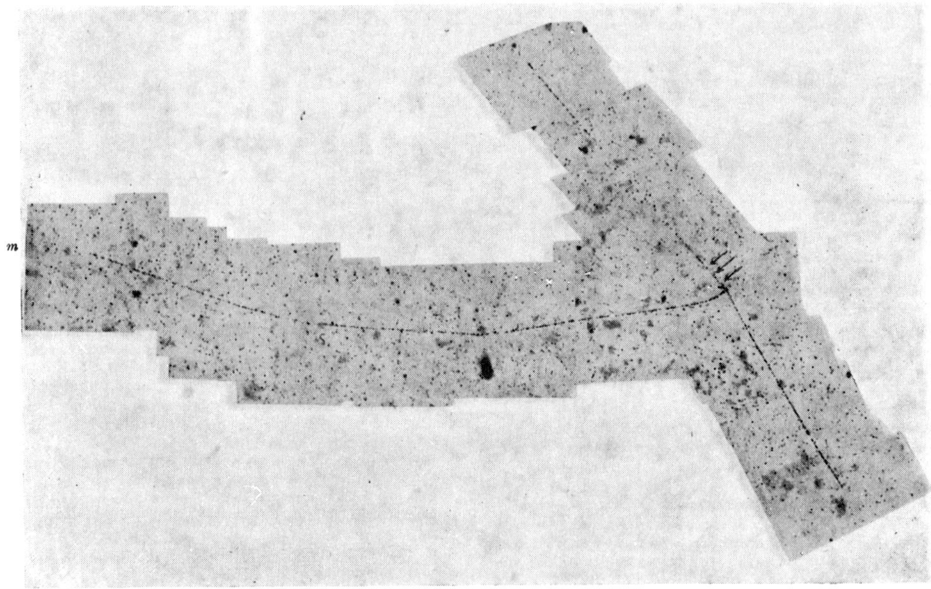

Fig. 1. COOKE × 45 'FLUORITE' OBJECTIVE, × 10 EYE-PIECE, WITH LEITZ PHOTOMICROGRAPHIC ATTACHMENT. *C*2 EMULSION LOADED WITH LITHIUM

The tracks of three disintegration particles can be distinguished, but only the particle of short-range remains in the emulsion

in the immediate neighbourhood of the disintegrating nucleus and becomes less as we recede from it.

We have now observed six of these events among a total of eight hundred stars. The probability, in any one case, that a charged particle, unrelated to the star, has, by chance, come to the end of its range within 1 micron of the disintegrating nucleus, is less than 1 in 10^5. We must therefore conclude that the particle entered the nucleus and produced a disintegration with the emission of heavy particles. Similar conclusions can be drawn from an inspection of the other photographs in Figs. 2 and 3.

The characteristics of the tracks which allow us to infer the direction of motion of the particles also lead to the conclusion that the particles were either at the end of their range or very near it when they entered the nucleus. In all cases the particles enter the emulsion from the glass or at the surface.

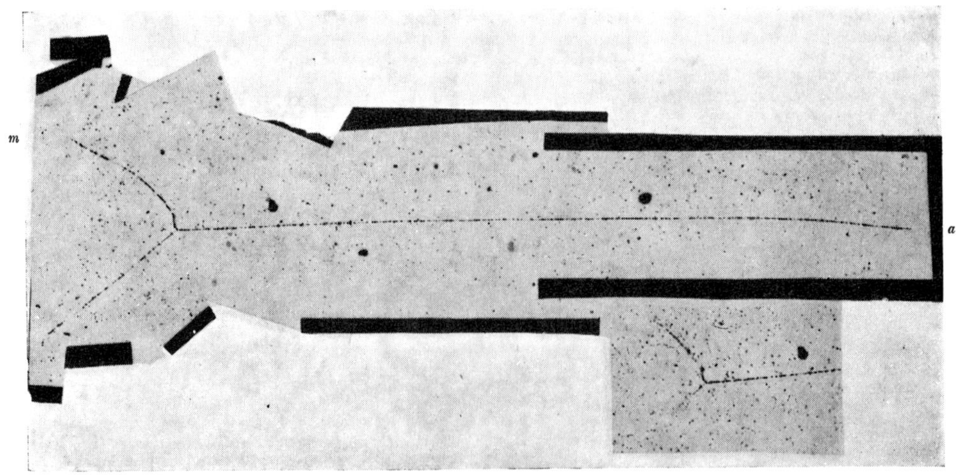

Fig. 2. COOKE × 45 'FLUORITE' OBJECTIVE, × 10 EYE-PIECE: DIRECT PHOTOGRAPHY WITH 'TELE-PANTO' PROJECTION MICROSCOPE. THE LONG TRACK *a* WHICH ENDS IN THE EMULSION IS PRODUCED BY A PARTICLE WITH A CHARGE EQUAL TO THAT OF THE PROTON. *C*2 EMULSION LOADED WITH LITHIUM

Fig. 3. Cooke × 95 'Achromat' objective with × 4 eye-piece. Direct photographs with projection microscope: $C2$ emulsion loaded with boron

Observations on the tracks of the slow particles indicated that the Coulomb scattering is more frequent than is to be expected if the particles are protons. Further, in moving along the trajectory, the increase in the grain density in the track, on approaching the centre of the star, is found to take place more rapidly than if the particles were protons. Both these qualitative observations suggested that the particles are of small mass, but more definite evidence is given by grain counts. Mr. Muirhead, in this Laboratory, has made a quantitative study of this subject, which is analogous to the problem of drop-counting in work with the expansion chamber. He has determined the variation of the grain-density along the tracks of protons in the emulsion in order to predict the distribution of grain density to be expected for particles with the same charge as a proton but with different values of the mass. A comparison of his results with the actual distribution of grains in the tracks of the particles producing the disintegration enables an estimate to be made of the mass of each particle. The values so obtained range from 100 m_e to 230 m_e, where m_e is the mass of the electron.

As in the case of drop-counting in experiments with Wilson chambers, any conclusion based on grain counting must be accepted with reserve. The latent image produced by charged particles in the 'Nuclear Research' emulsion is subject to fading, and although this effect is less marked in the types $B2$ and $C2$ than in $E1$ and $C1$, it is very serious in the usual exposures of six months duration. With this in mind we limited our exposures to six weeks; but even so the 'oldest' tracks obtained are subject to some fading. This will lead to an under-estimate of the mass of the particles, and Dr. Wassermann in this Laboratory therefore made observations by an alternative method based on measurements of the frequency of the Coulomb scattering of the particles.

Measurements on the tracks of the slow particles indicated that the number of deflexions which they suffer before being brought to rest is approximately three times greater than that of a proton of the same range. The expected frequency of scattering of particles of different mass, if they are subject only to Coulomb forces in collisions with nuclei of the atoms composing the emulsions, can be calculated. The observed frequency of scattering can thus be shown to correspond to a mass of the particle of 350 $m_e \pm 100$.

The estimates of the mass of the particles by both methods are subject to large statistical errors due to the small number of tracks available for measurement, and they give no basis for assuming that we are dealing with particles of different types. They show conclusively, however, that the particles have a mass less than one fifth of that of a proton.

During the last few years, following the pioneer experiments of Rasetti, a great deal of evidence has been accumulated concerning the fate of mesons, which can be summarized as follows. Positive mesons decay with the emission of an electron which can be detected by delayed coincidence experiments. On the other hand, in accordance with the theoretical predictions of Araki and Tomonaga, negative mesons at the end of their range do not suffer such a β-decay but have a high probability of being captured by nuclei. This process has previously been supposed to involve excitation of the nucleus and subsequent emission of β-particles.

In view of the evidence provided by the present experiments, we believe that the particles producing the observed disintegrations are negative mesons. We consider our photographs as evidence of the fact that the entry of the particle into the nucleus produces excitation, which is followed by the emission

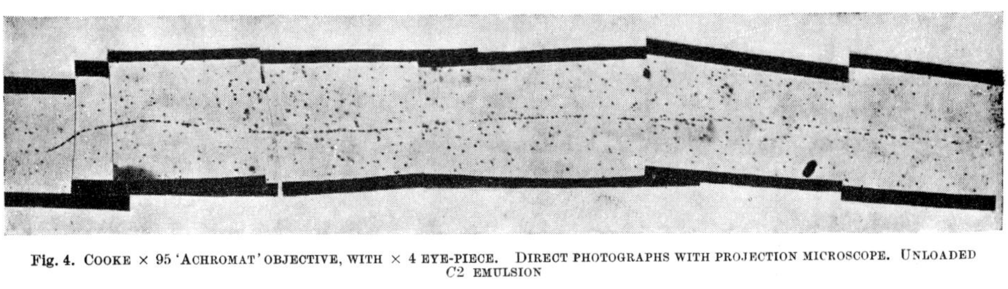

Fig. 4. Cooke × 95 'Achromat' objective, with × 4 eye-piece. Direct photographs with projection microscope. Unloaded $C2$ emulsion

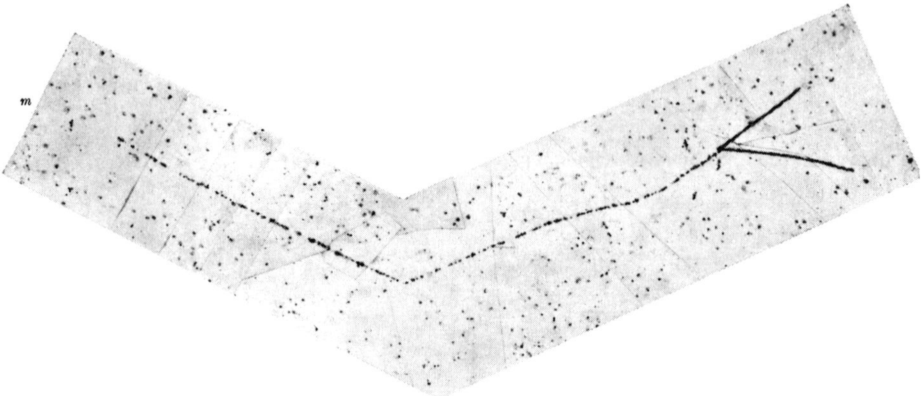

Fig. 5. COOKE × 45 'FLUORITE' OBJECTIVE, × 4 EYE-PIECE. TWO α-PARTICLES AND A PARTICLE OF LOW RANGE, POSSIBLY AN ELECTRON, ARE EJECTED IN THE DISINTEGRATION. $C2$ EMULSION LOADED WITH BORON

of heavy particles. In support of the general picture of the fate of the positive and negative mesons, we find examples of tracks which end in the emulsion and which we identify as positive mesons, an example being shown in Fig. 4. Such tracks are distinguished by inspection, from the much larger number of disintegration protons ending in the emulsion, by the frequency of the small-angle scattering which they display, and up to the present ten examples have been found. Grain counts in every case lead to an estimate of the mass of the particle equal to or less than 230 m_e.

In addition to the determination of the mass of the particle, a second quantity of great interest is the energy released in the disintegration. In the particular disintegration shown in Fig. 1, all but one of the ejected particles pass out of the emulsion before reaching the end of their range. Further, because we regard the momentum of the particle producing the disintegration as very small, the observed directions of ejection of the disintegration particles make it necessary to assume, at least in some cases, that neutrons or other particles unrecorded by the emulsion have also been emitted. In these circumstances we cannot make a precise estimate of the total release of energy in any one of the disintegrations of this type which we have observed, and thus make a comparison of this quantity with the rest-energy of the particles producing the disintegration.

These considerations suggest that the most favourable conditions of observation for the determination of the mass of the particles will be provided in work with thick emulsions loaded with the light elements. It is possible, for example, that reactions such as those represented by the equations

$$\mathrm{Li}^6_3 + m^0_{-1} \rightarrow \mathrm{H}^3_1 + \mathrm{H}^3_1 \text{ and } \mathrm{B}^{10}_5 + m^0_{-1} \rightarrow \mathrm{He}^5_2 + \mathrm{He}^5_2$$

may take place. Reactions of this type would be particularly easy to identify since they would lead to the ejection of two particles of equal range, moving in opposite directions from the disintegrating nucleus.

We are greatly indebted to the director of the Observatory of the Pic du Midi, Prof. J. Baillaud; to Dr. M. Hugon and Prof. Max Cosyns, who made it possible for us to obtain the exposures; to Mr. H. Muirhead and Dr. G. D. Wassermann, for making available to us the results of their work on the determination of the mass of charged particles; and to Mr. C. M. G. Lattes and Mr. P. H. Fowler for a number of discussions.

Note added in proof. Since this article was communicated, D. H. Perkins has published (*Nature*, January 25, p. 126) a photograph of an event similar to those we have discussed, and his conclusions are

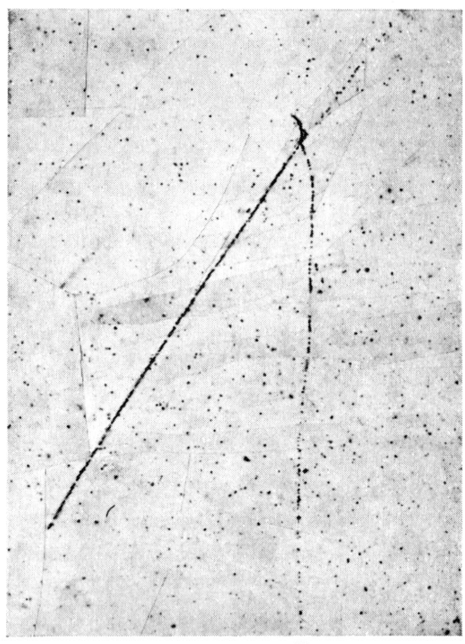

Fig. 6. COOKE × 95 'ACHROMAT' OBJECTIVE, × 4 EYE-PIECE. A FAST PARTICLE, POSSIBLY A PROTON, AN α-PARTICLE AND TWO HEAVY 'SPLINTERS' ARE EJECTED BY THE DISINTEGRATING NUCLEUS. UNLOADED $C2$ EMULSION

substantially identical with our own. The observed difference in the grain spacing of the meson tracks, in the *B*1 and *C*2 emulsions employed in the two experiments, is in good accord with expectations based on the known recording properties of the two types. The agreement between the results of observers in two different laboratories, working entirely independently with different experimental material, is a definite proof of the reliability of the photographic method in its present stage of development.

We have recently completed mosaics of two more of the six disintegrations referred to above, and reproductions of them are given in Figs. 5 and 6. We have also observed a number of disintegrations in which particles are emitted which are scattered more frequently than a proton of the same range, but which are more heavily ionizing than a meson of mass 240 m_e.

REFERENCES

Zhdanov, Perfilov and Deissenrod, *Phys. Rev.*, **65**, 202 (1944). For other references see Schapiro, *Rev. Mod. Phys.*, **13**, 58 (1941).

Tomonaga, S., and Araki, *Phys. Rev.*, **58**, 90 (1940).

Rasetti, *Phys. Rev.*, **59**, 706 (1941); **60**, 109 (1941).

Bernardini, Conversi, Pancini, Scrocco and Wick, *Phys. Rev.*, **60** 535 (1941); **68**, 109 (1945).

Conversi, Pancini and Piccioni, *Phys. Rev.*, **68**, 232 (1945).

OBITUARIES

Dr. D. Jordan Lloyd

DOROTHY JORDAN LLOYD was born in 1889, the daughter and grand-daughter of two prominent surgeons in Birmingham. It would appear natural that her early studies at Newnham College were in zoology, later physiological and then physico-chemical studies of muscle. During the First World War, her scientific pursuits were diversified, and under the auspices of the Medical Research Committee she was one of a team which studied substitutes for components of culture media for bacteria which were then in short supply. Then her brilliant short paper on the causes and prevention of ropiness in bread was most appropriate to war-time, and indicated the versatility of her mind. Under the stimulating influences of Gowland Hopkins and particularly of W. B. Hardy, she was led to the long-sustained studies of swelling in colloidal systems, which when carried over to the 'structured' proteins in animal skins gained for her the firmly established international reputation as the authority on the chemistry of leather which she held at her death on November 21, 1946. Thus, zoology in Part II of the Tripos followed by the M.A.(Cantab.), research in physiology followed by the D.Sc.(Lond.), after some interval the fellowship of the Institute of Chemistry in 1922, authorship of a standard work, "Chemistry of the Proteins" (first edit., 1926; second, with Ann Shore, 1938) and the directorship of the British Leather Manufacturers' Research Association (1927–46) are the signposts of an exceptional career.

The output of Dr. Jordan Lloyd's published work both in pure and applied chemistry is impressive. Returning after the First World War to her main line of research, she published her earliest papers under the auspices of the Medical Research Committee on the swelling of gelatin; but most of her work on this subject, and indeed by far the greater part of her scientific publications, was published after she joined the staff of the British Leather Manufacturers' Research Association in 1921. Her early work followed on that of Procter by establishing the swelling curves of gelatin as affected by hydrogen ion concentration, and demonstrating the effect of salts on the swelling. Later she showed how lyotropic effects are superimposed on the swelling effects of the Gibbs–Donnan theory, and illustrated the temperature effects which that theory predicts. She emphasized the importance of the balance between osmotic pressure and cohesion in controlling the degree of swelling. This work was extended to a study of the swelling of protein fibres (collagen, silk, hair) in aqueous systems, in which not only the water uptake but also the changes in length and thickness of the fibres were followed; these changes are of importance in giving a more detailed picture of the effects of swelling on the individual fibres and on the interwoven structure of hides. These studies of fibres showed the effect of hydrogen ion concentration on swelling, in so far as it produces salt formation by the protein, and also illustrated certain lyotropic effects. The weakening of the structure of the fibres which was observed in certain conditions was linked with the study of the molecular structure of fibrous proteins.

A number of Dr. Jordan Lloyd's papers deal with the closely related subjects of the chemical make-up of the fibrous proteins, and its bearing on their structure, water relationships and swelling properties. She carried out experiments on the amount of water in gelatin which is 'bound', that is, associated with specific groups in the protein molecules, and is to a considerable extent independent of osmotic effects. In addition to these studies of the molecular structure of fibrous proteins, she investigated the fibrous structure of hides and skins, and showed how the swelling properties of collagen fibres are affected by the sheath or network ('reticulin') which surrounds them and can be broken by strong swelling or weakened at very low and very high *p*H values.

The above work dealt with proteins in aqueous systems. In later developments Dr. Jordan Lloyd turned to a study of the effects of non-aqueous liquids (organic solvents and acids), using changes in volume, dimensions and extensibility of fibrous proteins in these liquids, to throw light on the nature of the cross-bonds which hold the protein molecules to one another, and on the orientation of these molecules in the fibres.

Her keen sense of the applicability of this mass of scientific results to the processes of the industry was coupled with a resolution and an unfailing affability towards industrialists and their technicians which was an immense advantage to the Research Association. The long series of her lectures, well attended by its members and their technical personnel, and based on the work in progress in the Association's laboratories, were symptomatic of her hold on the imaginations of those in the industry, and illustrated her wide grasp of the application of science to it. In the industry the measurement of hydrogen ion concentration and buffer index became usual, particularly in heavy-leather tanning; the use of the microscope and the correlation of photomicrographs with composition and use of tan liquors led to control of quality in leather in general and also in its production for special purposes—a control not hitherto attained; the disposition of the tanned fibres in leather is now related to its various measurable physical properties, and hence to quality and serviceability—such have been developed largely by her leadership and enthusiasm, which have contributed

in nodules are catalysed by a compound such as hæmoglobin, which has the remarkable property of oxygenation and is therefore an excellent oxygen carrier but a very inefficient catalyst. Although one cannot dismiss the possibility that under certain conditions hæmoglobin may act as an oxido-reduction catalyst, no evidence is so far available that this applies to hæmoglobin of root nodules.

On the other hand, the fact that this hæmoglobin is present in root nodules of every leguminous plant, that the pigment is formed only in nodules produced by an 'efficient' strain of *Rhizobium*, that the pigment is localized only within the large cells containing symbiotic organisms and that nitrogen fixation by nodules is strongly inhibited by small concentration of carbon monoxide, leave very little doubt that the activity of hæmoglobin is linked with the process of symbiotic nitrogen fixation. The presence of hæmoglobin in root nodules, therefore, will have to be taken into consideration in all further studies of nitrogen fixation by leguminous plants.

We are very grateful to Dr. H. G. Thornton for the strains 505 and 507 of *Rhizobium* used in this investigation.

1 Kubo, H., *Acta Phytochim.*, **11**, 195 (1939).
2 Burris, R. M., and Haas, E., *J. Biol. Chem.*, **155**, 227 (1944).
3 Keilin, D., and Wang, Y. L., *Nature*, **155**, 227 (1945).
4 Virtanen, A. I., *Nature*, **155**, 747 (1945).
5 Virtanen, A. I., and Laine, T., *Nature*, **157**, 25 (1946).
6 Keilin, D., *Proc. Roy. Soc.*, B, **113**, 393 (1933).
7 Keilin, D., and Wang, Y. L., *Biochem. J.*, **40**, 855 (1946).
8 Wilson, P. W., "The Biochemistry of Symbiotic Nitrogen Fixation" (Madison: Univ. Wisconsin Press, 1940).
9 Letsche, E., *Z. Physiol. Chem.*, **80**, 412 (1912).
10 Lipschitz, W., and Weber, J., *Z. physiol. Chem.*, **132**, 251 (1924).
11 Roche, J., *Bull. Soc. Chim. Biol.*, **8**, 98 (1926).

Processes involving charged mesons

By Dr. C. M. G. LATTES, H. MUIRHEAD, Dr. G. P. S. OCCHIALINI and Dr. C. F. POWELL

H. H. Wills Physical Laboratory, University of Bristol

In recent investigations with the photographic method[1,2], it has been shown that slow charged particles of small mass, present as a component of the cosmic radiation at high altitudes, can enter nuclei and produce disintegrations with the emission of heavy particles. It is convenient to apply the term 'meson' to any particle with a mass intermediate between that of a proton and an electron. In continuing our experiments we have found evidence of mesons which, at the end of their range, produce secondary mesons. We have also observed transmutations in which slow mesons are ejected from disintegrating nuclei. Several features of these processes remain to be elucidated, but we present the following account of the experiments because the results appear to bear closely on the important problem of developing a satisfactory meson theory of nuclear forces.

In identifying the tracks of mesons we employ the method of grain-counting. The method allows us, in principle[3], to determine the mass of a particle which comes to the end of its range in the emulsion, provided that we are correct in assuming that its charge is of magnitude $|e|$. We define the 'grain-density' in a track as the number of grains per unit length of the trajectory. Knowing the range-energy curve for the emulsion[4], we can make observations on the tracks of fast protons to determine a calibration curve showing the relation between the grain-density in a track and the rate of loss of energy of the particle producing it. With this curve, the observed distribution of grains along the track of a meson allows us to deduce the total loss of energy of the particle in the emulsion. The energy taken in conjunction with the observed range of the particle then gives a measure of its mass.

We have found that the above method gives satisfactory results when, in test experiments, it is applied to the determination of the mass of protons by observations on plates developed immediately after exposure. The errors in the observed values, based on grain-counts along individual tracks, are only a little greater than those corresponding to the statistical fluctuations associated with the finite number of grains in a track. As we have previously emphasized, however, serious errors arise when the method is applied to the plates exposed for several weeks to the cosmic rays[2]. These errors are due mainly to the fading of the latent image in the time elapsing between the passage of the particle and the development of the plate.

We have attempted to allow for fading by determining a calibration curve for each individual plate by grain-counts on the tracks of a number of protons, chosen at random from those originating in 'stars'. Such a calibration curve corresponds to an average value of the fading of the tracks in the plate. While we thus obtain improved mean values for the mass of particles of the same type, as shown by test measurements on the tracks of protons other than those used in making the calibration, the individual values are subject to wide variations. In no case, however, have mass determinations by grain-counts of particles, judged to be protons from the frequency of the small-angle scattering, given values exceeding 2,400 m_e or less than 1,300 m_e.

In these circumstances it is not possible to place serious reliance on the masses of individual mesons determined by grain-counts; and we employ the method, in the present experiments, only to distinguish the track of a meson from that of a proton. In searching a plate, an experienced observer quickly learns to recognize the track of a meson by inspection, provided that its range in the emulsion exceeds 100μ. Nevertheless, we regard it as established that a particular track was produced by a meson only if both the grain-density and the frequency of the Coulomb scattering correspond to the values characteristic of a particle of small mass. We have considered the possibility that as a result of a rare combination of circumstances we might, in spite of the above precautions, wrongly attribute the track of a proton to a meson of mass less than 400 m_e. It is difficult to give a numerical estimate of the probability of making such an error, but we believe it to be very small.

Secondary Mesons

We have now made an analysis of the tracks of sixty-five mesons which come to the end of their range in the emulsion. Of these, forty show no evidence for the production of a secondary particle. The remaining twenty-five lead to the production of secondary particles. Fifteen of them produce disintegrations with the emission of two or more heavy particles, and from each of the remaining ten we

Reprinted with permission from *Nature* **159** 694–7 (1947)
© 1983 Macmillan Journals Limited

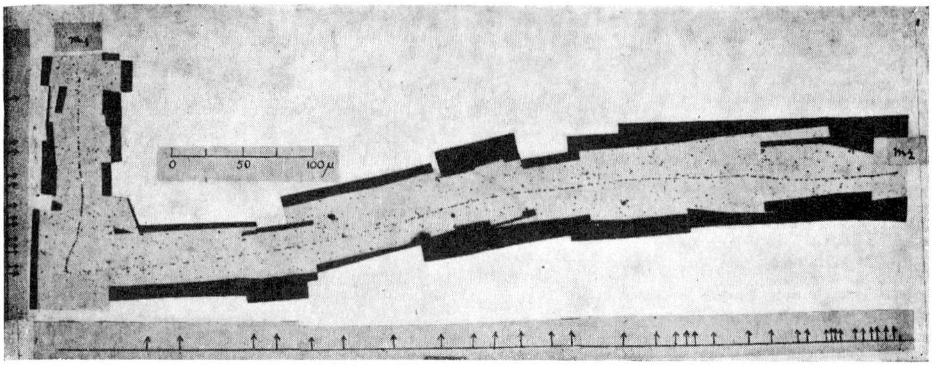

Fig. 1. OBSERVATION BY MRS. I. ROBERTS. PHOTOMICROGRAPH WITH COOKE × 45 'FLUORITE' OBJECTIVE. ILFORD 'NUCLEAR RESEARCH', BORON-LOADED $C2$ EMULSION. m_1 IS THE PRIMARY AND m_2 THE SECONDARY MESON. THE ARROWS, IN THIS AND THE FOLLOWING PHOTOGRAPHS, INDICATE POINTS WHERE CHANGES IN DIRECTION GREATER THAN 2° OCCUR, AS OBSERVED UNDER THE MICROSCOPE. ALL THE PHOTOGRAPHS ARE COMPLETELY UNRETOUCHED

observe a single secondary particle. Of these latter events, the secondary particle is in four cases a hydrogen or heavier nucleus; in four other cases the identification is uncertain, and in the last two cases it is a second meson.

Fig. 1 is a reproduction of a mosaic of photomicrographs which shows that a particle, m_1, has come to the end of its range in the emulsion. The frequent points of scattering and the rapid change of grain-density towards the end of the range show that the track was produced by a meson. It will be seen from the figure that the track of a second particle, m_2, starts from the point where the first one ends, and that the second track also has all the characteristics of that of a particle of small mass. A similar event is shown in Fig. 2. In each case the chance that the observation corresponds to a chance juxtaposition of two tracks from unrelated events is less than 1 in 10^9.

Grain-counts indicate that the masses of the primary particles in Figs. 1 and 2 are 350 ± 80 and $330 \pm 50\ m_e$, respectively; and of the secondary particle in Fig. 1, $330 \pm 50\ m_e$, the limits of error corresponding only to the standard deviations associated with the finite numbers of grains in the different tracks. All these values are deduced from calibration curves corresponding to an average value of the fading in the plate, and they will be too high if the track was produced late in the exposure, and too low if early. We may assume, however, that the two-component tracks in each event were produced in quick succession and were therefore subject to the same degree of fading. In these circumstances the measurements indicate that if there is a difference in mass between a primary and a secondary meson, it is unlikely that it is of magnitude greater than $100\ m_e$. The evidence provided by Fig. 2 is not so complete because the secondary particle passes out of the emulsion, but the variation in the grain density in the track indicates that it was then near the end of its range. We conclude that the secondary mesons were ejected with nearly equal energy.

We have attempted to interpret these two events in terms of an interaction of the primary meson with a nucleus in the emulsion which leads to the ejection of a second meson of the same mass as the first. Any reaction of the type represented by the equations

$$A_Z^N + \mu^0_{-1} \rightarrow B^N_{Z-1} + \mu^0_{+1} \text{ or } A_Z^N + \mu^0_{+1} \rightarrow C^N_{Z+1} + \mu^0_{-1}, \qquad (1)$$

in which A represents any stable nucleus known to be present in the emulsion, involves an absorption

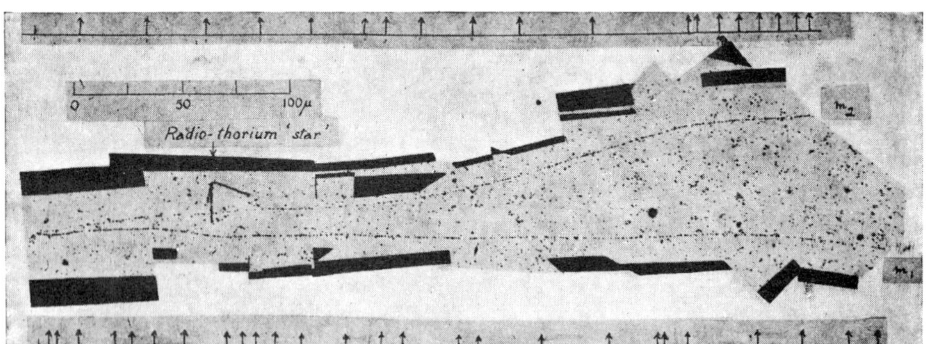

Fig. 2. OBSERVATION BY MISS M. KURZ. COOKE × 45 'FLUORITE' OBJECTIVE. ILFORD 'NUCLEAR RESEARCH' EMULSION, TYPE $C2$, BORON-LOADED. THE SECONDARY MESON, m_2, LEAVES THE EMULSION

of energy, in contradiction with the fact that the secondary meson is observed to have an energy of about 2 MeV.

A second process, represented by the equation

$$\mathrm{Ag}_{47} + \mu^{0}_{-1} \rightarrow X_Z + Y_{47-Z} + \mu^{0}_{+1}, \qquad (2)$$

in which X and Y represent two nuclei of approximately equal charge number, may be energetically possible, but the chance of it occurring in conditions where the total energy of the two recoiling nuclei is of the order of only a few million electron-volts is remote. It is therefore possible that our photographs indicate the existence of mesons of different mass[5,6,7]. The evidence provided by grain counts is not inconsistent with such an assumption. We have no direct evidence of the signs of the charges carried by the two mesons, except that the one secondary meson which comes to the end of its range in the emulsion does not lead to a disintegration with the emission of heavy particles. If, however, we assume that the transmutation corresponds to the interaction of the primary meson with a light nucleus, of a type represented by the equation

$$\mathrm{C}^{12}_{6} + \mu^{0}_{-1} \rightarrow \mathrm{Be}^{12}_{4} + \mu^{0}_{+1}, \qquad (3)$$

the difference in mass of the two mesons must be of the order of 60 m_e, according to estimates of the mass of the beryllium nucleus.

The only meson theory, to our knowledge, which assumes the existence of mesons of different mass is that of Schwinger[8]. It is visualized[9] that a negative vector meson should have a very short life and should lead to the production of a pseudo-scalar meson of the same charge but lower mass, together with a quantum of radiation. It will therefore be of great interest to determine whether the secondary meson, in transmutations of the type we have observed, is always emitted with the same energy. If this is so, we must assume that we are dealing with a more fundamental type of process than one involving particular nuclei such as is represented in equation (3). If, as an example of such a process, we assume that the momentum of the secondary meson appearing in our experiments is equal and opposite to that of an emitted photon, the total release of energy in the transmutation is of the order of 25 MeV.

In recent communications[10,11] very radical conclusions have been drawn from the results of observations on the delayed coincidences produced by positive and negative mesons in interactions with light and heavy nuclei[12,13]. It is assumed that a negative meson, at the end of its range, falls into a K orbit around a nucleus. In the case of a heavy nucleus, it is then captured, giving rise to a disintegration with the emission of heavy particles. With a light nucleus, on the other hand, it is regarded as suffering β-decay before being captured, so that, like a positive meson, it can produce a delayed coincidence. The conclusion is drawn that the nuclear forces are smaller by several orders of magnitude than has been assumed hitherto. Since our observations indicate a new mode of decay of mesons, it is possible that they may contribute to the solution of these difficulties.

Emission of Mesons from Nuclei

Fig. 3 shows a mosaic of photomicrographs of a disintegration in which six tracks can be distinguished radiating from a common centre. The letters at the edge of the mosaic indicate whether a particular track passes out of the surface of the emulsion, *s*, into the glass, *g*, or ends in the emulsion, *e*. The grain-density in tracks *a* and *c* indicate that the time between the occurrence of the disintegration and the development of the plate was sufficiently short to avoid serious fading of the latent image.

The track marked *f* suffers frequent changes in direction due to scattering, and there is a very rapid change in the grain-density in moving along the trajectory. These two features, taken together, make it certain that the track was produced by a light particle, and grain counts give an estimate for the mass of 375 ± 70 m_e[14].

We have now observed a total of 1,600 disintegration 'stars', in each of which three or more charged particles are ejected from a nucleus. Of these, 170 correspond to the liberation of an amount of energy equal to, or greater than, that in the 'star' represented in Fig. 4, but only in two cases can we identify an

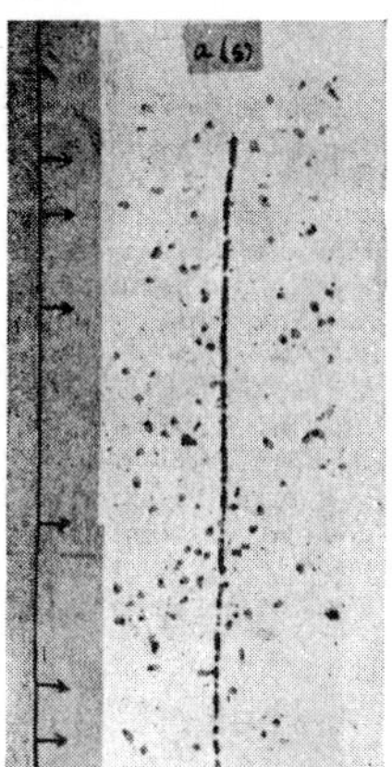
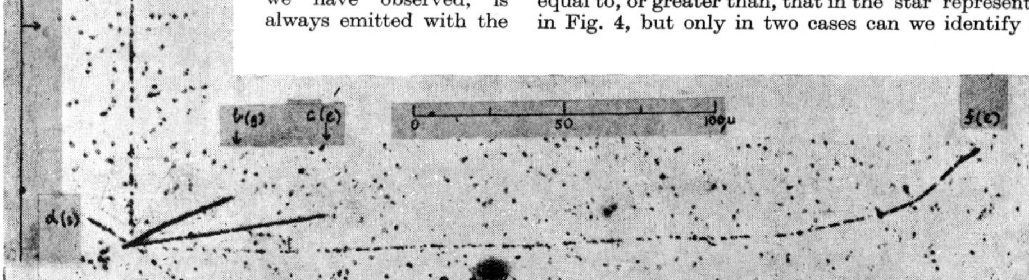

Fig. 3. Observation by Mrs. I. Roberts. Photomicrograph with Cooke × 45 'Fluorite' objective. Ilford 'Nuclear Research', boron-loaded C2 emulsion. The track (*b*) dips steeply and its apparent grain density is greater than the true value through foreshortening. Both (*b*) and (*c*) were probably produced by α-particles

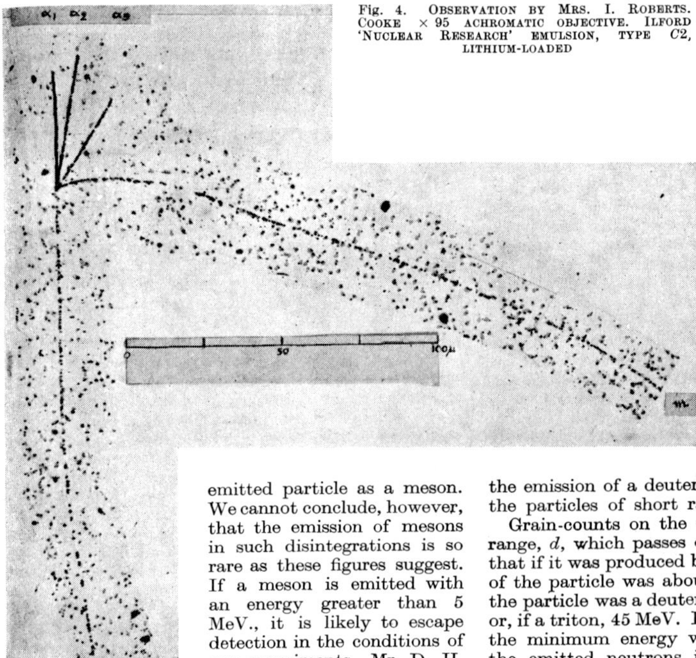

Fig. 4. Observation by Mrs. I. Roberts. Cooke × 95 achromatic objective. Ilford 'Nuclear Research' emulsion, type *C*2, lithium-loaded

emitted particle as a meson. We cannot conclude, however, that the emission of mesons in such disintegrations is so rare as these figures suggest. If a meson is emitted with an energy greater than 5 MeV., it is likely to escape detection in the conditions of our experiments. Mr. D. H. Perkins, of the Imperial College of Science and Technology, has shown that, in the B_1 emulsion, the grain-density in the track of a meson becomes very small at energies greater than 2 MeV., and we must anticipate a similar result in the C_2 emulsion at higher energies. Our observations are therefore not inconsistent with the view that the ejection of mesons is a common feature of the disintegration of nuclei by primary particles of great energy, and that the present instance, in which the velocity of ejection has been exceptionally low so that an identification of the particle has been possible, is a rare example. It is possible that the example of meson production recently described[15] is due to a similar process, produced by a primary particle of higher energy, in which some of the heavier fragments emitted on the disintegration have escaped detection because of the depth inside the lead plate at which the event occurred.

The disintegration shown in Fig. 3 may be the representative of a type, common in the high atmosphere with particles of great energy. In the present instance the energy of the primary particle must have been of at least 200 MeV., and, if its mass was equal to or less than that of a proton, it would not have been recorded by the emulsion.

Disintegrations Produced by Mesons

The observation of the transmutations of nuclei by charged mesons has led to the suggestion of a method for determining the mass of these particles based on observations of the total energy released in the disintegration[1,2]. In attempting to apply the method, we meet the difficulty of identifying the particular type of nucleus undergoing disintegration and of taking account of any ejected neutrons which will not be recorded by the emulsion. A photograph of such a disintegration which, at first sight, appears to allow us to draw definite conclusions, is shown in Fig. 4. In the photograph, the tracks of four heavy particles can be distinguished, of which the short tracks α_1, α_2 and α_3 end in the emulsion; α_1 and α_2 were certainly produced by α-particles, and grain-counts show that α_3 is due to a proton. The observations are therefore consistent with the equation

$$N^{14}_{7} + \mu^{0}_{-1} \rightarrow 2He^{4}_{2} + H^{1}_{1} + H^{1}_{1} + 4n^{1}_{0}; \quad (4)$$

or, less probably, to a similar equation involving the emission of a deuteron or a triton in addition to the particles of short range.

Grain-counts on the track of the particle of long range, *d*, which passes out of the emulsion, indicate that if it was produced by a proton, the initial energy of the particle was about 15 MeV. Alternatively, if the particle was a deuteron, its energy was 30 MeV.; or, if a triton, 45 MeV. In any case, we can determine the minimum energy which must be attributed to the emitted neutrons if momentum is to be conserved in the disintegration. As a result, we find a minimum value for the mass of the primary meson of 240 m_e. The value determined by grain-counts is also 240 ± 50 m_e.

In view of the recent results of experiments on delayed coincidences, referred to previously[12,13], such results must, for the present, be accepted with great reserve. We must expect the liberation of an amount of energy of magnitude 100 MeV. in any nucleus to lead to the ejection of several particles, some of which may be neutrons. There is therefore no firm basis for assuming that the disintegration represented in Fig. 4 corresponds to the disintegration of a nucleus of nitrogen rather than one of silver or bromine. Indeed, the delayed coincidence experiments suggest that the second assumption is the more probable. When a sufficient number of observations with loaded plates has been accumulated, it may be possible to draw more definite conclusions from observed regularities in the modes of disintegration of particular types of nuclei.

A detailed account of the experiments will be published elsewhere.

1 Perkins, D. H., *Nature*, **159**, 126 (1947).
2 Occhialini and Powell, *Nature*, **159**, 186 (1947).
3 Bose and Choudhuri, *Nature*, **148**, 259 (1941); **149**, 302 (1942).
4 Lattes, Fowler and Cuer, *Nature*, **159**, 301 (1947).
5 Hughes, *Phys. Rev.*, **69**, 371 (1946).
6 Leprince-Ringuet and L'Héritier, *J. Phys. et Rad.*, **7**, 65 (1946).
7 Bethe, *Phys. Rev.*, **70**, 821 (1947).
8 Schwinger, *Phys. Rev.*, **61**, 387 (1942).
9 Wentzel, *Rev. Mod. Phys.*, **19**, 4 (1947).
10 Fermi, Teller and Weisskopf, *Phys. Rev.*, **71**, 314 (1947).
11 Wheeler, *Phys. Rev.*, **71**, 320 (1947).
12 Conversi, Pancini and Piccioni, *Phys. Rev.*, **71**, 209 (1947).
13 Sigurgeirsson and Yamakawa, *Phys. Rev.*, **71**, 319 (1947).
14 Zhdanov, *Akad. Nauk. Odtel. Bull. Ser. Phys.*, **3**, 734 (1938); **4**, 272 (1939). *C.R. (U.S.S.R.)*, **28**, 110 (1940).
15 Rochester, Butler and Runcorn, *Nature*, **159**, 227 (1947).

since they can only be produced by neutrons by (*n*, 2*n*) reactions requiring very fast neutrons.

Radioactive isotopes produced by (*n,p*) reactions would require chemical separation; and when the processes have been established at the Harwell station, routine separation will be carried out by the Radiochemical Centre of the Ministry of Supply. The allocation of radioactive isotopes will be carried out by a sub-committee of the Advisory Council for the Radiochemical Centre. This sub-committee, under the chairmanship of Dr. J. D. Cockcroft, includes representatives of user interests—the Medical and Agricultural Research Councils and the Department of Scientific and Industrial Research, which will represent the universities and industry.

Radioactive isotopes from piles in the United States have been available to American users only since 1946, and a large number of applications to medical, biological and industrial research have already been found, while the future has been declared by Dr. G. Seaborg to hold unlimited possibilities.

J. D. Cockcroft

Observations on the tracks of slow mesons in photographic emulsions: Part 1

By C. M. G. LATTES, Dr. G. P. S. OCCHIALINI and Dr. C. F. POWELL

H. H. Wills Physical Laboratory, University of Bristol

INTRODUCTION. In recent experiments, it has been shown that charged mesons, brought to rest in photographic emulsions, sometimes lead to the production of secondary mesons. We have now extended these observations by examining plates exposed in the Bolivian Andes at a height of 5,500 m., and have found, in all, forty examples of the process leading to the production of secondary mesons. In eleven of these, the secondary particle is brought to rest in the emulsion so that its range can be determined. In Part 1 of this article, the measurements made on these tracks are described, and it is shown that they provide evidence for the existence of mesons of different mass. In Part 2, we present further evidence on the production of mesons, which allows us to show that many of the observed mesons are locally generated in the 'explosive' disintegration of nuclei, and to discuss the relationship of the different types of mesons observed in photographic plates to the penetrating component of the cosmic radiation investigated in experiments with Wilson chambers and counters.

Part I. Existence of Mesons of Different Mass

As in the previous communications[1], we refer to any particle with a mass intermediate between that of a proton and an electron as a meson. It may be emphasized that, in using this term, we do not imply that the corresponding particle necessarily has a strong interaction with nucleons, or that it is closely associated with the forces responsible for the cohesion of nuclei.

We have now observed a total of 644 meson tracks which end in the emulsion of our plates. 451 of these were found, in plates of various types, exposed at an altitude of 2,800 m. at the Observatory of the Pic du Midi, in the Pyrenees; and 193 in similar plates exposed at 5,500 m. at Chacaltaya in the Bolivian Andes. The 451 tracks in the plates exposed at an altitude of 2,800 m. were observed in the examination of 5 c.c. emulsion. This corresponds to the arrival of about 1·5 mesons per c.c. per day, a figure which represents a lower limit, for the tracks of some mesons may be lost through fading, and through failure to observe tracks of very short range. The true number will thus be somewhat higher. In any event, the value is of the same order of magnitude as that we should expect to observe in delayed coincidence experiments at a height of 2,800 m., basing our estimates on the observations obtained in similar experiments at sea-level, and making reasonable assumptions about the increase in the number of slow mesons with altitude. It is therefore certain that the mesons we observe are a common constituent of the cosmic radiation.

Photomicrographs of two of the new examples of secondary mesons, Nos. III and IV, are shown in Figs. 1 and 2. Table 1 gives details of the characteristics of all events of this type observed up to the time of writing, in which the secondary particle comes to the end of its range in the emulsion.

Table 1

Event No.	Range in emulsion in microns of Primary meson	Secondary meson
I	133	613
II	84	565
III	1040	621
IV	133	591
V	117	638
VI	49	595
VII	460	616
VIII	900	610
IX	239	666
X	256	637
XI	81	590

Mean range 614 ± 8 μ. Straggling coefficient $\sqrt{\Sigma\Delta_i^2/n}$ = 4·3 per cent, where $\Delta_i = R_i - \bar{R}$, R_i being the range of a secondary meson, and $\bar{R}$ the mean value for n particles of this type.

The distribution in range of the secondary particles is shown in Fig. 3. The values refer to the lengths of the projections of the actual trajectories of the particles on a plane parallel to the surface of the emulsion. The true ranges cannot, however, be very different from the values given, for each track is inclined at only a small angle to the plane of the emulsion over the greater part of its length. In addition to the results for the secondary mesons which stop in the emulsion, and which are represented in Fig. 3 by black squares, the length of a number of tracks from the same process, which pass out of the emulsion when near the end of their range, are represented by open squares.

The μ-Decay of Mesons

Two important conclusions follow from these measurements. Our observations show that the directions of ejection of the secondary mesons are orientated at random. We can therefore calculate the probability that the trajectory of a secondary

* This article contains a summary of the main features of a number of lectures given, one at Manchester on June 18 and four at the Conference on Cosmic Rays and Nuclear Physics, organised by Prof. W. Heitler, at the Dublin Institute of Advanced Studies, July 5–12. A complete account of the observations, and of the conclusions which follow from them, will be published elsewhere.

Reprinted with permission from *Nature* **160** 453–6 (1947)
© 1983 Macmillan Journals Limited

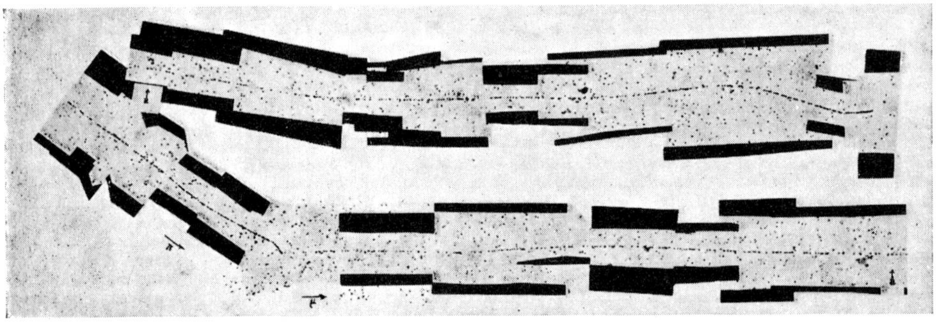

Fig. 1. OBSERVATION BY MRS. I. POWELL. COOKE × 95 ACHROMATIC OBJECTIVE; *C*2 ILFORD NUCLEAR RESEARCH EMULSION LOADED WITH BORON. THE TRACK OF THE μ-MESON IS GIVEN IN TWO PARTS, THE POINT OF JUNCTION BEING INDICATED BY *a* AND AN ARROW

meson, produced in a process of the type which we observe, will remain within the emulsion, of thickness 50 μ, for a distance greater than 500 μ. If we assume, as a first approximation, that the trajectories are rectilinear, we obtain a value for the probability of 1 in 20. The marked Coulomb scattering of mesons in the Nuclear Research emulsions will, in fact, increase the probability of 'escape'. The six events which we observe in plates exposed at 2,800 m., in which the secondary particle remains in the emulsion for a distance greater than 500 μ, therefore correspond to the occurrence in the emulsion of 120 $\pm$ 50 events of this particular type. Our observations, therefore, prove that the production of a secondary meson is a common mode of decay of a considerable fraction of those mesons which come to the end of their range in the emulsion.

Second, there is remarkable consistency between the values of the range of the secondary mesons, the variation among the individual values being similar to that to be expected from 'straggling', if the particles are always ejected with the same velocity. We can therefore conclude that the secondary mesons are all of the same mass and that they are emitted with constant kinetic energy.

If mesons of lower range are sometimes emitted in an alternative type of process, they must occur much less frequently than those which we have observed; for the geometrical conditions, and the greater average grain-density in the tracks, would provide much more favourable conditions for their detection. In fact, we have found no such mesons of shorter range. We cannot, however, be certain that mesons of greater range are not sometimes produced. Both the lower ionization in the beginning of the trajectory, and the even more unfavourable conditions of detection associated with the greater lengths of the tracks, would make such a group, or groups, difficult to observe. Because of the large fraction of the mesons which, as we have seen, can be attributed to the observed process, it is reasonable to assume that alternative modes of decay, if they exist, are much less frequent than that which we have observed. There is, therefore, good evidence for the production of a single homogeneous group of secondary mesons, constant in mass and kinetic energy. This strongly suggests a fundamental process, and not one involving an interaction of a primary meson with a particular type of nucleus in the emulsion. It is convenient to refer to this process

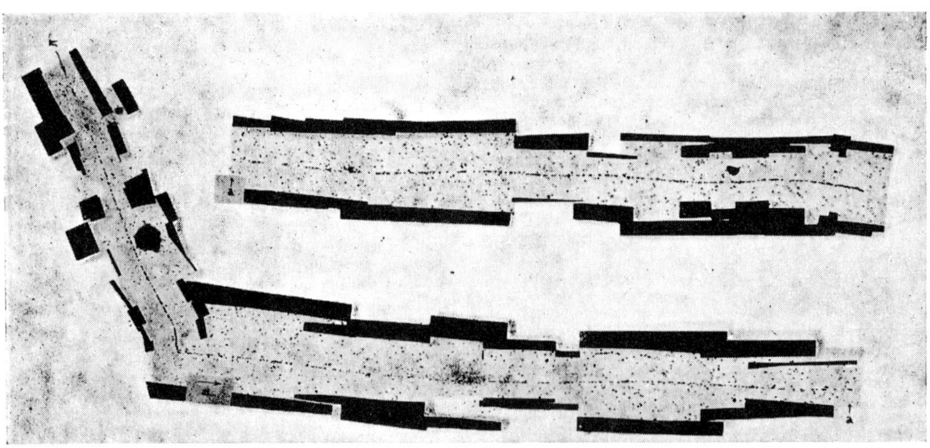

Fig. 2. COOKE × 95 ACHROMATIC OBJECTIVE. *C*2 ILFORD NUCLEAR RESEARCH EMULSION LOADED WITH BORON

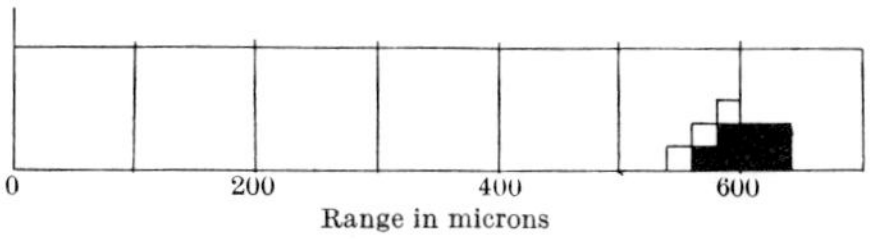

Fig. 3. Distribution in range of ten secondary mesons. Those marked ■ stop in the emulsion; the three marked □ leave the emulsion when near the end of their range. Mean range of secondary mesons, 606 microns. The results for events Nos. VIII to XI are not included in the figure

in what follows as the μ-decay. We represent the primary mesons by the symbol π, and the secondary by μ. Up to the present, we have no evidence from which to deduce the sign of the electric charge of these particles. In every case in which they have been observed to come to the end of their range in the emulsion, the particles appear to stop without entering nuclei to produce disintegrations with the emission of heavy particles.

Knowing the range-energy relation for protons in the emulsion, the energy of ejection of the secondary mesons can be deduced from their observed range, if a value of the mass of the particles is assumed. The values thus calculated for various masses are shown in Table 2.

Table 2

Mass in m_e	100	150	200	250	300
Energy in MeV.	3·0	3·6	4·1	4·5	4·85

No established range-energy relation is available for protons of energies above 13 MeV., and it has therefore been necessary to rely on an extrapolation of the relation established for low energies. We estimate that the energies given in Table 2 are correct to within 10 per cent.

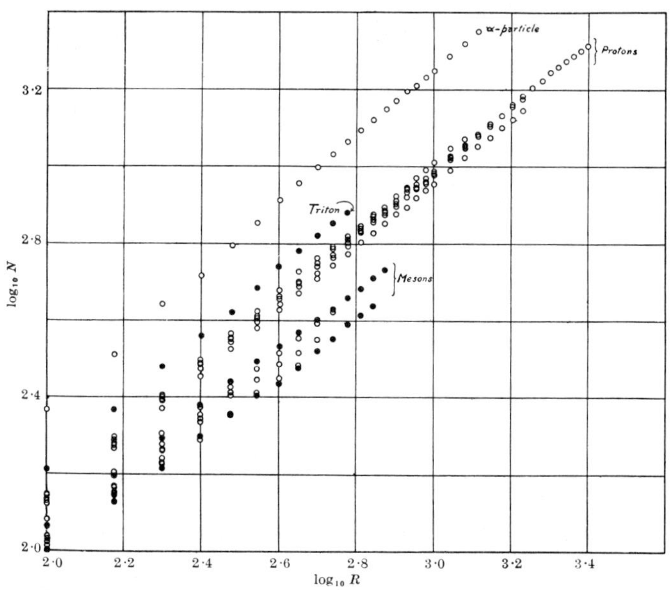

Fig. 4. N is total number of grains in track of residual range R (scale-divisions). 1 scale-division = 0·85 microns

Evidence of a Difference in Mass of π- and μ-Mesons

It has been pointed out[1] that it is difficult to account for the μ-decay in terms of an interaction of the primary meson with the nucleus of an atom in the emulsion leading to the production of an energetic meson of the same mass as the first. It was therefore suggested that the observations indicate the existence of mesons of different mass. Since the argument in support of this view relied entirely on the principle of the conservation of energy, a search was made for processes which were capable of yielding the necessary release of energy, irrespective of their plausibility on other grounds. Dr. F. C. Frank has re-examined such possibilities in much more detail, and his conclusions are given in an article to follow. His analysis shows that it is very difficult to account for our observations, either in terms of a nuclear disintegration, or of a 'building-up' process in which, through an assumed combination of a negative meson with a hydrogen nucleus, protons are enabled to enter stable nuclei of the light elements with the release of binding energy. We have now found it possible to reinforce this general argument for the existence of mesons of different mass with evidence based on grain-counts.

We have emphasized repeatedly[1] that it is necessary to observe great caution in drawing conclusions about the mass of particles from grain-counts. The main source of error in such determinations arises from the fugitive nature of the latent image produced in the silver halide granules by the passage of fast particles. In the case of the μ-decay process, however, an important simplification occurs. It is reasonable to assume that the two meson tracks are formed in quick succession, and are subject to the same degree of fading. Secondly, the complete double track in such an event is contained in a very small volume of the emulsion, and the processing conditions are therefore identical for both tracks, apart from the variation of the degree of development with depth. These features ensure that we are provided with very favourable conditions in which to determine the ratio of the masses of the π- and μ-mesons, in some of these events.

In determining the grain density in a track, we count the number of individual grains in successive intervals of length 50 μ along the trajectory, the observation being made with optical equipment giving large magnification ($\times$ 2,000), and the highest available resolving power. Typical results for protons and mesons are shown in Fig. 4. These results were obtained from observations on the tracks in a single plate, and it will be seen that there is satisfactory resolution between the curves for particles of different types. The 'spread' in the results for different particles of the same type can be attributed to the different degrees of fading associated with the different times of passage of the particles through the emulsion during an exposure of six weeks.

Applying these methods to the examples of the μ-decay process, in which the secondary mesons come to the end of their range in the emulsion, it is found that in every case the line representing the observations on the primary meson lies above that for the secondary particle. We can therefore conclude that there is a significant difference in the grain-density in the tracks of the primary and secondary mesons, and therefore a difference in the mass of the particles. This conclusion depends, of course, on the assumption that the π- and μ-particles carry equal charges. The grain-density at the ends of the tracks, of particles of both types, are consistent with the view that the charges are of magnitude $|e|$.

A more precise comparison of the masses of the π- and μ-mesons can only be made in those cases in which the length of the track of the primary meson in the emulsion is of the order of 600 μ. The probability of such a favourable event is rather small, and the only examples we have hitherto observed are those listed as Nos. III and VIII in Table 1. A mosaic of micrographs of a part only of the first of these events is reproduced in Fig. 1, for the length of the track of the π-meson in the emulsion exceeds 1,000 μ. The logarithms of the numbers of grains in the tracks of the primary and secondary mesons in this event are plotted against the logarithm of the residual range in Fig. 5. By comparing the residual ranges at which the grain-densities in the two tracks have the same value, we can deduce the ratio of the masses. We thus obtain the result $m_\pi/m_\mu = 2 \cdot 0$. Similar measurements on event No. VIII give the value 1·8. In considering the significance which can be attached to this result, it must be noticed that in addition to the standard deviations in the number of grains counted, there are other possible sources of error. Difficulties arise, for example, from the fact that the emulsions do not consist of a completely uniform distribution of silver halide grains. 'Islands' exist, in which the concentration of grains is significantly higher, or significantly lower, than the average values, the variations being much greater than those associated with random fluctuations. The measurements on the other examples of μ-decay are much less reliable on account of the restricted range of the π-mesons in the emulsion; but they give results lower than the above values. We think it unlikely, however, that the true ratio is as low as 1·5.

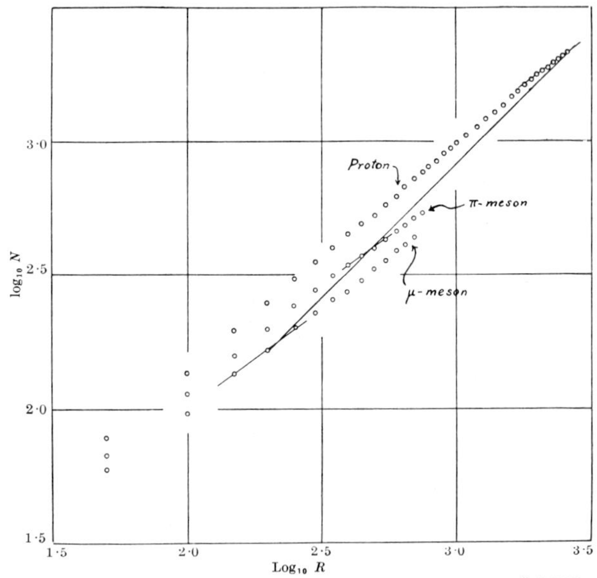

Fig. 5. N IS TOTAL NUMBER OF GRAINS IN TRACK OF RESIDUAL RANGE R (SCALE-DIVISIONS). 1 SCALE-DIVISION = 0·85 MICRONS
THE 45°-LINE CUTS THE CURVES OF THE MESONS AND PROTON IN THE REGION OF THE SAME GRAIN DENSITY

The above result has an important bearing on the interpretation of the μ-decay process. Let us assume that it corresponds to the spontaneous decay of the heavier π-meson, in which the momentum of the μ-meson is equal and opposite to that of an emitted photon. For any assumed value of the mass of the μ-meson, we can calculate the energy of ejection of the particle from its observed range, and thus determine its momentum. The momentum, and hence the energy of the emitted photon, is thus defined; the mass of the π-meson follows from the relation

$$c^2 m_\pi = c^2 m_\mu + E_\mu + h\nu.$$

It can thus be shown that the ratio m_π/m_μ is less than 1·45 for any assumed value of m_μ in the range from 100 to 300 m_e, m_e being the mass of the electron (see Table 3). A similar result is obtained if it is assumed that a particle of low mass, such as an electron or a neutrino, is ejected in the opposite direction to the μ-meson.

TABLE 3

Assumed mass m_μ	E (MeV.)	$h\nu$ (MeV.)	m_π	$m_\pi/m_\mu \pm 3$ per cent
100 m	3·0	17	140 m_e	1·40
150	3·6	23	203	1·35
200	4·1	29	264	1·32
250	4·5	34	325	1·30
300	4·85	39	387	1·29

On the other hand, if it is assumed that the momentum balance in the μ-decay is obtained by the emission of a neutral particle of mass equal to the μ-meson mass, the calculated ratio is about 2·1 : 1.

Our preliminary measurements appear to indicate, therefore, that the emission of the secondary meson cannot be regarded as due to a spontaneous decay of the primary particle, in which the momentum balance is provided by a photon, or by a particle of small rest-mass. On the other hand, the results are consistent with the view that a neutral particle of approximately the same rest-mass as the μ-meson is emitted. A final conclusion may become possible when further examples of the μ-decay, giving favourable conditions for grain-counts, have been discovered.

[1] *Nature*, **159**, 93, 186, 694 (1947).

Flying across wind at 120 m.p.h. at a height of about 50 ft., with wind speeds of 6–10 m.p.h., an effective swathe width of 70 yards was obtained, approximately 50 acres per minute being sprayed. The cost of dressing about 100 square miles was 2*s*. an acre for the D.D.T. and 1*s*. per acre for the air-craft services. Six dressings were given, so the total cost was 18*s*. an acre. Up to six aircraft were used at a time, succeeding ones following the leader at intervals and taking alignment at the edge of the visible spray left by the one in front. Flying so close to the ground involves some definite risk of accident, and to meet this point it is suggested that a height of 75 ft. might be used without materially affecting the lie of the D.D.T.

Elaborate tests were made of the actual spread of the D.D.T. in the bush country used for the experiment. Where there were breaks in the ground because of small ravines, the spread was uneven, and D.D.T. grenades placed by hand were used to supplement the spraying at such places.

Many factors have, of course, to be taken into consideration in assessing the value of the results obtained. So far there has been no indication that there is serious danger that useful forms of life would be permanently exterminated.

The spraying has resulted in a significant drop in the incidence of the tsetse flies in the area treated. The catches in the Harris traps after the spraying were exceedingly low. The flies were not eradicated, but the numbers that survived were small.

This is a pilot experiment, and an enormous amount of work will be required before the relative value of this or any other method of control can be assessed. It has provided certain definite and significant results. Where the fly was a nuisance to individuals working in or passing through the area, they are now rarely noticed. It is considered that with such a considerable reduction in the numbers, the ordinary causes of mortality of the tsetse fly—and these are considerable because of the hazards involved in its life-history and of its slow breeding-rate—may lead to its disappearance. On the other hand, the flies may be able again to attain sufficient concentration to menace the health of the livestock.

Sufficient has been done in Africa to justify a tremendously increased scale of operations against the tsetse fly. The amount of money spent on research on methods of controlling the fly since Africa has been developed by the white race has been small indeed—pitifully small—in comparison with the harm it causes and the extent to which it limits the utilization of land potentially valuable for husbandry and other purposes. Here is a problem which requires costly large-scale research. If very large tracts of useful agricultural land could be made available as a result of experimentation, the money would be well spent. It is, of course, to be understood that, hand in hand with successful tsetse fly control in tropical Africa, measures for dealing with prevalent social and agricultural systems that lead to over-stocking and consequent soil erosion would require attention. Not one of the least important problems would be the provision of adequate water supplies for livestock and efficient use of such water.

The film showed very clearly the main features of the work done, and Dr. du Toit discussed the many questions involved in consideration of this gigantic problem. South Africa is making a very determined effort to ensure that progress is made, and this pioneer work will serve as a stimulus to those interested in the development of tropical Africa. All concerned are to be congratulated on a piece of work which is a very important landmark in the history of parasite control.

Observations on the tracks of slow mesons in photographic emulsions: Part 2

By C. M. G. LATTES, Dr. G. P. S. OCCHIALINI
AND
Dr. C. F. POWELL
H. H. Wills Physical Laboratory, University of Bristol

Part 2: Origin of the Slow Mesons

In Part 1 of the present article*, we showed that two types of mesons exist, and it was suggested that the heavier, π-mesons, decay to produce the lighter, μ-mesons. In this second part, we discuss the origin of the slow mesons observed in photographic emulsions, and their relation to the mesons forming the penetrating component of the cosmic rays, of which evidence is provided by experiments with Wilson chambers and counters. We also give photomicrographs which show that some slow mesons, ejected from nuclei during 'explosive disintegrations', can enter nuclei and produce a second disintegration. Most of the observations on plates exposed at 5,500 m. were made with boron-loaded emulsions and, for the most part, we have confined our analysis to the results obtained with plates of this type. The loading material has an important influence on the rate of fading of the latent image, but by the above procedure we can compare the results of experiments at different altitudes.

Disintegrations Produced by Mesons

Of the total of 644 mesons observed to come to the end of their range in our emulsions, 145 give rise to secondary particles. In 40 cases, the track of the secondary particle has characteristics which suggest that it was produced by a meson. We therefore regard these events as due to the decay of π-mesons although, in several instances, the length of the track of the secondary particle is too short for the identification to be certain. In the remaining 105 cases, the mesons produce disintegrations with the emission of heavy particles. We have previously published photomicrographs of such disintegrations, and Fig. 6 shows an example in which a single proton is ejected. Although we have no evidence that they are all of the same type, it will be convenient to refer, provisionally, to all mesons producing nuclear disintegrations, when moving at low velocity, as σ-mesons. Since the specifically nuclear interaction between mesons and nucleons may be assumed to be of short range, and since the Coulomb repulsion would tend to prevent a slow positively charged meson from approaching a nucleus, we regard the observation of the disintegrations produced by σ-mesons as providing strong evidence that they are negatively charged. We show in a later paragraph that there are indications that π-mesons and some, at least, of the σ-mesons

* *Nature*, **160**, 453 (Oct. 4, 1947).

Reprinted with permission from *Nature* **160** 486–92 (1947)
© 1983 Macmillan Journals Limited

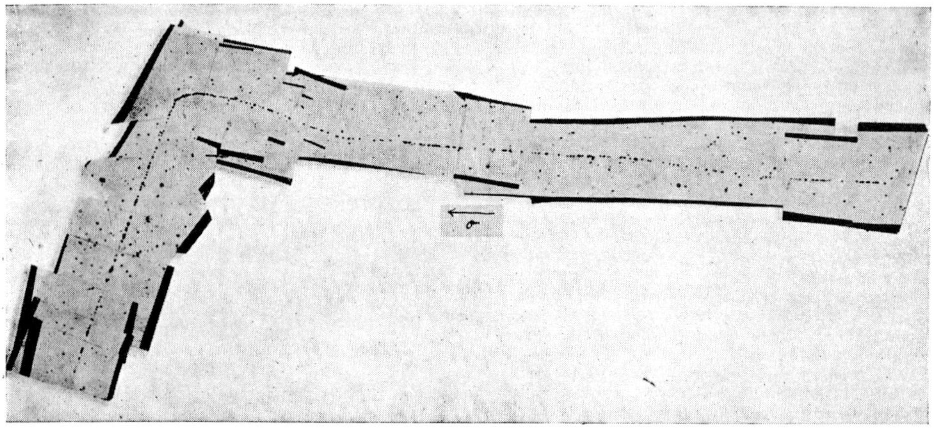

Fig. 6. A MESON, σ, COMES TO THE END OF ITS RANGE AND LEADS TO A DISINTEGRATION WITH THE EMISSION OF A SINGLE HEAVY CHARGED PARTICLE, PROBABLY A PROTON. PHOTOMICROGRAPHS WITH COOKE × 95 ACHROMAT

are, respectively, positively and negatively charged particles of the same type.

Fig. 7 shows the relative frequency with which different numbers of heavy charged particles are ejected in the disintegrations produced by mesons. Prof. W. Heitler has pointed out to us that the observed events probably do not represent all the disintegrations actually produced by slow mesons in the area of the plates examined. In some cases, for example, a single fast neutron may result from the interaction of a negative meson with a proton in a nucleus. Further, for reasons similar to those discussed below, it is likely that we sometimes fail to observe a single fast proton ejected from a nucleus because of the large angle of dip of the track relative to the plane of the emulsion.

It is interesting to compare the distribution shown in Fig. 7 with the results obtained by Gardner[2] in

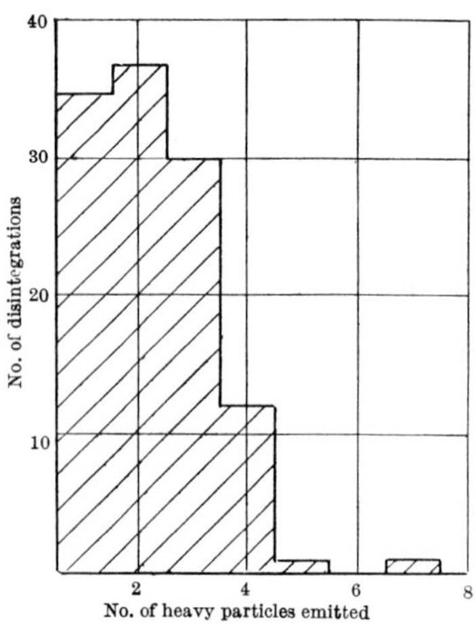

Fig. 7. RELATIVE FREQUENCY OF THE DISINTEGRATIONS PRODUCED BY MESONS, IN WHICH DIFFERENT NUMBERS OF HEAVY CHARGED PARTICLES ARE EMITTED

studying the disintegrations, produced in photographic emulsions by 200 MeV. deuterons generated by the 184-in. synchro-cyclotron. In these experiments, disintegrations are commonly observed in which two, three or four charged particles are produced; but in only a very small fraction of the cases are five charged particles emitted. Our results are therefore consistent, in this respect, with the view that the absorption of a σ-meson sometimes leads to the liberation in a nucleus, of an amount of energy corresponding to a rest-mass of the particle of the order of 400 m_e.

Total Number of Meson Tracks in the Emulsion

In discussing the origin of the mesons observed in our experiments, it is important to determine the rate of production of the different types and, for this purpose, we cannot use the observed numbers of tracks without correction. If a track 'dips' at a large angle, and if the length of its projection on the plane of the emulsion is less than 50 μ, it is likely either to escape observation, or to provide insufficient evidence to justify it being identified as due to a meson. A certain fraction of the mesons will therefore not be counted. This 'loss' of tracks will not, however, affect the observations of different types of mesons equally. In the case of the σ-mesons, for example, the observation of the track is made easier by the existence of the secondary particles resulting from the disintegration, which direct the attention of the observer to the event. Similarly, in the case of those π-mesons which lead to μ-mesons of which the track remains in the emulsion, the observation of the track of the μ-meson ensures that the track of the parent π-meson is observed, even although it is very short.

To overcome these difficulties, we adopt the following procedure. We plot, in Fig. 8, the distribution in length of the projections, on the surface of the emulsion, of the tracks produced by mesons which come to the end of their range without giving rise to secondary heavy particles. The results are taken from observations on boron-loaded plates, 50 μ thick, exposed at the Pic du Midi.

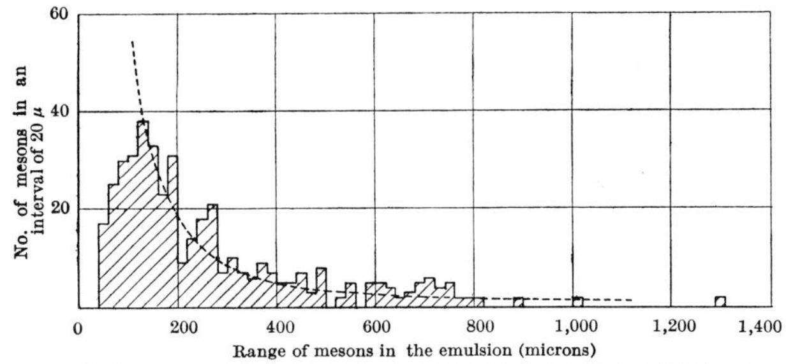

Fig. 8. HISTOGRAM SHOWING THE DISTRIBUTION OF THE OBSERVED LENGTHS OF THE PROJECTIONS OF MESON TRACKS ON THE SURFACE OF THE EMULSION. THE DOTTED LINE SHOWS THE EXPECTED DISTRIBUTION ASSUMING THE TRACKS TO BE RECTILINEAR

Simple geometrical considerations allow us to compare the distribution shown in Fig. 8 with that to be expected if all tracks, without exception, were observed in the given volume of the emulsion. We assume that the tracks are rectilinear and the ranges of the particles indefinitely great; and that the end points of the tracks, and the orientation of the trajectories, are distributed at random. It is then easily shown that the numbers of tracks, $N(t)\delta t$, of which the projected range lies in the interval from t to $t + \delta t$, is given by the relation:

$$N(t) \,.\, \delta t = \frac{\Sigma N_0}{d}\left(1 - \frac{t}{(d^2+t^2)^{1/2}}\right)\delta t, \quad . \quad (1)$$

where ΣN_0 is the total number of tracks in the observed area, and d the thickness of the emulsion. Further, the number of tracks, ΣN_t, of projected range greater than t, is related to the total number, ΣN_0, by the equation

$$\Sigma N_t = \Sigma N_0 \,[\,(d^2 + t^2)^{1/2} - t]. \quad . \quad . \quad (2)$$

From equation (2) we can calculate the quantity ΣN_0, from the observed values of ΣN_t, corresponding to the measurements represented in Fig. 8, and the results are shown in Table 4.

TABLE 4. BORON-LOADED EMULSION, TYPE C2. MESON TRACKS WITHOUT SECONDARY HEAVY PARTICLES. EXPOSURE 40 DAYS AT THE PIC DU MIDI. TOTAL VOLUME OF EMULSION EXAMINED 3·5 C.C.

$t(\mu)$	ΣN_t	ΣN_0	Number of mesons per c.c. per diem
400	42·5	679 ± 104	
300	62	748 ± 95	
200	96·5	783 ± 77	
100	185·5	795 ± 58	5·7
80	199·5	709 ± 49	
60	212	569 ± 41	
40	224	465 ± 31	

The dotted line in Fig. 8 is calculated, from equation (1), corresponding to a total of 795 mesons ending in the emulsion. It will be seen, from this figure and from Table 4, that the results are consistent with the assumption that no meson tracks of this class are lost of which the projected range is greater than 100 μ. On the other hand, with tracks of shorter range, there is a serious 'loss'. We can therefore assume that the total number of tracks in the emulsion can be determined from the observations on the longer tracks, for which t is greater than 100 μ, and we use the result $\Sigma N_0 = 795$, obtained from the value of ΣN_{100}, as the most reliable. This number includes the tracks of π-mesons of which we fail to distinguish the track of the secondary meson, and of any σ-mesons which produce disintegrations with no recognizable secondary particle.

From similar observations on the σ-mesons which lead to observed disintegrations, we conclude that the total number of such particles in the measured area of the emulsion is 122 ± 20, and, therefore, the total number of slow mesons of all types 917 ± 70.

Number of π-Mesons

The analysis described in the previous paragraph can also be applied to determine the total number of π-mesons in the emulsion. Thus in the boron plates exposed at the Pic du Midi, we have observed six π-meson tracks, in which the length of the track of the secondary μ-meson exceeds 500 μ. Applying equation (2), it is found that this figure corresponds to the occurrence in the measured area of the emulsion of 120 ± 50 events of this class. In many of these events the track of the μ-meson will be very short, and we shall have failed to observe it and that of its associated π-meson. The number of events of which we infer the occurrence must be compared, therefore, with the total number of mesons, 917; for, as we have seen, the observation of the μ-meson track, in those cases in which its length exceeds 500 μ, ensures that we shall observe the track of the parent π-meson, however short its range in the emulsion.

A second estimate of the number of π-mesons can be obtained from the observed events in which the secondary μ-meson has a restricted range in the emulsion. The result so obtained is 115 ± 20, a value which does not differ significantly from that obtained by the first method. Combining the two results, we obtain the value for the number of π-mesons, 117 ± 20.

Relative Numbers of Slow Mesons of Different Types at 2,800 and 5,500 m.

Of the 917 slow mesons, of which we infer the occurrence in the measured area of plates exposed at the Pic du Midi, we have seen that 122 ± 20 must be attributed to σ-mesons, and 117 ± 20 to π-mesons. An approximately equal number of μ-mesons will end in the emulsion from the decay of π-mesons brought to rest in the glass or other matter in the immediate neighbourhood of the emulsion. These results are summarized in Table 5, and show that we can account for approximately one third of the observed mesons in terms of π- and σ-mesons brought to rest in the assembly of plates, and of the μ-mesons resulting from the decay of the former. The remaining tracks, apart from a small number which may be due to σ-mesons producing disintegrations with the emission of neutrons, remain to be accounted for.

Very similar results have been obtained from an examination of the plates exposed at 5,500 m., and the corresponding figures are included in Table 5. The results indicate that there is no significant difference, at the two altitudes, in the distribution of the observed mesons among the different types. There were differences in the conditions of the exposures at the two altitudes; and different time intervals elapsed between the ends of an exposure and the development of the plates, in the two cases. We cannot, therefore, rely on the absolute intensities given in the table.

TABLE 5. SUMMARY OF RESULTS OBTAINED IN AN EXAMINATION OF BORON-LOADED EMULSIONS EXPOSED FOR 40 DAYS AT THE PIC DU MIDI, 2,800 M.; AND AT COCHABAMBA, BOLIVIA, AT 5,500 M. AT 2,800 M., EMULSION THICKNESS, 50 μ; VOLUME OF EMULSION EXAMINED, 3·5 C.C. AT 5,500 M., EMULSION THICKNESS, 100 μ; VOLUME OF EMULSION EXAMINED, 1·1 C.C.

	At 2,800 m.		At 5,500 m.	
	Estimated total	Number per c.c. per diem	Estimated total	Number per c.c. per diem
Mesons of all types	917 ± 70	6·5 ± 0·5	269 ± 25	6·1 ± 0·6
(*a*) π-mesons	117 ± 20	0·8 ± 0·2	40 ± 20	0·9 ± 0·5
(*b*) σ-mesons	122 ± 20	0·9 ± 0·2	46 ± 9	1·0 ± 0·2
(*c*) μ-mesons arising from the decay of π-mesons at rest	117 ± 20	0·8 ± 0·2	40 ± 20	0·9 ± 0·5
(*d*) Number unaccounted for	561 ± 100	4·0 ± 0·7	143 ± 40	3·3 ± 0·9
(*e*) Number of nuclear disintegrations giving 5 or more charged particles		10·5		
(*f*) Total of π- + σ-mesons observed		1·7 ± 0·5		
(*g*) Estimated number of slow mesons originating in disintegrations in plate assembly		0·6 ± 0·3		

Emission of Mesons from Disintegrating Nuclei

A mosaic of photomicrographs showing the ejection of a meson from a disintegrating nucleus was given in a previous communication. We have now observed five events of this type, including two in which the ejected meson, at the end of its range, leads to the disintegration of a nucleus. Photomicrographs of these two events are shown in Figs. 9 and 10, the number of charged particles from the secondary 'stars', *S*, being four and two, respectively. In the nomenclature of the present paper, these two events prove that σ-mesons can be generated in the disintegration of nuclei by cosmic ray particles of great energy.

We have also observed that one of the ejected mesons, at the end of its range, leads to the production of a single fast secondary particle. In this case, the length of the trajectory of the secondary particle in the emulsion, before it passes into the glass, is only about 150 μ. We cannot definitely state, therefore, that the secondary particle is a μ-meson, and not a fast proton, although the grain density in the track is consistent with such an assumption, and we believe it to be the correct interpretation of the event. If this is so, the observation corresponds to the generation of π-mesons in processes associated with the explosive disintegration of nuclei. For decisive evidence on this point we must await further observations. If our present view is correct, we shall expect to observe the decay of an ejected π-meson, in which the secondary particle stops in the emulsion after traversing a distance of about 600 μ.

Of the five mesons which we have observed to be ejected in the explosive disintegration of nuclei, two come to the end of their range without giving rise to secondary particles of which we can distinguish the tracks. We have seen that σ- and, possibly, π-mesons can be generated during explosive disintegrations; the important question arises whether other types, μ-mesons for example, can also be produced in such processes. We consider that our evidence is not sufficient to provide an answer to this question. The two ejected mesons under consideration come to the end of their range close to the surface of the emulsion. If they were π-mesons, the probability of observing the track of the associated μ-meson would be, in the case of each event, less than 0·5.

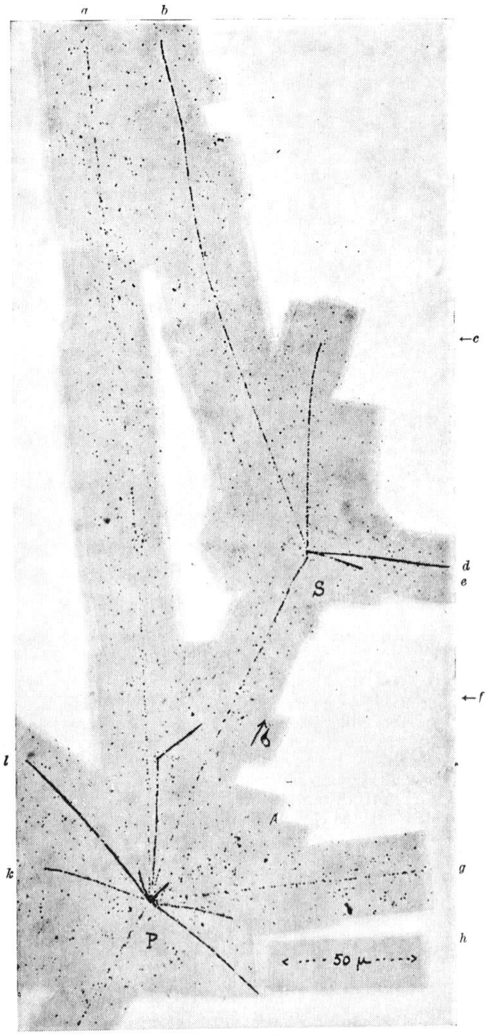

Fig. 9. MOSAIC OF PHOTOMICROGRAPHS WITH COOKE × 95 ACHROMAT; ILFORD 'NUCLEAR RESEARCH' EMULSION TYPE *C2*, BORON LOADED. A MESON, σ, EJECTED DURING THE EXPLOSIVE DISINTEGRATION OF A NUCLEUS, *P*, COMES TO THE END OF ITS RANGE AND PRODUCES A SECOND DISINTEGRATION, *S*. THE TRACKS LABELLED '*b*', '*c*' AND '*k*' ARE DUE TO PROTONS WHICH REACHED THE END OF THEIR RANGE IN THE EMULSION. MOST OF THE OTHER TRACKS END IN ONE OF THE SURFACES OF THE EMULSION

The creation of a meson requires an amount of energy equal to, let us assume, 100 MeV. As we have previously emphasized, the probability of such a particle being ejected with a kinetic energy of only

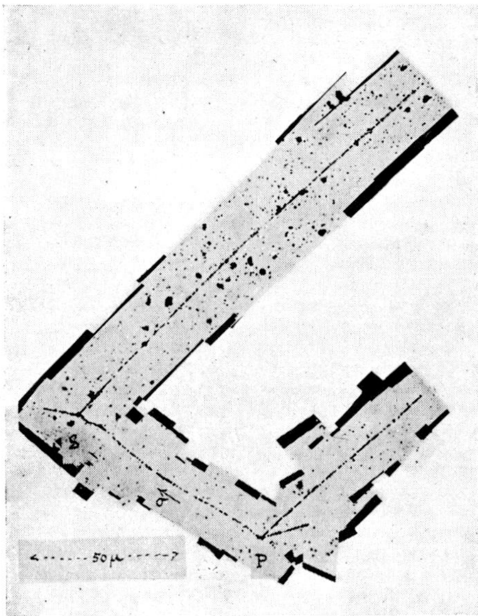

Fig. 10. MOSAIC OF PHOTOMICROGRAPHS WITH COOKE × 95 ACHROMAT; ILFORD 'NUCLEAR RESEARCH' EMULSION TYPE *C*2, BORON LOADED. A MESON, *σ*, EJECTED DURING THE EXPLOSIVE DISINTEGRATION OF A NUCLEUS, *P*, COMES TO THE END OF ITS RANGE AND PRODUCES A SECOND DISINTEGRATION, *S*

a few MeV. must be very small. Further, we shall expect to identify such an ejected meson only if it happens to be emitted in a direction nearly parallel to the surface of the emulsion. The new observations, therefore, lend support to the view that the ejection of mesons is commonly associated with the explosive disintegration of nuclei. The probability of observing two slow mesons ejected from the same nucleus must be very small, even if, as is often supposed, the production of several mesons, by an interaction of a fast particle with a single nucleus, is of frequent occurrence.

Origin of the Slow Mesons

We have now observed four thousand examples of the explosive disintegration of nuclei in plates exposed at an altitude of 2,800 m., which correspond to the liberation of a greater amount of energy, as judged by the numbers of the ejected charged particles and their ranges, than occurs in some of the similar events which we observe to be accompanied by the emission of slow mesons. In a mass of solid material such as is constituted by a box of photographic plates, nuclear explosions will be produced which will be accompanied by the emission of mesons. A certain fraction of these mesons will be brought to rest in the emulsion of the photographic plate. It is of great interest to determine what proportion of the mesons we observe can be attributed to such processes.

We may assume as a first approximation that those of the ejected mesons will stop in the solid material the ranges of which are less than the linear dimensions of the mass. Further, that the ends of the meson tracks will be uniformly distributed at random throughout the solid material, their number per unit volume being equal to that of the parent nuclear explosions. We estimate that the energy of a meson able to traverse the solid material constituted by the assembly of our plates during exposure is of the order of 50 MeV. Ejected mesons with a kinetic energy less than this value will contribute to the mesons stopping in the emulsion or in the glass; those with greater energy may be regarded as escaping.

In plates exposed at the Pic du Midi, we have observed four disintegrations in which the ejected meson has an energy between 1 and 3 MeV. Allowing for geometrical factors, we can estimate that this number corresponds to the occurrence, in the volume of the emulsion which we have examined, of at least twelve events of this type, the others having escaped detection because of the large angles of dip of the tracks of the mesons. If we assume that the values of the energy of the ejected mesons are distributed at random in the interval from 0 to 50 MeV., all values of the energy being equally probable, then the estimated numbers of mesons with energies between 1 and 3 MeV. correspond to a frequency of occurrence in our emulsions, of mesons at the end of their range, of $0{\cdot}6 \pm 0{\cdot}3$ per c.c. per diem. This result depends on estimates of quantities which are not known with precision, but the agreement, in order of magnitude, with the numbers of π- and σ-mesons is striking. It gives strong support to the view that a considerable fraction of the observed mesons of these types arises in locally produced disintegrations. The observations are not, however, sufficiently precise to allow us to determine the fraction of the π- and σ-mesons which must be attributed to similar processes in neighbouring matter, located within a few metres of the plates during exposure, a point of great interest in connexion with the determination of the life-time of these mesons.

Interpretation of the Observations

In view of the complexity of the phenomena, and the limited numbers of tracks of the different types of mesons which we have observed, we cannot draw any final conclusions about many important questions suggested by our observations. We have, however, considered a number of possible interpretations of the present experiments and of those made with Wilson chambers and counters. The following working hypothesis is the simplest, from among those we have considered, which allows us to account for all the main experimental results without serious contradictions.

We assume that the π-mesons, and most of the σ-mesons, are, respectively, positive and negative particles of the same type, which are produced in processes associated with the explosive disintegration of nuclei, such as we have observed. The positive π-mesons suffer μ-decay and give rise to μ-mesons, for the fate of which our experiments give no evidence. On the other hand, the negative π-mesons, which appear in our experiments as σ-mesons, are captured by nuclei to produce disintegrations with the emission of heavy particles.

The above hypothesis allows us to account for the approximate equality of the numbers of π- and σ-mesons which we have deduced from our observations, and for the fact that the relative frequency of occurrence of the different types shows no significant change with increasing altitude. Further, since particles observed to be produced by the decay of

π-mesons at rest are regarded as being positively charged, we shall not expect them, in conformity with the observations, to react with nuclei and produce disintegrations with the emission of heavy particles.

It remains to account for those of the mesons which are not π- or σ-mesons, and which cannot be attributed to the μ-mesons arising from the decay of π-mesons stopping in the plate assembly. We attribute them to μ-mesons, produced by the decay in flight of positive and negative π-mesons which have been generated in nuclear processes taking place at a distance from the plate assembly. To this extent, our observations can be interpreted in terms of a simple and self-consistent scheme, similar, in several respects, to those visualized by other workers[3].

If the above view is correct, we may attribute the penetrating component observed at sea-level to particles of the same type as the μ-mesons, produced by energetic π-mesons which have been formed in the high atmosphere, and which have decayed in flight. These μ-mesons may be assumed to be both positively and negatively charged. If we accept the general conclusions which have been drawn from delayed coincidence experiments, we should expect those of the negative μ-mesons which stop in a silver halide grain to be captured by a heavy nucleus. On the other hand, those stopping in the gelatine will be expected to suffer β-decay. It is not possible to make a precise determination of the relative probability of these two processes, since it involves a knowledge of the atomic stopping power of the various constituents of the emulsion in a range of energies, $\sim$ 20 keV., in which no measurements have been made. Further, Dr. H. Fröhlich has pointed out to us that the range-energy curve will be different for positive and negative mesons, since the latter do not capture and lose electrons in their passage through the emulsion. Rough estimates indicate, however, that there is about twice the probability of a meson stopping in the gelatine rather than in a silver bromide grain. The mesons in line (*d*) of Table 5 would, on these assumptions, represent about five sixths of such μ-mesons stopping in the emulsion, the other one sixth interacting with nuclei. If we assume that these latter mesons, like the negative π-mesons, lead to disintegrations with the emission of heavy particles, they will appear among the σ-mesons. No disintegrations produced by slow mesons have been observed, however, in expansion chambers operating at sea-level. It is possible, therefore, that the interaction of negative μ-mesons with nuclei leads to a different type of process from that represented by the σ-meson disintegrations.

If the above view is accepted, it remains to be explained why, in spite of their assumed short life-time, it is possible for the negative π-mesons to suffer nuclear capture in both gelatine and silver bromide, whereas the negative μ-mesons, assumed to be longer-lived, suffer β-decay before being captured by light nuclei. We can only resolve this difficulty if we assume that the time taken for a negative meson to approach a nucleus is less than the life-time of both types of mesons. At present, the magnitude of this time interval is still under discussion. Assuming, however, that the close approach to the nucleus can be made before the decay of the meson, whatever its type, it is necessary to assume that the interaction of μ-mesons with nucleons is much weaker than that of π-mesons. If the genetical relationship which we have visualized between the different types of particles is correct—the π-mesons decaying to produce μ-mesons, which in turn emit the β-particles responsible for the delayed coincidences—it follows from the counter experiments that the sum of the life-times of the π- and μ-mesons must be 2×10^{-6} sec. The only available information about the individual values of the decay-constants of the two types is derived from the fact that the times of flight of the π- and μ-mesons in our emulsions are of the order of 10^{-11} sec. The life-time for both types must, accordingly, be greater than this value.

Let us, therefore, as an example of an alternative hypothesis, assume that the π-mesons have a life-time of 2×10^{-6} sec., and that the μ-mesons are much shorter-lived. The majority of the cosmic ray mesons observed in counter and Wilson chamber experiments at sea-level must then be attributed to π-mesons. In this case, to be consistent with the delayed coincidence experiments, we must assume that those negative π-mesons which stop in the gelatine suffer μ-decay, whereas those stopping in silver bromide grains produce disintegrations. This would lead us to expect about five times as many π- as σ-mesons. Further, on this hypothesis, some negative π-mesons will suffer μ-decay to produce negative μ-mesons which might react with nuclei. To account for the observed absence of disintegrations produced by the eleven of our observed mesons which undoubtedly originate by μ-decay, some of which we shall now expect to be negatively charged, we can attribute to them either a sufficiently short life-time or assume that they do not interact with nucleons.

If the above difficulties are resolved, two serious objections to this second hypothesis remain. In the first place, in contradiction with experience, we should expect Wilson chamber experiments to have given some evidence for the μ-decay of mesons. Secondly, we can now make no contribution towards solving the contradiction between the weak interaction with nucleons of the mesons observed at sea-level, as shown by the delayed coincidence and other experiments, and the fact of the observed generation of the σ-mesons. In view of these numerous difficulties, we can reject this second hypothesis in favour of the first assumption that the π-mesons are short-lived and interact strongly with nucleons.

We have already emphasized that no final conclusions can be drawn at the present stage of our knowledge. It is, indeed, quite possible that the phenomena are more complicated than we have visualized. It is important, therefore, to extend the observations in order to reduce the statistical errors, and to make determinations of the masses of the various types of mesons by all available methods. In this connexion it may be noticed that the large numbers of tracks which are now available for measurement allow us to apply statistical methods. Thus the π- and μ-mesons probably constitute unique classes; and it is therefore possible to make a statistical study of the scattering of the particles in their passage through the emulsion in order to determine their mass. It is possible that the σ-mesons, which, we suggest, are mainly composed of negative π-mesons, may contain some negative μ-mesons. In this case also, however, a statistical study of the small-angle scattering should allow us to decide this question and to prove whether or not the majority of the σ-mesons have the same mass as the π-mesons. Such investigations are in progress, together with experiments to determine the life-time of the π- and σ-mesons.

Summary

We may summarize our results in the following terms. We regard our observations as establishing:

(*a*), that two types of mesons exist, of different mass, which we refer to as π- and μ-mesons;

(*b*), taken in conjunction with the observations of D. H. Perkins, that some slow mesons can enter nuclei to produce disintegrations with the emission of heavy particles; we regard these mesons as negatively charged and refer to them, provisionally, as σ-mesons; and

(*c*), that σ-, and possibly π-mesons, can be produced in processes associated with explosive disintegration of nuclei.

Our observations suggest:

(*d*), that π-, and a large proportion of the σ-mesons are, respectively, positively and negatively charged particles of the same type, which interact strongly with nucleons, the negatively charged mesons being captured by both light and heavy nuclei to produce disintegrations with the emission of heavy particles;

(*e*), that the heavier π-mesons suffer spontaneous decay with the emission of the lighter μ-mesons; and that in the decay process the momentum of the μ-meson is balanced by that of a neutral particle of approximately the same mass.

In addition, the main features of the present experimental results, and those obtained in delayed coincidence and Wilson chamber experiments, can be explained on the assumption that:

(*f*), the greater part of the mesons observed at sea-level are μ-mesons formed by the decay in flight of π-mesons; and

(*g*), that positive and negative π-mesons are short-lived, with a mean life-time in the interval from 10^{-6} to 10^{-11} sec.

We have pleasure in acknowledging our indebtedness to Prof. W. Heitler, Prof. C. Møller, Dr. H. Fröhlich, Prof. L. Jánossy and other members of the Dublin colloquia, held in July, for a number of discussions which were of great value to us in the formulation of our ideas. Our thanks are also due to Prof. J. Baillaud and Dr. M. Hugon of l'Observatoire du Pic du Midi, and to M. René Varin, cultural attaché of the French Embassy in London, for further assistance in securing exposures; and to Prof. Ismael Escobar, director of the Bolivian Meteorological Service, for assistance during the course of the expedition to the Andes.

[6] Gardner, E., *Phys. Rev.* (in the press).

[7] Bethe and Marshak, *Phys. Rev.* (in the press). We are indebted to these authors for a typescript copy of an article entitled "The Two-Meson Hypothesis". Møller and Pais have also considered the possibility of genetical relationships between different types of particles of intermediate mass.

PROGRESS IN NUCLEAR PHYSICS

CONFERENCE AT HARWELL

A CONFERENCE on nuclear physics was held at the Atomic Energy Research Establishment (Ministry of Supply) during September 18–19. Opening the first session, the director of the Establishment, Dr. J. D. Cockcroft said that he was glad to welcome so many distinguished guests to the first nuclear physics conference at Harwell. The conference celebrated both the start up of the first British pile, and the restarting of nuclear physics research in Britain after the War. One object of the Establishment is to provide research facilities which cannot be provided in the universities, and it is hoped that it will soon be possible for visitors from universities to work at Harwell. Another object is the provision of radioactive and enriched stable isotopes for use in research. Harwell has also taken the initiative in the development of electronic equipment for nuclear physics, and hopes to be able increasingly to make this available to universities.

Prof. H. R. Staub gave an account of the comparison of the magnetic moments of the proton and neutron carried out at Stanford University. The determinations were made by the methods of Rabi for the proton, and Alvarez and Bloch for the neutron, and were conducted close together in a large uniform magnetic field, thus avoiding the absolute measurement of the field, which gave the large experimental error in the original neutron determination. Two blocks of strongly magnetized iron were used as polarizer and analyser for the neutron beam. At normal saturation values of field strength, the degree of polarization produced was far less than that predicted by theory; but at much higher field strengths (order of 5,000 oersted) theoretical predictions of intensity were fulfilled, the effect varying as the square of the thickness of material used. In conditions of high permeability and low field strength, the ferromagnetic domains are discrete and not quite parallel, so that in travelling between them a neutron may suffer appreciable rotation in the transverse field: only for high field strengths were the transverse fields between domains sufficiently small. The grain size of the domains can be deduced from the polarization effect.

The original value for the magnetic moment of the neutron (1·93) was equal to the difference between those of the proton and deuteron, which conflicts with Rabi's discovery of the quadripole moment of the deuteron and Schwinger's estimate that 3·9 per cent of deuterons are normally in the $^3D_1$ state. The new result by Staub and Bloch is (−) $1{\cdot}9112 \pm 0{\cdot}0017$, which agrees excellently with that obtained by Wayne and Roberts ($1{\cdot}9107 \pm 0{\cdot}0003$) and with Schwinger's predictions on the deuteron.

Dr. K. T. Bainbridge opened his lecture on current physics work at Harvard by referring to a reversible transition which has been observed by Bridgman in cæsium at a pressure of 50,000 atmospheres. Street at Harvard and Shockley at the Bell Telephone Laboratories are working on the characteristics of silver chloride crystals for use as β- or γ-counters. Crystals 2 cm. thick have been used for measuring the migration velocities of electrons: a value of 40–50 cm./sec./volt/cm. is obtained, and trapping is avoided above 600 volts/cm. The energy loss of radium gamma rays is 11 eV. per ion pair. With a crystal 3 mm. thick, a response time of 1/30 µsec. has been obtained. The life-time of the crystal counters is estimated as 100,000 counts.

Attempts have been made to detect the positive particles which have been observed by cloud chamber techniques to occur to the extent of one particle in 5,000 from uranium X_2 and radium E. With a fringing field semi-circular magnetic spectrometer and 20-millicurie P32 source, it was established that if such positive particles exist they are less frequent than 1 in 75,000 disintegrations.

Dr. Bainbridge also described the new Harvard frequency-modulated cyclotron now under construction, which is to have a 95-in. magnet giving a field of 18,000 gauss.

Evidence for the existence of new unstable elementary particles

By Dr. G. D. ROCHESTER

AND

Dr. C. C. BUTLER

Physical Laboratories, University, Manchester

Among some fifty counter-controlled cloud-chamber photographs of penetrating showers which we have obtained during the past year as part of an investigation of the nature of penetrating particles occurring in cosmic ray showers under lead, there are two photographs containing forked tracks of a very striking character. These photographs have been selected from five thousand photographs taken in an effective time of operation of 1,500 hours. On the basis of the analysis given below we believe that one of the forked tracks, shown in Fig. 1 (tracks *a* and *b*), represents the spontaneous transformation in the gas of the chamber of a new type of uncharged elementary particle into lighter charged particles, and that the other, shown in Fig. 2 (tracks *a* and *b*), represents similarly the transformation of a new type of charged particle into two light particles, one of which is charged and the other uncharged.

The experimental data for the two forks are given in Table 1 ; H is the value of the magnetic field, α the angle between the tracks, p and Δp the measured momentum and the estimated error. The signs of the particles are given in the last column of the table, a plus sign indicating that the particle is positive if moving down in the chamber. Careful re-projection of the stereoscopic photographs has shown that each pair of tracks is copunctal. Moreover, both tracks occur in the middle of the chamber in a region of uniform illumination, the presence of background fog surrounding the tracks indicating good condensation conditions.

Though the two forks differ in many important respects, they have at least two essential features in common : first, each consists of a two-pronged fork with the apex in the gas ; and secondly, in neither case is there any sign of a track due to a third ionizing particle. Further, very few events at all similar to these forks have been observed in the 3-cm. lead plate, whereas if the forks were due to any type of collision process one would have expected several hundred times as many as in the gas. This argument indicates, therefore, that the tracks cannot be due to a collision process but must be due to some type of spontaneous process for which the probability depends on the distance travelled and not on the amount of matter traversed.

Table 1. Experimental data

Photo-graph	H (gauss)	α (deg.)	Track	p (eV./c.)	Δp (eV./c.)	Sign
1	3500	66·6	*a*	$3{\cdot}4 \times 10^8$	$1{\cdot}0 \times 10^8$	+
			b	$3{\cdot}5 \times 10^8$	$1{\cdot}5 \times 10^8$	−
2	7200	161·1	*a*	$6{\cdot}0 \times 10^8$	$3{\cdot}0 \times 10^8$	+
			b	$7{\cdot}7 \times 10^8$	$1{\cdot}0 \times 10^8$	+

This conclusion can be supported by detailed arguments. For example, if either forked track were due to the deflexion of a charged particle by collision with a nucleus, the transfer of momentum would be so large as to produce an easily visible recoil track. Then, again, the attempt to account for Fig. 2 by a collision process meets with the difficulty that the incident particle is deflected through 19° in a single collision in the gas and only 2·4° in traversing 3 cm. of lead—a most unlikely event. One specific collision process, that of electron pair production by a high-energy photon in the field of the nucleus, can be excluded on two grounds : the observed angle between the tracks would only be a fraction of a degree, for example, 0·1° for Fig. 1, and a large amount of electronic component should have accompanied the photon, as in each case a lead plate is close above the fork.

We conclude, therefore, that the two forked tracks do not represent collision processes, but do represent spontaneous transformations. They represent a type of process with which we are already familiar in the decay of the meson into an electron and an assumed neutrino, and the presumed decay of the heavy meson recently discovered by Lattes, Occhialini and Powell[1].

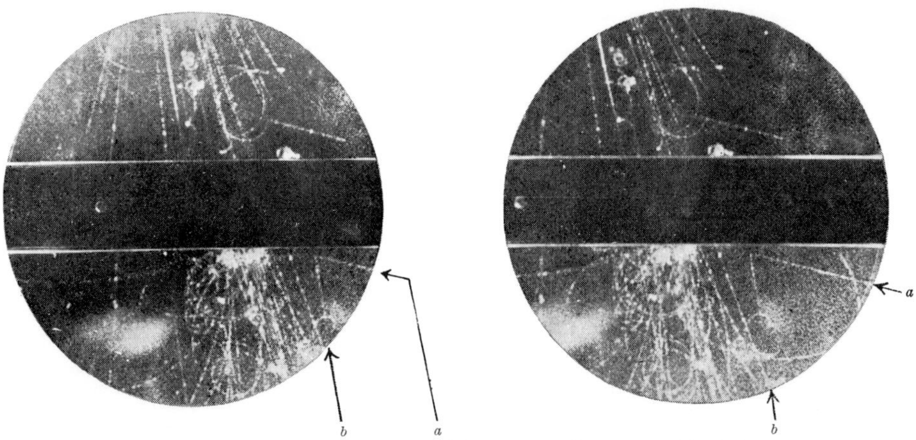

Fig. 1. Stereoscopic photographs showing an unusual fork (*a b*) in the gas. The direction of the magnetic field is such that a positive particle coming downwards is deviated in an anticlockwise direction

Reprinted with permission from *Nature* **160** 855–7 (1947)
© 1983 Macmillan Journals Limited

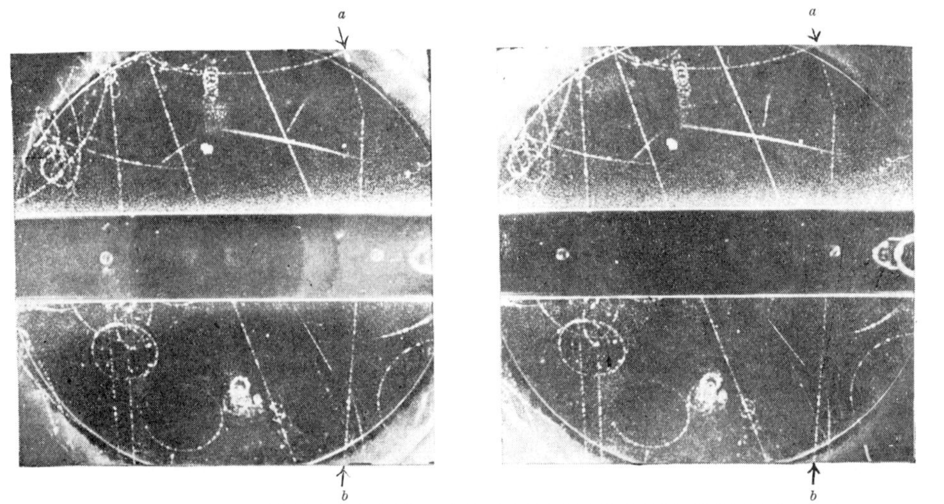

Fig. 2. STEREOSCOPIC PHOTOGRAPHS SHOWING AN UNUSUAL FORK (a b). THE DIRECTION OF THE MAGNETIC FIELD IS SUCH THAT A POSITIVE PARTICLE COMING DOWNWARDS IS DEVIATED IN A CLOCKWISE DIRECTION

The Masses of the Incident Particles

Let us assume that a particle of mass M and initial momentum P is transformed spontaneously into two particles of masses m_1 and m_2, momenta p_1 and p_2 at angles of θ and φ with the direction of the incident particle. Then the following relations must hold :

$$\sqrt{M^2c^4+P^2c^2} = \sqrt{m_1^2c^4+p_1^2c^2} + \sqrt{m_2^2c^4+p_2^2c^2} \quad (1)$$

$$P = p_1 \cos\theta + p_2 \cos\varphi \quad (2)$$

$$p_1 \sin\theta = p_2 \sin\varphi. \quad (3)$$

These general relations may be used to obtain the mass of the incident particle as a function of the assumed masses of the secondary particles.

The value of M must be greater than that obtained by taking the rest masses of the secondary particles as small compared with their momenta ; thus the minimum value $M_{\min}$ is given by the following equation :

$$M_{\min}c^2 = c\sqrt{(p_1 + p_2)^2 - P^2}. \quad (4)$$

Applying this equation to the forked track of Fig. 1, after calculating P from the observed values of p_1 and p_2, it is found that $M_{\min}$ is $(770 \pm 200)m$, where m is the mass of the electron. The application of equation (4) to the forked track of Fig. 2, however, after calculating p_2 from the observed values of P and p_1, shows that $M_{\min} = (1{,}700 \pm 150)m$. This value of the mass would require an ionization for the incident particle of twice minimum, which is inconsistent with the observed ionization. We are therefore justified in assuming that the real value of P is greater than the observed value which, as indicated in Table 1, has a large error. If larger values of P are assumed, then $M_{\min}$ is reduced in value. The lowest value of $M_{\min}$ is $(980 \pm 150)m$ if P is $14{\cdot}5 \times 10^8$ eV./c. Beyond this value of P the mass increases slowly with increasing momentum. No choice of incident momentum will bring the mass of the incident particle below 980 m.

In the special case where the incident particle disintegrates transversely into two particles of equal mass m_0, giving a symmetrical fork, equation (1) reduces to the following expression,

$$\frac{M}{m} = \frac{2m_0}{m}\left(1 + \frac{p^2c^2}{m_0^2c^4} \cdot \sin^2\theta\right)^{1/2}, \quad (5)$$

where p is the momentum of each of the secondary particles. Some typical results for different assumed secondary particles, calculated from equation (5), are given in Table 2. On the reasonable assumption that the secondary particles are light or heavy mesons, that is, with masses of $200m$ or $400m$, we find that the incident particle in each photograph has a mass of the order of $1{,}000m$.

TABLE 2. MASS OF INCIDENT PARTICLE AS A FUNCTION OF MASS OF SECONDARY PARTICLE

Photograph	Assumed secondary particle m_0/m	Momentum of observed secondary particle (eV./c.)	Incident particle M/m
1	0	$3{\cdot}5 \times 10^8 \pm 1{\cdot}0 \times 10^8$	770 ± 200
	200	,,	870 ± 200
	400	,,	1110 ± 150
	1837	,,	3750 ± 50
2	0	$7{\cdot}7 \times 10^8 \pm 1{\cdot}0 \times 10^8$	980 ± 150
	200	,,	1080 ± 100
	400	,,	1280 ± 100
	1837	,,	3820 ± 50

Upper values of the masses of the incident particles may also be obtained from the values of the ionization and the momenta. Thus for each of the observed particles in Fig. 1, the ionization is indistinguishable from that of a very fast particle. We conclude, therefore, that $\beta = v/c \gg 0{\cdot}7$. Since the momentum of the incident particle may be found from the observed momenta of the secondary particles, we can apply equation (1) to calculate M. In this way we find $M/m \ll 1{,}600$. Again, since the ionization of the incident particle in Fig. 2 is light, $\beta \gg 0{\cdot}7$, from which it can be shown that $M/m \ll 1200$. This last result, however, must be taken with caution because of the uncertainty in the measured value of the momentum of the incident particle.

One further general comment may be made. This is that the observation of two spontaneous disintegrations in such a small number of penetrating showers suggests that the life-time of the unstable particles is much less than the life-time of the ordinary meson. An approximate value of this life-time may be derived as follows. The probability of an unstable particle of life-time τ_0 decaying in a short distance D is given by

$$p = \frac{D(1 - \beta^2)^{1/2}}{\tau_0 \, c \, \beta}. \tag{6}$$

Since the total number of penetrating particles in the penetrating showers so far observed is certainly less than 50, we must assume that the number of our new unstable particles is unlikely to have been greater than 50. Since one particle of each type has been observed to decay, we can therefore put $p \approx 0{\cdot}02$. Setting $D \approx 30$ cm., and $\beta = 0{\cdot}7$, we find from equation (6) that $\tau_0 = 5{\cdot}0 \times 10^{-8}$ sec.

We shall now discuss possible alternative explanations of the two forks.

Photograph 1. We must examine the alternative possibility of Photograph 1 representing the spontaneous disintegration of a charged particle, coming up from below the chamber, into a charged and an uncharged particle. If we apply the argument which led to equation (4) to this process, it is readily seen that the incident particle would have a minimum mass of 1,280m. Thus the photograph cannot be explained by the decay of a back-scattered ordinary meson. Bearing in mind the general direction of the other particles in the shower, it is thought that assumption of the disintegration of a neutral particle moving downwards into a pair of particles of about equal mass is more probable. Further, it can be stated with some confidence that the observed ionizing particles are unlikely to be protons because the ionization of a proton of momentum $3{\cdot}5 \times 10^8$ eV./c. would be more than four times the observed ionization.

Photograph 2. In this case we must examine the possibility of the photograph representing the spontaneous decay of a neutral particle coming from the right-hand side of the chamber into two charged particles. The result of applying equation (4) to this process is to show that the minimum mass of the neutral particle would be about 3,000m. In view of the fact that the direction of the neutral particle would have to be very different from the direction of the main part of the shower, it is thought that the original assumption of the decay of a charged particle into a charged penetrating particle and an assumed neutral particle is the more probable.

We conclude from all the evidence that Photograph 1 represents the decay of a neutral particle, the mass of which is unlikely to be less than 770m or greater than 1,600m, into the two observed charged particles. Similarly, Photograph 2 represents the disintegration of a charged particle of mass greater than 980m and less than that of a proton into an observed penetrating particle and a neutral particle. It may be noted that no neutral particle of mass 1,000m has yet been observed; a charged particle of mass 990m ± 12 per cent has, however, been observed by Leprince-Ringuet and L'héritier[2].

Peculiar cloud-chamber photographs taken by Jánossy, Rochester and Broadbent[3] and by Daudin[4] may be other examples of Photograph 2.

It is a pleasure to record our thanks to Prof. P. M. S. Blackett for the keen interest he has taken in this investigation and for the benefit of numerous stimulating discussions. We also wish to acknowledge the help given us by Prof. L. Rosenfeld, Mr. J. Hamilton and Mr. H. Y. Tzu of the Department of Theoretical Physics, University of Manchester. We are indebted to Mr. S. K. Runcorn for his assistance in running the cloud chamber in the early stages of the work.

[1] Lattes, C. M. G., Occhialini, G. P. S., and Powell, C. F., *Nature*, **160**, 453, 486 (1947).

[2] Leprince-Ringuet, L., and L'héritier, M., *J. Phys. Radium.* (Sér. 8), **7**, 66, 69 (1946). Bethe, H. A., *Phys. Rev.*, **70**, 821 (1946).

[3] Jánossy, L., Rochester, G. D., and Broadbent, D., *Nature*, **155**, 142 (1945). (Fig. 2. Track at lower left-hand side of the photograph.)

[4] Daudin, J., *Annales de Physique*, 11e Série, **19** (Avril–Juin), 1944 (Planche IV, Cliché 16).

THE GRASSHOPPER PROBLEM IN NORTH AMERICA

By Dr. B. P. UVAROV, C.M.G.

Anti-Locust Research Centre, British Museum (Natural History)

OFFICIAL figures compiled by the Division of Grasshopper Control of the United States Bureau of Entomology show that in the ten-year period 1936–46, crops to the value of 424,563,614 dollars were destroyed by grasshoppers, and 26,088,211 dollars was spent on control measures. No similar statistics are available for Canada, where the general situation with regard to grasshoppers is similar. The annual figures for the same period show considerable fluctuation, apparently dependent on natural variations in grasshopper populations; but they do not suggest any general downward trend which might be interpreted as a cumulative result of persistent artificial control measures. It must be stressed that these measures are unquestionably very successful so far as the protection of standing crops of the year is concerned, and the organisation of control, based on close co-operation of Federal and State agencies and the farmers, leaves little to be desired. Both the organisation and the technique of artificial control are being continually perfected, and it is estimated that, on the average, each dollar spent on grasshopper control serves to save crops to the value of 35 dollars. This is a remarkable achievement, and a large share of credit for it belongs to the Division of Grasshopper Control, admirably organised and run by Dr. Claude Wakeland and his staff. Nevertheless, it is clear that the protection of crops by direct control methods cannot be relaxed, and efforts on an ever-increasing scale will have to continue indefinitely. The policy of direct control is a palliative which cannot lead towards a lasting solution of the grasshopper problem in the North American continent.

This state of affairs has been realized by entomologists of the United States and Canada for some time, and their views were forcibly expressed in the following resolution adopted at a session of the Committee on Grasshopper Research at Lincoln, Nebraska, in 1946: "Millions of dollars have been appropriated and spent during the past twenty years by state, provincial, federal and dominion agencies for materials and labour for the applied control of grasshoppers. During the same period, the amount of money spent on research for a solution of the grasshopper problem has been woefully inadequate.

Observations with electron-sensitive plates exposed to cosmic radiation: Part 1

By Miss R. BROWN, U. CAMERINI, P. H. FOWLER, H. MUIRHEAD and Prof. C. F. POWELL
H. H. Wills Physical Laboratory, University of Bristol

and D. M. RITSON
Clarendon Laboratory, Oxford

WE have recently made observations with the new Kodak nuclear research emulsions, described by Dr. R. W. Berriman[1], which record the tracks of particles of charge *e*, even at minimum ionization. The new plates represent a technical advance of great importance; they allow us to obtain a much deeper insight into nuclear processes than was possible with the older emulsions, and they greatly extend the field of application of the photographic method to physical and biological problems.

In Part 1 of the present paper, some of the more important processes involving charged mesons, which can be observed when the new plates are exposed to the cosmic radiation, are described. They include examples of disintegrations produced by π^--particles; and the decay of μ-particles. In about 10 per cent of the cases, the track of the particle produced by the decay of a μ-meson has a length in the emulsion greater than 1,000 μ. By studying the Coulomb scattering of these particles in their passage through the emulsion, estimates of their energy can be obtained.

A striking feature of the exposed plates is the large number of tracks with a grain-density corresponding to that of particles near minimum ionization. These tracks provide a very simple test for 'fading', for the variation of the degree of development with depth, and for the sensitivity and other characteristics of the emulsions.

In Part 2, an event is described which provides further evidence for the existence of mesons of greater mass than the π-particles[2]. A particle, of which the mass, as determined by grain-counts, is about 1,000 m_e, comes to rest and leads to the emission of three charged particles, one of which reaches the end of its range in the emulsion and produces a disintegration. Evidence is given which suggests that the event corresponds to the spontaneous disintegration of the heavy particle. Assuming that the three charged particles observed are the only products of decay, the event can be analysed by applying the principles of the conservation of mass-energy and of momentum. It is then found that the observations are consistent, to within the errors of the measurements, with the assumption that the three secondary particles are π- or μ-particles.

Before the correctness of this interpretation can be taken as established, it will be necessary to observe further examples of the same process. We give this preliminary account of the observations, however, because the number of particles of this type, compared with that of π- and μ-particles, appears to be small at moderate altitudes (11,000 ft.). Further, if our assumptions are correct, only a small proportion of such events will, for purely geometrical reasons, permit a detailed analysis of the type we have been able to make. Taken in conjunction with the present observations, however, other events of a similar character, although less suitable for analysis, would provide important information supporting or disproving the correctness of our suggestions.

PART I. DECAY OF μ-MESONS

Experimental

The plates employed in the present experiments were experimental Kodak emulsions, type *N.T.*4, kindly supplied by Dr. R. W. Berriman. They were transported to the Jungfraujoch immediately after manufacture and some were placed beneath a block of lead 10 cm. thick for periods of from eight to sixteen days, others being left uncovered. The lead block was employed in order to reduce the intensity of the soft component; but it appears to have been an unnecessary precaution, for the number of tracks in the unshielded plates is not inconveniently large. The plates were developed at the Jungfraujoch immediately after exposure. Different plates were processed in different baths in an attempt to determine the most favourable conditions of development, and there are corresponding differences in the grain-density of the tracks of similar particles in different plates.

Because of the short exposure, the number of events per unit area is relatively small, but the greatly increased grain-density in the tracks, due to the higher sensitivity of the emulsion, makes it possible to search plates under the microscope with a much lower magnification than that previously employed. Thus, in searching for mesons, we have employed a $\times$ 10 objective and $\times$ 6 eyepieces. With a binocular microscope, this equipment gives an overall magnification of only $\times$ 90. It is thus possible to increase the speed of search of the plates, and in an examination of 300 cm.^{2} of the new plates, many examples of familiar processes involving charged mesons have been found.

Disintegrations Produced by π^--Particles

The observations include thirty examples of disintegrations produced by π^--particles (see Fig. 1), and thirty examples of π^+-particles which suffer μ-decay. A notable feature of the new plates is the large number of tracks of slow electrons which can be clearly seen under the microscope. In Fig. 1, for example, two electrons appear to be associated with the star produced by the nuclear capture of a π^--particle.

The tracks of electrons originate only rarely from the point at the end of the range of π^--particles where the disintegration occurs (see Fig. 2). Thus of thirty disintegrations produced by π^--particles, only three show the tracks of associated electrons. This result appears to have a bearing on the processes by which a π^--particle loses energy in falling to a *k*-orbit around a nucleus. In those cases in which electrons are observed, they may be attributed to the β-decay of the residual nucleus formed by the disintegration;

Reprinted with permission from *Nature* **163** 47–51 (1949)
© 1983 Macmillan Journals Limited

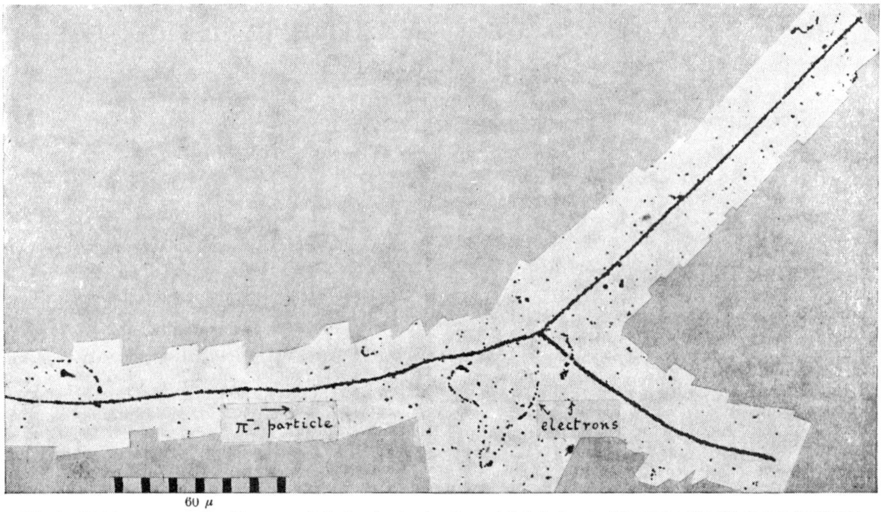

Fig. 1. Disintegration produced by π⁻-particle showing tracks of associated electrons. Observer: Mrs. W. J. van der Merwe

or possibly to the internal conversion of γ-rays. If the meson sometimes loses energy by the ejection of Auger electrons, the number of such electrons, of energy greater than 15 keV., must be small.

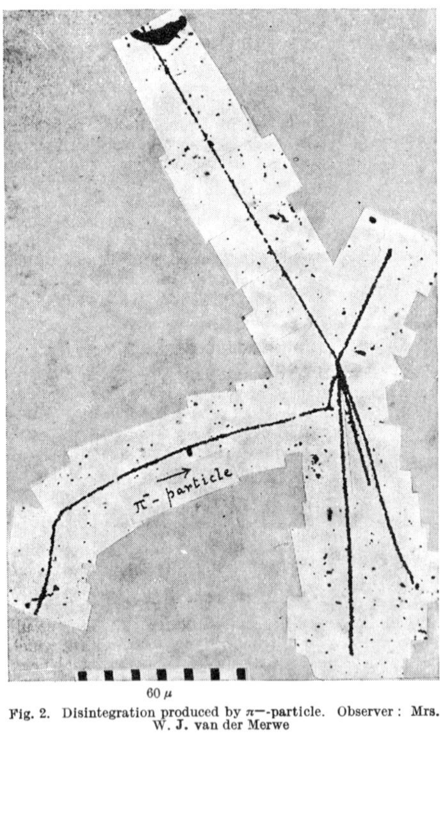

Fig. 2. Disintegration produced by π⁻-particle. Observer: Mrs. W. J. van der Merwe

Particles of Low Specific Ionization

In addition to the phenomena with which we are familiar, we can distinguish in the new plates the tracks of many particles of low specific ionization which pass across the emulsion, from the surface to the glass, without suffering a deviation, through scattering, of more than a few degrees. These may be attributed to fast particles, mostly electrons and mesons, with a specific ionization near the minimum value for a particle of charge *e*. In order to confirm this view, grain-counts were made on twenty-five tracks of this type, of length greater than 1,000 μ, chosen at random from a particular plate, No. ***K*** **47.** The results of this study are represented in Fig. 3, where the relative frequency of occurrence of tracks with different values of the average grain-density is shown. The plate was one of those exposed under the lead block.

It will be seen from Fig. 3 that there is a sharply defined group, with a mean grain-density of 0·22 grains per micron. We regard this group as due to particles with a specific ionization near the minimum value. The observations can be explained in terms of the slow variation with energy of the ionization of particles of momentum greater than 2 *mc*, where *m* is the rest-mass.

The results represented in Fig. 3 prove that effects due to fading of the latent image are small, for fading will tend to reduce the grain-density. Had such effects been appreciable, a dispersed set of values for the grain-density of particles with minimum ionization would have been expected—corresponding to the

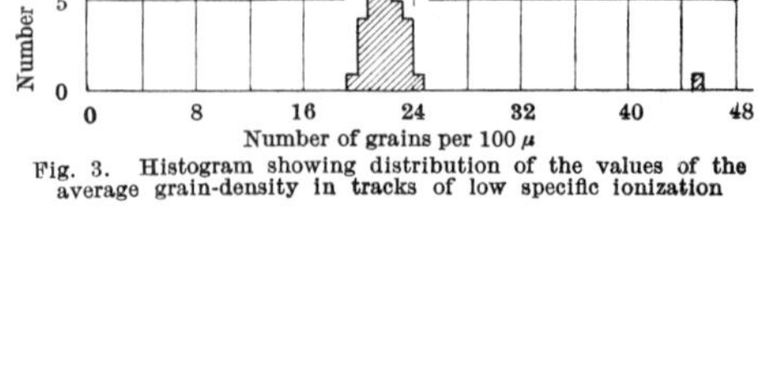

Fig. 3. Histogram showing distribution of the values of the average grain-density in tracks of low specific ionization

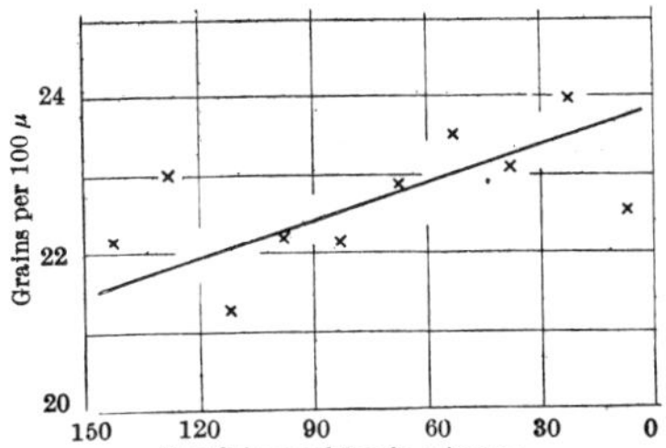

Fig. 4. Average values of the grain-density, in elements of the tracks of 25 particles of low specific ionization, at different depths in the emulsion

presence of tracks of different 'ages'—instead of the sharply defined group actually found.

The tracks under consideration pass right through the emulsion. They can therefore be studied in order to determine the variation of the degree of development with depth. The grain-density along the tracks has been measured as a function of the depth in the emulsion, and the mean results for twenty-five tracks are represented in Fig. 4. It will be seen that the grain-density corresponding to minimum ionization varies with depth, in the particular plate under examination, from 23·5 per 100 μ at the surface, to 21·5 at the glass.

It follows from this result that the changes in the grain-density in a track, due to a variation of the degree of development with depth, are less than 8 per cent. Further, a correction can be made, to allow for the small variations observed, in the case of tracks of low specific ionization which are of particular interest. Long tracks of small grain-density are a common feature in electron-sensitive plates exposed to the cosmic radiation, and the above methods may therefore be useful as a standard procedure for determining, (*a*), the degree of fading in a given exposure; (*b*), the degree of development with depth; and (*c*), the grain-density corresponding to the minimum value of the specific ionization of a particle of charge *e*.

A second conclusion follows from the results represented in Fig. 3. The magnitude of the minimum value of the specific ionization of a particle is proportional to the square of its charge. The sharply defined peak, shown in Fig. 3, proves that the great majority of the fast particles recorded by the emulsion have charges which are equal, to within about 5 per cent. This makes it almost certain that the charge in question is of magnitude *e*.

Decay of μ-Mesons

The tracks of 100 μ^+- or μ^--particles, which reach the end of their range in the emulsion and emit a single charged particle of low specific ionization, have now been observed. A mosaic of photomicrographs of such a disintegration is reproduced in Fig. 5. Events of this type can be easily distinguished from those in which a π-meson decays to produce a μ-meson, because of the much greater grain-density in the track of the latter particle, when it is formed by the decay of a π-meson at rest. Further, the particles appearing at the end of the range of the μ-mesons, if ejected in a direction nearly parallel to the surface, commonly have a range of several thousand microns before leaving the emulsion. Such particles are very little scattered, and they have a grain-density nearly equal to that corresponding to minimum ionization. This observation shows the particles to have the electronic charge.

Fig. 6 shows a mosaic of photographs of an event in which a π-particle stops in the emulsion and emits a μ-particle of range approximately 600 μ. The particle of low specific ionization which originates from the end of the track of the μ-particle can be clearly seen. In eight other events of this type which we have observed, the secondary particles all have characteristics indistinguishable from those of the particles arising at the end of the range of the μ-mesons which enter the emulsion after originating at points outside it. Taken in conjunction with evidence obtained in previous experiments, the new observations provide additional weight for the assumption that the μ-mesons are identical with some at least of the ordinary mesons of the penetrating component of the cosmic radiation; and that the π- and μ-mesons carry electric charges of magnitude *e*.

About 10 per cent of the μ-mesons, which stop in the emulsion after being created at points outside it, give rise to a particle of which the track in the emulsion has a length greater than 1,000 μ. We have considered the possibility of determining the values of the momentum of these particles by making observations on the change in their directions of motion in the emulsion, produced by a magnetic field of 50,000 gauss. A simple calculation shows, however, that the deflexions would be of the same order of magnitude as those to be expected from scattering. Observations were therefore made on the scattering of the particles in the absence of a magnetic field, the method employed being similar to that described previously. The following values are typical of those obtained by this method, and show the order of accuracy which can be obtained in favourable cases: 15 ± 3, 25 ± 5, 38 ± 6, 42 ± 2, 44 ± 4, 48 ± 6 MeV./c. The limits correspond to the probable errors associated with the statistical fluctuations.

It is reasonable to suppose that, in the course of a few months, we shall find some hundreds of tracks of

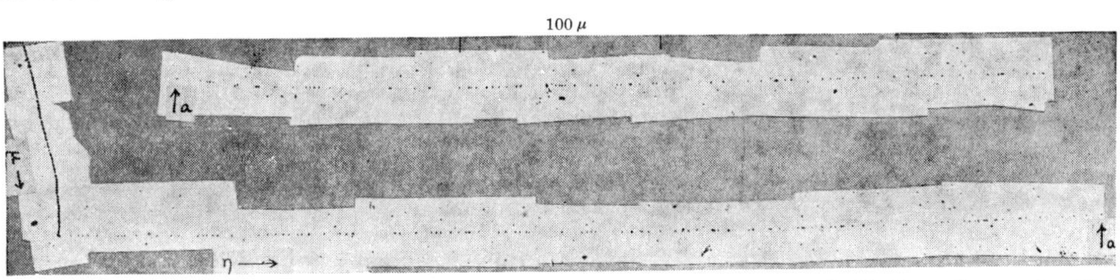

Fig. 5. Characteristic example of the track of a particle of low specific ionization, produced by the decay of a μ-meson. The track of the secondary particle is given in two parts which should join at the points marked *a*. Observer: Mrs. B. A. Moore

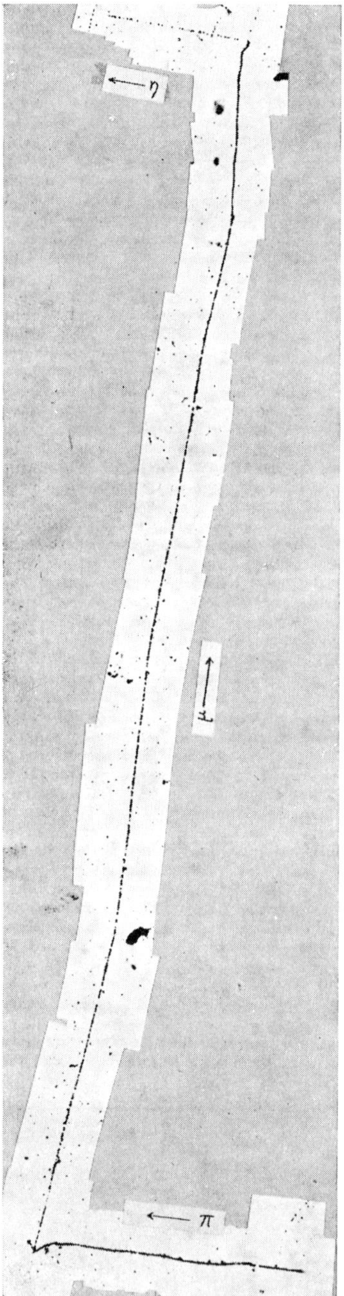

Fig. 6. Successive decay $\pi \rightarrow \mu \rightarrow \eta$. The particle, η, arising from the decay of the μ-meson, is commonly assumed to be an electron, but this assumption remains to be finally established. A few grains in the track of a second particle of low specific ionization can be seen near the end of the range of the μ-meson, but this particle is not associated with the main event. The tracks of several δ-rays produced by the μ-meson, one of which occurred near the point at which the π^+-particle decayed, can be distinguished. In every example observed hitherto, the particle η leaves the emulsion.
Observer: Miss R. Brown

such particles which are suitable for measurement, and that it will be possible to make a detailed study of their distribution in momentum; the values obtained on the first twenty tracks with a range greater than 1,000 microns suggest that the momentum of the decay particle is not constant.

Of the μ-particles which are observed to stop in the emulsion, about 40 per cent give rise to no secondary particles of which the tracks can be distinguished. These may be assumed to be either μ^--particles which have been captured by silver or bromine atoms, and which therefore interact with nuclei before they have had time to decay; or, to be events in which the decay particle escapes observation because of its direction of ejection. In order to estimate the number of events which can be attributed to the latter process, the lengths of the projections of the tracks of the decay particles, on the surface of the emulsion, have been measured. From the distribution of the values thus obtained, the total number of μ-mesons producing secondaries can be estimated. The problem is purely geometrical, and the approach is identical with that employed previously in the case of the μ-decay of π-mesons[5]. We have thus found that of the μ-mesons observed, 65 ± 10 per cent emitted secondaries, and 35 ± 10 per cent gave none. This result appears to be consistent with previous estimates of the proportions of the μ^--particles which stop in gelatine, and in silver bromide.

We have suggested that the observations on the disintegrations produced by π^--particles indicate that Auger electrons of energy greater than 15 keV. are only rarely produced during the 'atomic capture' of a negative meson. A similar conclusion follows from a study of the tracks of the μ-particles. Of the hundred events in which we can distinguish the track of the fast particle arising from the decay of a μ-meson, there is no case in which the track of an electron of short range, arising from the point of disintegration, can be distinguished. It is reasonable to suppose that in about a third of these events we are observing the spontaneous decay of a negative meson. It follows that for the μ^--particles also, the ejection of Auger electrons, of energy greater than 15 keV., during the process of atomic capture, is a rare phenomenon. A closer study of this problem would appear to be desirable; for in the case of μ^--particles which decay spontaneously, no confusion due to the β-decay of a residual nucleus, following meson capture, can occur.

Showers of Penetrating Particles

In examining the 'stars' produced in the new emulsions, we have observed eight cases in which five or more charged particles emerge which produce tracks having a grain-density near the value corresponding to minimum ionization. In each case, these particles moved downwards in directions making only small angles with the vertical, so that they are confined to a narrow cone. They are accompanied by protons and α-particles of low energy which were ejected in directions orientated at random.

We suggest that each of these events corresponds to the interaction of a fast particle with a nucleus of silver or bromine, which led to the production of several penetrating particles, the heavy particles being afterwards 'evaporated' from the nucleus as a result of its increased 'temperature'; that the events represent the same type of process for which evidence has been given in recently published photographs showing the production of showers of penetrating particles obtained in experiments with the expansion chamber (Hazen, Jánossy, Rochester and other authors). The processes observed are reminiscent of those visualized by Hamilton, Heitler and Peng, in their theory of the production of mesons by the interaction of fast nucleons with nuclei.

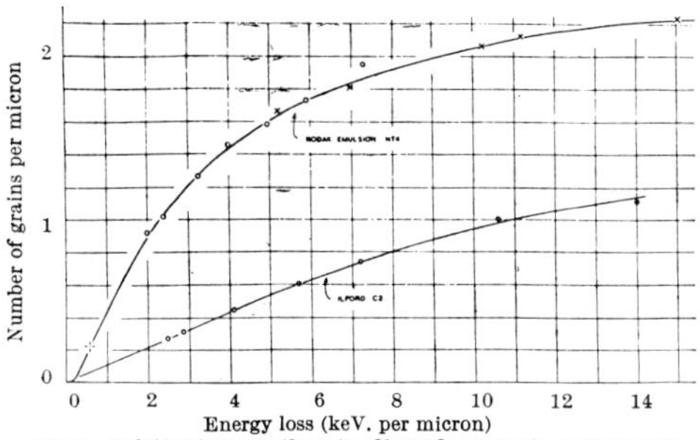

Fig. 7. Relation between the rate of loss of energy of a particle and the grain-density in the track. Points marked —o—o have been deduced from observations on the tracks of μ-particles of range greater than 2,000 μ; and those marked —×—×, from similar measurements with protons

Relation Between Grain-Density and Specific Ionization

It is a matter of great technical importance to be able to infer the specific ionization of a charged particle from the observed grain-density in its track in an emulsion. We have seen that curves of the type represented in Fig. 3 allow us to determine the grain-density in the tracks of particles of charge *e*, corresponding to minimum ionization. Further, from observations on the tracks of μ-mesons and protons of great range, observed in the same plate, we can determine the grain-density corresponding to a wide range of values of the specific ionization. In making such determinations, the rate of loss of energy corresponding to a given value of the residual range of a particle of known mass can be determined from the range-energy curve for protons. In this way, we have obtained the results represented in Fig. 7. The point corresponding to minimum ionization in this figure is shown thus +, the ionization having been determined in this case from the formulæ developed by Bloch, assuming the atomic composition of the emulsion to be identical with that of the Ilford plates.

Included in Fig. 7 is a similar curve for an Ilford *C* 2 emulsion. It will be seen that, as a result of the increased sensitivity of the Kodak *N.T.*4 emulsion, there is a reduction in discriminating power for tracks with a specific ionization of the order of 6 keV. per micron. In the Kodak plates, a track, in regions of high ionization, appears as an almost continuous succession of developed grains which cannot be individually resolved. This has the result that for determining the mass of particles of mass approximately 1,000 m_e, tracks of much greater length must be available for measurement than is the case with less sensitive emulsions.

In contrast with the difficulties met in the case of tracks with a high specific ionization, the calibration curve shown in Fig. 7 allows us to make estimates of the rate of loss of energy for weakly ionizing particles, a technical advance of great importance in relation to the subjects discussed in Part 2 of the present paper.

(To be continued)

[1] Berriman, *Nature*, **162**, 992 (1948).
[2] Leprince-Ringuet, *C.R. Acad. Sci., Paris*, **226**, 1897 (1948). Rochester and Butler, *Nature*, **160**, 855 (1947). Bradt and Peters, Report to the Bristol Symposium, 1948 (in the press). Alichanian, Alichanov and Weissenberg, *J. of Exp. and Theoret. Phys., U.S.S.R.*, **18**, 301 (1948); and other references.
[3] Camerini, Muirhead, Powell and Ritson, *Nature*, **162**, 433 (1948).
[4] Goldschmidt-Clermont, King, Muirhead and Ritson, *Proc. Phys. Soc.*, **61**, 138 (1948).
[5] Lattes, Occhialini and Powell, *Proc. Phys. Soc.*, **61**, 173 (1948).

PROBLEMS AND POLICY IN PLANT BREEDING

THE second annual Conference of Plant Breeders under the auspices of the Agricultural Research Council was held during November 11–12 at the John Innes Horticultural Institution. Sixty-four members attended, including representatives of English, Scottish and Welsh plant-breeding stations, some overseas visitors, and geneticists from the universities. Prof. F. T. Brooks represented the Agricultural Research Council. Dr. C. D. Darlington was in the chair.

Demonstrations

(1) The Genetics Department of the John Innes Horticultural Institution demonstrated genetical problems in seed-growing, rogues in tomatoes, the spontaneous and induced mutation of the incompatibility gene, *Phaseolus* hybrids and the effects of selection in *Drosophila*.

(2) The Pomology Department showed new mildew-resistant tomatoes, incompatibility in pears, the breeding of sweet corn and 'bolters' in potatoes.

(3) The Cytology Department showed iso-chromosomes and heterochromatin in plants. Chromosome numbers were used to reveal the origin and development of the cultivated daffodil since 1890. In earthworms meiosis was shown in a tetraploid species —a novelty in animals—and the relations of chromosome numbers to species were illustrated in this group.

(4) The Garden Department demonstrated soil sterilization and artificial illumination as aids to the breeder.

PROBLEMS

Quantitative Inheritance and Selection

Prof. K. Mather, in the opening talk, gave an outline of the methods available in the study of quantitative inheritance. The Mendelian method of experiment, he pointed out, can be applied only to differences caused by genes of large effect. Biometrical methods can be used with all differences, including quantitative variation which is controlled by polygenes. Mendel's principles apply to nearly all heritable variation in higher organisms, and therefore have been used as the basis for biometrical analysis.

On this basis methods had been developed of measuring polygenic variation; of partitioning such variation into non-heritable, heritable and fixable, and heritable but depending on dominance and therefore unfixable; of testing for residual genic interaction and for linkage; and of finding the number of effective units of segregation. By these means the magnitude and speed of progress under selection could be predicted in some circumstances, and the progress of inbreeding could be measured experimentally.

Most of the variability was normally hidden, and its exposure to the action of selection depended on crossing, segregation and recombination. These processes were controlled in Nature by the natural breeding system and by the frequency of chiasma formation.

Control of the breeding system is a regular tool of plant breeding; and the control of recombination by adjustment of the environment may become another.

Correlated responses in characters other than those under direct selection, Prof. Mather suggested, might also be useful as a means of splitting up stabilized

Observations with electron-sensitive plates exposed to cosmic radiation: Part 2

By Miss R. BROWN, U. CAMERINI, P. H. FOWLER, H. MUIRHEAD
and Prof. C. F. POWELL
H. H. Wills Physical Laboratory, University of Bristol

and D. M. RITSON
Clarendon Laboratory, Oxford

PART 2. FURTHER EVIDENCE FOR THE EXISTENCE OF UNSTABLE CHARGED PARTICLES, OF MASS $\sim$ 1,000 m_e, AND OBSERVATIONS ON THEIR MODE OF DECAY

ONE of the first events found in the examination of electron-sensitive plates exposed at the Jungfraujoch is represented in the mosaic of photomicrographs shown in Fig. 8. There are two centres, A and B, from which the tracks of charged particles diverge, and these are joined by a common track, t. Because of the short duration of the exposure, and the small number of disintegrations occurring in the plate, the chance that the observation corresponds to a fortuitous juxtaposition of the tracks of unrelated events is very small—of the order 1 in 10^7. It is therefore reasonable to exclude it as a serious possibility. Further observations in support of this assumption are presented in a later paragraph.

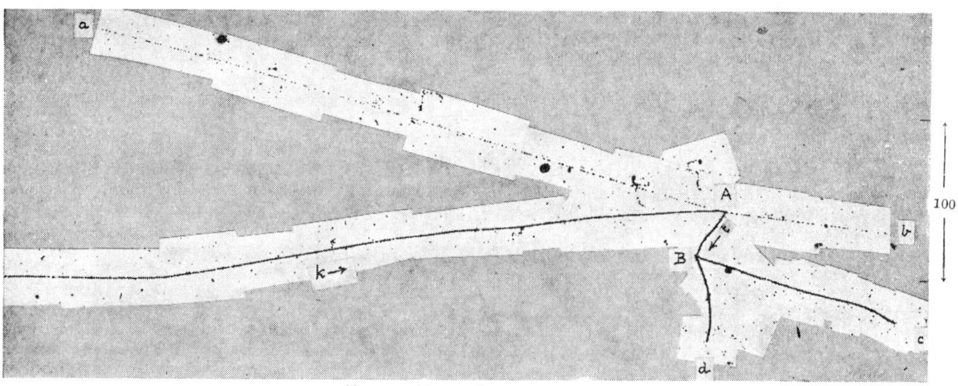

Observer: Mrs. W. J. van der Merwe
Fig. 8

An inspection of the track k shows that the particle producing it approached the centre of disintegration A. The range of the particle in the emulsion exceeds 3,000 μ, and there is continuous increase in the grain-density along the track in approaching A. Near A, the grain-density is indistinguishable from that of particles of charge e, recorded in the same plate, near the end of their range.

The evidence for the direction of motion of the particle based on grain-counts is supported by observations on the small-angle deviations in the track due to Coulomb scattering. These deviations are most frequent near A, and the scattering is less marked at points remote from it.

From these observations, it is reasonable to conclude that the particle k approached the point A; that it carried the elementary electronic charge; and that it had reached, or was near, the end of its range at the point A. We therefore assume that the particle k initiated the train of events represented by the tracks radiating from A and B. It follows that the particle producing track t originated in star A, and produced the disintegration B. In order to analyse the event, we first attempted to determine the mass of the particle k.

Mass Determinations by Grain-Counts

About a year ago, experiments were made in this Laboratory to determine the ratio, m_π/m_μ, of the masses of π- and μ-mesons, by the method of grain-counting[5], and by studying the small-angle scattering of the particles in their passage through the emulsion[4]. The values obtained by the two methods were $m_\pi/m_\mu = 1{\cdot}65 \pm 0{\cdot}11$, and $m_\pi/m_\mu = 1{\cdot}35 \pm 0{\cdot}10$*, respectively. Recent experiments at Berkeley[6] suggest that the true value is $1{\cdot}33 \pm 0{\cdot}02$, a result which throws serious doubt on the reliability of the method based on grain-counts. Because of the advantage of this method, and of the important conclusions which have been based on it, experiments were made to determine the conditions in which reliable results can be obtained.

In the first experiments[5], the two most serious experimental difficulties arose from the fading of the latent image and from the variation of the degree of development with depth. This made it necessary to

* Continued from page 51.

* For the following reasons, the limits of error quoted above, in the determination of m_π/m_μ by observations on scattering, are less than those given in ref. 4. Previously, values for the mass of the different types of mesons, classified phenomenologically, were given separately. It is now known, however, that at least the majority of the σ-mesons are π^--particles; and the ρ-mesons, μ^+- and μ^--particles. The different results can therefore be combined to give a value for m_π/m_μ with a greater statistical weight.

Reprinted with permission from *Nature* **163** 82–7 (1949)
© 1983 Macmillan Journals Limited

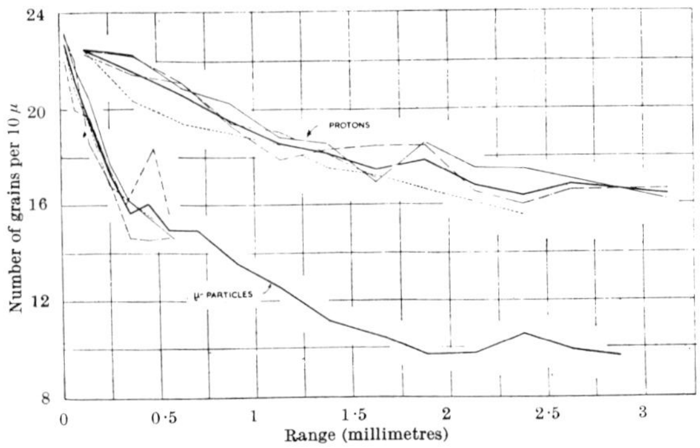

Fig. 9

work only with tracks formed contemporaneously; to compare the grain-density along the tracks of the π- and μ-mesons of the same pair. As a result, the tracks of the π-mesons available for measurement were, in most cases, shorter than 400 μ. In continuing the experiments, much more favourable conditions were obtained by using short exposures, so that the effects of fading were negligible; and by developing the plates by the method employed by Dilworth, Occhialini and Payne[7], which gives a nearly uniform degree of development with depth.

In the plates obtained by these methods, it is legitimate to compare the grain-density in the tracks of unrelated particles. Further, it is now known that at least the majority, and possibly all, the mesons which produce 'stars' are π^--particles[6,8]; and that most of the ρ-mesons are μ^+- and μ^--particles. In determining m_π and m_μ, we have therefore made measurements on the tracks of π^+- and π^--, μ^+- and μ^--particles, of length greater than 1,000 μ, comparing the results with those of similar measurements made on the tracks of protons. In these conditions, we have found $m_\pi/m_\mu = 1{\cdot}33 \pm 0{\cdot}05$. A detailed account of the observations will be published elsewhere; but, for the purpose of the present paper, it is sufficient to note that the results appear to be in good accord with those obtained by other methods. We conclude that, using the Ilford *C*2 emulsion in the new conditions, reliable information can be obtained.

We have seen that the conditions of uniform development and absence of fading have been achieved in the present experiments with the new Kodak emulsions, and we therefore attempted to measure the mass of particles by similar methods to those employed with the Ilford plates. The results obtained in observations on the tracks of four protons and four μ-particles, occurring in the same plate, are represented in Fig. 9. In this figure, the number of grains per unit length in the tracks is plotted for different values of the residual range; and the mean values, for tracks of the same type, are indicated by the full lines. The ratio of the masses of the two types of particles can be deduced by making a comparison of the values of the residual range at which the grain densities have the same value. The result thus obtained is $m_\mu = 220 \pm 20\ m_e$.

Using similar methods, we have made estimates of the mass of the particle, k, and the measurements are represented in Fig. 10. This figure shows the mean values of the grain-density in the tracks of the four μ-mesons and four protons, together with the corresponding results for the particle k. All the tracks under consideration occurred in the same plate.

Table 1 shows the values of the mass of the particle, k, as determined from these results, by making a comparison of the grain-density in the track of the particle with the mean curve for protons. The values thus obtained are all independent and the mean is $m_k = 1{,}080 \pm 160\ m_e$.

TABLE 1. Determination of the ratio, m_P/m_k, of the mass of a proton to that of particle, k, by grain-counting

	Individual independent values						
m_P/m_k	1·77	1·88	1·49	1·64	2·17	1·79	1·32
	1·71	1·66	1·27	1·69	2·13	1·55	

Mean value: 1·70; $m_k = 1{,}080 \pm 160\ m_e$

The limits of error given above have been deduced in the following manner: We have compared the grain-density in the tracks of the four individual protons with the mean curve for the same particles—(see Fig. 9)—and have thus obtained a number of independent values for the apparent mass of each of these particles. The distribution in these values allows us to calculate the 'probable error' associated with the mass as determined from the observations on any one track, expressed as a percentage of the apparent mass of the particle. It is then assumed that the 'probable' percentage error in the calculated mass of the particle k has the same value.

We have also determined the mass m_k by studying the small-angle scattering of the particle, by the methods recently described[4], and the result thus obtained is $m_k = 1{,}800 \pm 400\ m_e$. If the true mass of the particle is 1,080 m_e, the chance that the value obtained by observations on scattering shall be equal to, or greater than, 1,800 m_e is one in four. Because of the large statistical fluctuations associated with

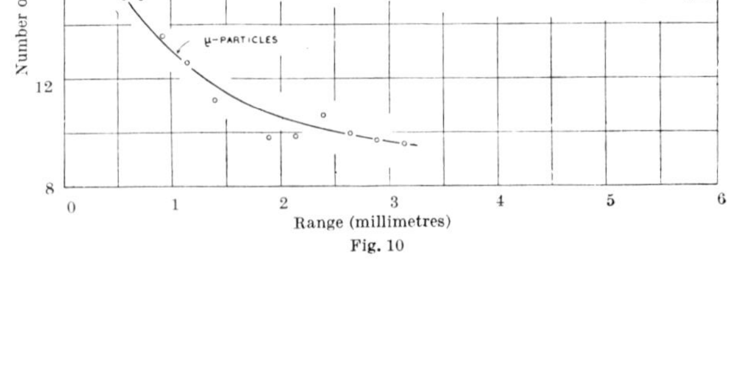

Fig. 10

the observations in the scattering experiments, we give more weight to the measurements by grain-counting. It appears certain, from these observations, that the true value of m_k lies between 700 and 1,800 m_e, and we think it highly probable that it is substantially less than that of the proton. Thus every individual point representing the grain-density in the track k, at a particular value of the residual range, lies below the corresponding points for each of the four protons.

Disintegration '*B*'

The tracks, c and d, of the two particles emitted from point B are characteristic of protons or heavier particles, and we regard them as due to a disintegration produced by the particle t. This particle was frequently scattered in passing through the emulsion and was therefore of low velocity; and the evidence is consistent with the assumption that it had reached the end of its range at the point B.

The only known slow charged particle which is capable of producing a disintegration of the type represented by star B is a π^--particle[6,8]. We therefore assume that a negative meson of mass 286 m_e was created at the point A, and reached the end of its range to produce the disintegration B.

Transmutation '*A*'

In order to interpret the transmutation A, we first made a detailed examination of the tracks of the emitted particles. Of the two tracks a and b, the former has a length in the emulsion of more than 2,000 μ, and ends in the surface, whereas b ends in the glass and is 116 μ long. The grain-densities in the two tracks are equal to within the limits defined by the statistical fluctuations. The average grain-density in the long track a is 49·0 grains per 100 μ; that is, 2·17 times the value characteristic of minimum ionization for a particle of charge e. Unless we admit the existence of fractional values of the electronic charge, we must conclude that the particles producing the tracks a and b both carried charges of magnitude e.

In order to determine the possible values for the energy of the particles producing tracks a and b, we have calculated the variation with energy of the specific ionization of a particle of charge e, from the formula of Halpern and Hall[9], assuming the atomic composition of the emulsion to be identical with that of the Ilford *C*2 plates. This formula is a modification of that of Bloch[10]; it applies to particles moving in a solid medium and gives results in good agreement with experiment for particles of low energy. The results are shown in Fig. 11, where the specific ionization is plotted as a function of the quantity E/m, where E is the energy and m the mass of the particle, both quantities being measured in MeV. From Fig. 11, we have determined the possible values of the energy of the particles, a, b, corresponding to the observed grain-density in the tracks, assuming them to be protons, π-mesons, μ-mesons or electrons. The resulting values are tabulated in Table 2.

TABLE 2. Values of the energy and momentum of the particle producing track a, as deduced from the observed grain-density and scattering, making various assumptions concerning the mass of the particle

Assumed particle		proton	π-meson	μ-meson	electron
Energy in MeV.	(*a*) below minimum ionization	235±95	37±13	27±11	0·13±0·05
	(*b*) above minimum ionization				> 1,000
Momentum MeV./*c*	(*a*) below minimum ionization	700±160	109±22	80±15	0·4±0·1
	(*b*) above minimum ionization				> 1,000
	(*c*) from scattering observations	245±40	113±18	100±16	68±11
	(*d*) from momentum balance	98±5	98±5	98±5	98±5

There are two possible interpretations of the transmutation produced at A by the particle k. We can assume, either that the particle was captured by a nucleus, or that it decayed spontaneously. From the measured values of the mass of the particle, it would be possible, from the point of view of the conservation of mass and energy, to admit that, at the end of its range in the emulsion, it was captured by a nucleus and led to the ejection of two energetic protons and a π^--particle. It appears almost certain, however, that the release in a nucleus of such a large amount of energy would lead to the 'evaporation' of many nucleons, a process commonly observed in plates exposed to the cosmic radiation; and that two protons of great energy would be only two components of a 'many-pronged' star. (It may be noticed that we cannot assume that the particle k was captured by one of the rare nuclei of heavy hydrogen, present in the gelatine. In such an interaction, the algebraic sum of the charges on the two initial particles is 0 or $2e$, whereas that of the product particles is e or $3e$.) We shall see later that there are other objections to the hypothesis that the tracks a and b were produced by protons, or heavier nuclei of charge e.

It follows from the above considerations that if we are to describe the transmutation in terms of particles of which the existence is already established, we must attribute the tracks a and b either to electrons, to μ-mesons or to π-mesons. Considering the first of

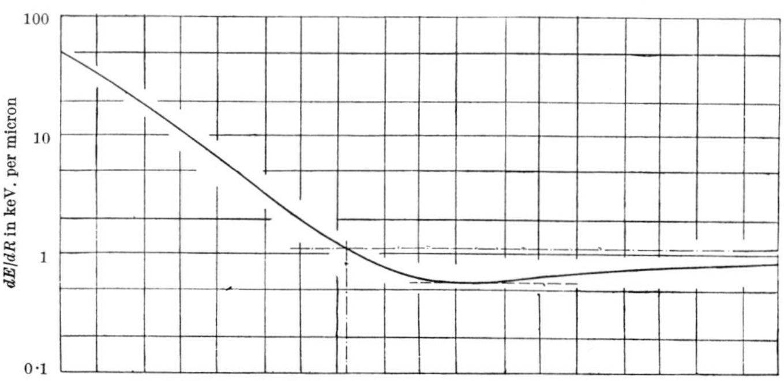

Fig. 11. Variation of the rate of loss of energy of a particle of charge $|e|$ as a function of the quantity E/m, where E is the kinetic energy and m the mass of the particle, both quantities being measured in MeV.

these possibilities, we must assume the electrons to have had an energy value greater than that corresponding to minimum ionization, namely, greater than 1,000 MeV.; for with the alternative lower value corresponding to the observed ionization, 300 keV., the particle would have had a range in the emulsion of only about 100 μ, and would have been frequently scattered. The assumption that the particles *a* and *b* were electrons is therefore inconsistent with the conservation of energy and can be rejected. We are left with the alternatives that the tracks were produced either by π- or by μ-mesons.

If the particles *a* and *b* were mesons, we must assume, in order to conserve mass-energy, that their kinetic energies were 27 MeV. or 37 MeV., respectively, in the case of μ- or π-mesons (see Fig. 11). In either case, it appears to be very difficult to reconcile the observations with the assumption that the particles were emitted as a consequence of the liberation in a nucleus of the energy corresponding to the rest-mass of particle *k*. We are therefore led to examine the possibility of explaining the observations in terms of a spontaneous decay of this particle.

Assumption of a Spontaneous Decay of the *k*-Particle

In examining the possibility that the transmutation *A* corresponds to a spontaneous decay of the particle *k*, we require to know the relative directions of motion of the three ejected particles. For this purpose it is necessary to determine the shrinkage of the emulsion; the ratio, *S*, of the thickness of the emulsion during exposure to that after it had been developed, fixed and dried. We have measured this quantity by examining the tracks of α-particles, produced in the emulsion by uncontrolled radioactive contamination. Among such 'stars', it is possible to identify some, due to an original atom of radiothorium, from which an α-particle of thorium C′ was emitted. The shrinkage has been measured by determining the lengths of the projection of the corresponding tracks on the surface of the emulsion, and their apparent angles of 'dip'. The value of the 'shrinkage' thus found is $S = 2{\cdot}7 \pm 0{\cdot}1$. Knowing the value of *S*, the original orientation of a track in the emulsion, before processing, can be determined, in favourable cases, with a precision of the order of 1°, by observing the apparent angle of 'dip' of the particle, and the direction of its projection on the plane defined by the surface of the emulsion. Using these methods, the original directions of motion of the three particles *a*, *b* and *t* were found to be coplanar. The departure of the direction of motion of any one particle from the plane defined by the other two is less than 4°. The error in this determination is largely due to the fact that track *t* is of short range, and the particle producing it was of low velocity, and frequently scattered.

The values of the angles between the directions of motion of the particles in the common plane are shown in Fig. 12. The observed coplanarity makes it legitimate to assume that the three particles arise as a result of the spontaneous decay of the *k*-particle at the end of its range in the emulsion, and that they are the only product of its disintegration; that no neutral particles, which would escape observation, are emitted. It follows that the vector sum of the momenta of the three particles must be assumed to be equal to zero.

If we are correct in attributing the track *t* to a π⁻-particle, it follows from the observed range, 45 μ,

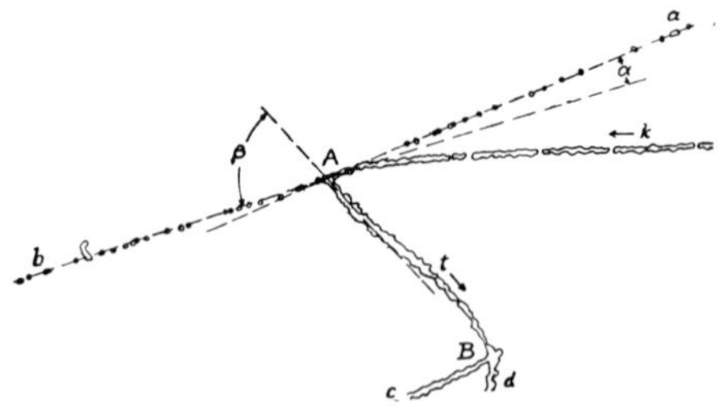

Fig. 12. Facsimile drawing of the event shown in Fig. 8, made with the projection microscope. The actual angles *α* and *β*, measured in the common plane of the three tracks, *a*, *b* and *t*, are: $\angle\alpha = 9{\cdot}8°$; $\angle\beta = 76{\cdot}6°$

that the kinetic energy of ejection was 1·04 MeV. The corresponding value of the momentum of the particle is 17·5 MeV./*c*. From the observed directions of motion, the momenta of the particles giving tracks *a* and *b* are then found to be 98 ± 5 and 104 ± 5 MeV./*c*, respectively. These values are to be compared with those corresponding to electrons or mesons listed in Table 2, which have been deduced from the observed grain-density in the tracks. We have seen that the values given in Table 2 for the momenta of the two particles, if they are assumed to be electrons, are many times too large. It follows that there is a wide departure from a momentum balance if the tracks *a* and *b* are assumed to be due to either electrons or protons. Further, the values of the momenta, as deduced from observations on the scattering of the particles, are inconsistent with those obtained from grain-counts, if the particles are assumed to have been either electrons or protons (see Table 2).

The agreement between the sets of values for mesons, however, is most remarkable, and gives strong support for the assumption of a spontaneous decay of the *k*-particle. Only a very rare combination of unrelated features, including the coplanarity of the tracks and the directions of motion of the particles in the common plane, the range of the particle *t*, and the specific ionization of the particles producing tracks *a* and *b*, could produce such an agreement between the estimated values of the momenta, if the result is fortuitous.

The values of the momenta of the particles producing tracks *a* and *b*, as determined by the three different methods, are consistent, within the errors of measurements, with the assumption of a spontaneous decay of the *k*-particle whether the product particles are assumed to be μ-mesons or π-mesons. We can apply a further test by calculating the values of the rest-mass of the particle *k* which corresponds to the two different assumptions, and the results are tabulated in Table 3.

TABLE 3. Estimates of the mass of particle *k* based on total release of mass and energy, for two assumed modes of decay

(i) $k \rightarrow \pi^- + \pi + \pi$

	Track '*a*'	Track '*b*'	Track '*t*'
Particle	π	π	π−
Rest-mass (m_e)	286	286	286
Energy in m_e	61	64	2

Total = m_k = 985 m_e

(ii) $k \rightarrow \pi^- + \mu + \mu$

	Track '*a*'	Track '*b*'	Track '*t*'
Particle	μ	μ	π−
Rest-mass (m_e)	212	212	286
Energy in m_e	76	81	2

Total = m_k = 869 m_e

In calculating the energy of the particles producing tracks *a* and *b*, it is assumed that the particle producing track *t* is a π−-particle, of momentum 17·5 MeV./*c*; knowing the relative directions of motion of the three ejected particles, the momenta of the other two particles are determined, and hence the energies corresponding to any assumed mass.

It will be seen from Table 3 that the assumption of two μ-mesons corresponds to a rest mass of the *k*-particle of 869 m_e; and for two π-mesons, 985 m_e. The assumption of different particles, one π- and one μ-meson, gives an intermediate value of approximately 925 m_e. In view of the error in the direct determination of m_k, the results are not decisive.

If the transmutation is to be interpreted in terms of particles of which the existence is already established, we are left with four possibilities for the nature of the particles producing tracks *a* and *b*. These are indicated schematically in Table 4.

TABLE 4. Comparison of the observed and calculated values of the grain-density in track *b*, for various assumptions regarding the nature of the particles producing the tracks *a* and *b*

	Length of track in microns	Number of grains	Grain-density	Assumed particles 1	2	3	4
Track *a*	2100	1025	49±1·5	π	μ	π	μ
Track *b*	116	59	51±6	π	μ	μ	π
Calculated grain-density in *b*				45	45	34	64

Values of the grain-density are given in grains per 100 μ.

For the following reasons, case 3, Table 4, is the most improbable. If track *a* is that of a π-meson, we can calculate the momentum and the grain-density to be expected in track *b*. We thus obtain the value of 34 grains per micron instead of 51·0 ± 6·0 as observed. For case 4, on the other hand, if *a* is a μ-meson, the calculated grain-density for track *b* is 64, a value which differs from that observed by an amount only twice that corresponding to the standard deviations. The observed grain-densities agree best with the assumption that the two particles are of the same type.

Observations on the scattering of the particle producing track *a* are in better accord with the assumption that it is a π-meson rather than a μ-meson (see Table 2); but the results are again indecisive. We may sum up this evidence, and that provided by the mass determinations by grain-counting, by saying that there is some support for the view that the three product-particles are π-mesons; but that the alternative possibilities of one π- and two μ-, or two π- and one μ-meson cannot be excluded.

Chance Juxtaposition of Unrelated Events

In the light of the analysis made in the preceding sections, we can now return to the original assumption that the event is not to be regarded as a fortuitous juxtaposition of tracks. The accuracy of the determination of the mass of the particle *k* does not allow us to exclude the possibility that it has a mass as great as that of a proton, although the observations by grain-counts render it very improbable. Suppose then that a proton, unrelated to the particles producing the other tracks, came to the end of its range at *A*. Even with this assumption, the event is still difficult to explain in conventional terms. Many examples of π^--particles ejected from stars have been observed in this Laboratory[8], but in the present instance the existence of a nuclear interaction in which two protons of great energy are emitted, unaccompanied by slow protons and α-particles, would remain to be explained. A similar difficulty is met if we assume that a particle producing one of the tracks, *a* or *b*, approached *A* and produced the transmutation.

If, alternatively, the tracks *c* and *d*, diverging from star *B*, represent an unrelated disintegration—produced, for example, by a γ-ray—we could then assume track *t* to be that of a proton. We are then left with the difficulties associated with the features peculiar to star *A*, which must now be assumed to have been produced by a slow, charged particle; difficulties which have already been discussed in a previous paragraph. These considerations give further support to the original assumption, that all the tracks shown in the mosaic represent a succession of associated processes.

Relation of the Present Results to Other Observations

If a particle with the elementary electronic charge suffers a spontaneous decay, the law of the conservation of charge demands that the number of emitted particles of charge *e* shall be odd. From this point of view, the sign of the charge of the original particle can have been either positive or negative. If the particles producing tracks *a* and *b* form a pair of opposite sign, then the original *k*-particle was negative. The only other alternative is that they were both positively charged, in which case the *k*-particle was also positive. It is therefore possible that our observations correspond to a mode of decay of positive particles of mass approximately 900 m_e, and that the observation by Leprince-Ringuet[2] demonstrates the fate of the corresponding negative particles—nuclear capture with the production of a 'star' and the ejection of a π^--particle.

Rochester and Butler[2] have published an expansion-chamber photograph which appears to be due to the spontaneous decay of a neutral particle of mass approximately 900 m_e into a pair of oppositely charged particles of rest-mass approximately 300 m_e. We have therefore considered the possibility that the decay process suggested by the present results can be regarded as taking place in two stages: the emission of a π^--particle of low energy, followed by the spontaneous decay of the resulting neutral particle. On this view, however, it would be necessary to assume that the neutral particle has a life-time of the order of 10^{-14} sec. Otherwise, in recoiling from the π-particle, it would move away from the original point of decay, and the two charged particles into which it became transformed would originate from a point separated from the beginning of the track of the π^--particle. It follows that we cannot identify such a postulated unstable neutral particle with that for which evidence is provided in the experiments of Rochester and Butler.

Finally, we have considered the possible relations of the present results to the particles of mass approximately 800 m_e referred to as τ-mesons, evidence for which has been recently reported by Bradt and Peters[2]. It is a remarkable feature of their experiments that their τ-mesons give rise to no recorded secondary particles at the end of their range. It appears to be possible that these particles also decay with the emission of three fast mesons, but that the transmutation usually takes place with a more equal partition of kinetic energy than in the case we have observed. It would then follow that in the Ilford *C*2 emulsion the disintegration products would commonly escape observation. If this view is correct, we must regard the event we have observed as representing a rare example of a common mode of decay of these mesons; an example which, by chance, has allowed a detailed analysis to be carried out. If so, the τ-meson of Bradt and Peters, when recorded by electron-sensitive emulsions, should show the tracks of three particles, of low specific ionization, and of which the directions of motion are co-planar.

We have pleasure in thanking Prof. von Muralt and members of the staff of the Jungfraujoch Forschungsstation for hospitality and assistance in obtaining the exposures; Dr. E. R. Davies and Dr. W. E. Berriman, of Messrs. Kodak, Ltd., for special photographic plates; Miss C. Dilworth and Dr. G. P. S. Occhialini for advice on development; Mr. W. O. Lock and Mr. J. H. Davies for assistance in making observations on the scattering of particles in the emulsion; and to the team of microscope observers of this Laboratory. We are indebted to Prof. N. F. Mott and other colleagues for a number of discussions on the processes associated with the capture of negative mesons by nuclei.

Note added in proof. Since completing this article, we have been informed by Dr. Peters that, in Ilford *C*2 emulsions exposed at 90,000 feet, he and Dr. Bradt have observed three events with the following characteristics. A particle, which they judge to be similar in mass to their τ-mesons, appears to come to rest and to lead to the emission of a particle of smaller mass, which, at the end of its range, produces a nuclear disintegration. The ranges of the secondary particles, in the three cases, are 20, 25 and 45 μ, respectively. The authors were not aware of our results when they suggested to us that their observations may correspond to the spontaneous decay of heavy mesons. According to their description, these events are precisely similar to those we should expect to observe in *C*2 emulsions as a result of the spontaneous decay of heavy particles of the type we have postulated; for any particles of low specific ionization will not be recorded by the Ilford plates. The observations of Peters and Bradt appear, therefore, to give further support for the assumption that the present observations are not due to a chance juxtaposition of tracks; and they suggest that it will be possible, in the near future, to find similar examples suitable for making a detailed analysis.

[1] Berriman, *Nature*, **162**, 992 (1948).
[2] Leprince-Ringuet, *C.R.*, **226**, 1897 (1948). Rochester and Butler, *Nature*, **160**, 855 (1947). Bradt and Peters, Report to the Bristol Symposium, 1948 (in the press). Alichanian, Alichanov and Weissenberg, *J. Exp. and Theoret. Phys., U.S.S.R.*, **18**, 301 (1948); and other references.
[3] Camerini, Muirhead, Powell and Ritson, *Nature*, **162**, 433 (1948).
[4] Goldschmidt-Clermont, King, Muirhead and Ritson, *Proc. Phys. Soc.*, **61**, 138 (1948).
[5] Lattes, Occhialini and Powell, *Proc. Phys. Soc.*, **61**, 173 (1948).
[6] Serber, Report of Solvay Conference for 1948.
[7] Dilworth, Occhialini and Payne, *Nature*, **162**, 102 (1948).
[8] Occhialini and Powell, *Nature*, **162**, 168 (1948).
[9] Halpern and Hall, *Phys. Rev.*, **73**, 477 (1948).
[10] Livingston and Bethe, *Rev. Mod. Phys.*, **9**, 263 (1937).
[11] Camerini and Lattes (private communication); see also Powell and Occhialini, "Nuclear Physics in Photographs", 112 (Oxford, 1947).

DEVELOPMENT OF LEATHER CHEMISTRY

THE first applications of scientific knowledge to the problems of leather manufacture may be said to have been made about 1890, and the initiation in the following year at the Yorkshire College (now University of Leeds) of a course of lectures devoted to the subject was an indication that a new field of technology had gained recognition. Five years later, through the foresight and generosity of local tanners and the Worshipful Company of Skinners, a special department was built to house the new subject, and H. R. Procter was appointed to fill the newly created chair in 1898.

The industry, which was being encouraged to examine its methods along scientific lines, was one dating from primitive times, and little advance had been made in methods of manufacture or in the principles underlying them since their inception. Moreover, and perhaps because of this, the industry as a whole was a conservative one, and early workers in the field had to contend with a rooted distrust of new ideas which took a long time to break down. It was without doubt fortunate for the future development of the subject that the man chosen to hold a key position at the outset was Procter, who deservedly earned the title of 'father of leather chemistry'. Working in a more leisurely age than the present, he did not publish his well-known theory of protein swelling until 1916; but during the intervening years he was laying the foundations firmly by his influence on the succession of students who passed through his hands, as well as by his own published work. Largely by reason of his efforts, the industry was of a sufficiently scientific turn of mind by 1920 to support the foundation of the British Leather Manufacturers Research Association, which is one of the oldest associations of its kind in Great Britain. The growth of knowledge in this field, and the amount of work which has been done, may be appreciated by a perusal of three publications issued to commemorate the twenty-fifth anniversary of this Association*, which is, in effect, a summary of almost all important investigations relating to leather carried out during the period throughout the world.

Since the materials used in leather manufacture are of many different types, both organic and inorganic, and as the processes employed involve to some extent the application of almost all branches of chemical knowledge, there can be few other industries in which the research worker must be familiar with so many different aspects of chemistry. Many problems have been elucidated by virtue of investigations in other subjects, probably the most fruitful contributions coming from biochemists, especially those working on proteins, as a result of which the composition and structure of the tanner's most important raw material, collagen, is more or less settled. From the realm of organic chemistry have come numerous studies on the constitution of the tannins which, from Fischer's early synthesis of penta-*m*-digalloyl-β-glucose and from Freudenberg's work on the condensed tannins to the work of present day, have materially contributed to the understanding of vegetable tanning. Information on these matters is, however, by no means complete, and recent research, especially on the important tannin constituent of mimosa bark, has directed the attention of leather chemists in particular to the importance of this line of investigation.

In this branch of their subject, as in many others, leather chemists have in many cases made notable contributions not only to the understanding of their own problems but also to chemical knowledge in general. It may not be widely known that Procter's work on the swelling of gelatine was promoted by the observation of swelling phenomena in hides during operations prior to tanning; and the discovery that chromium salts were valuable tanning agents led to a large amount of research on the properties of their solutions, with results of wide interest.

* Progress in Leather Science, 1920–1945. Vols. 1, 2, 3. Pp. xvi + 705. (London: British Leather Manufacturers Research Association, 1946–48.)

The properties of charged V-particles

By

R. Armenteros, K. H. Barker, C. C. Butler, A. Cachon and C. M. York
The Physical Laboratories, University of Manchester*

[Received March 11, 1952]

Abstract

Twenty-one examples of the decay of charged *V*-particles have been found among 25 000 counter-controlled cloud-chamber photographs obtained on the Pic-du-Midi (2 867 m). These events have been analysed and compared with the data obtained by other groups. The measurements show that, if the majority of the $V^{\pm}$-tracks are produced by one type of particle having a single mode of decay, there must be at least two neutral secondary particles, in addition to the charged secondary particle. The latter is probably a μ- or a π-meson and the combined mass of the neutral particles must be less than 1 000 m_e. Since the κ-particles, of mass about 1 200 m_e, which were discovered in photographic emulsions, also decay into a charged meson and at least two neutral particles, it is plausible to assume, provisionally, that the $V^{\pm}$-particles and the κ-particles are the same. On this assumption most of the available data concerning $V^{\pm}$- and κ-particles can be explained if the unstable particle has a mass of about 1 200 m_e and decays into a charged μ-meson, a π°-meson and a neutrino, or alternatively into a charged μ-meson and two neutrinos. As yet no direct evidence has been found for the existence of a π°-meson in the decay scheme. One negatively charged *V*-particle probably produced a neutral *V*-track and a charged meson.

§1. Introduction

The cloud-chamber work on the Pic-du-Midi (2 867 m), described by Armenteros *et al.* (1951 a and b), has been continued and a special study made of the charged *V*-particles. These particles have been found in small numbers in many cloud-chamber experiments, for example, by Rochester and Butler (1947), Seriff *et al.* (1951), Bridge and Annis (1951), Fretter (1951), Thompson *et al.* (1951), Leighton *et al.* (1951) and Astbury *et al.* (1951). Nearly all these workers have found a smaller number of charged than neutral *V*-tracks ; in most cases the $V^{\pm}$-tracks number about 15% of the V°-tracks. A tentative explanation of the phenomena observed on the Pic-du-Midi is given below, where the cloud-chamber

* Communicated by Professor P. M. S. Blackett, F.R.S.

SER. 7, VOL. 43, NO. 341.—JUNE 1952 2 S

Reprinted with permission from *Phil.Mag.* **43** 597–612 (1952)
© Taylor and Francis

events are compared with the decay process of κ-particles, discovered in photographic emulsions by O'Ceallaigh (1951). The lifetime and frequency of production of $V^{\pm}$-particles will be discussed in a later paper.

§ 2. Experimental Data

(i) *The Classification of the Data*

Twenty-five thousand pictures, obtained with the cloud chamber and electromagnet described by Armenteros *et al.* (1951 a and b), have been examined and forty-five examples of the sudden deflection of tracks in the gas of the cloud chamber, without the production of visible recoil tracks, have been found. It has already been shown by Armenteros *et al.* (1951 a) and by Butler (1951) that this type of track is only very rarely due to the production of a two-pronged star in the cloud-chamber gas or to the large-angle scattering of particles with momenta greater than 500 Mev/c. The tracks may, however, be divided into two categories, those due to (*a*) the large-angle elastic scattering of low-energy particles making nuclear collisions and (*b*) the decay-in-flight of π-mesons and particles heavier than the π-meson, known as charged V-particles.

The events of type (*a*) have been identified by using the momentum and ionization measurements ; two were slow mesons, one was a slow proton and eight were electrons. An argon recoil with a momentum of 75 Mev/c would be just visible in the chamber ; for each of the events of type (*a*) the recoil was too short to be visible. The decay of a slow, that is heavily-ionizing, π-meson can easily be identified but it is sometimes difficult to distinguish the decay of a fast π-meson from that of a $V^{\pm}$-particle, because both types of particle frequently produce minimum ionization tracks. However, the dynamics of the π-meson decay process can be used to ascertain which of these events can be interpreted by means of the scheme : $\pi \rightarrow \mu +$ neutrino.

The momentum component of the μ-meson transverse to the direction of the π-meson, namely p_T, cannot be greater than the momentum p^* of the μ-meson in the rest system of the π-meson. The value of p^* can be obtained, in terms of m_1 and m_2, the masses (in energy units) of the π- and μ-mesons respectively, by using the relativistic energy equation, $m_1 = (m_2{}^2 + p^{*2})^{\frac{1}{2}} + p^*$. Since $m_1 = 275\, m_e$ and $m_2 = 210\, m_e$, $p^* = 29$ Mev/c. In addition to this condition, the momentum P (in Mev/c) of the π-meson and the maximum permissible angle, $\phi_{\max}$, between the directions of the π- and μ-mesons are related by an equation which reduces to $\sin \phi_{\max} = 40/P$. The majority of the values of ϕ will be close to $\phi_{\max}$; if a particular track has a value of ϕ greater than the appropriate $\phi_{\max}$, then it cannot be due to the decay of a π-meson. If P cannot be measured, then the first condition alone is used. These two tests, based on the dynamics of the π-meson decay scheme, have been used to classify the 35 events in category (*b*): 14 can be explained as $\pi \rightarrow \mu$ decays leaving the remaining 21 to be explained as the decay of $V^{\pm}$-particles. Following this procedure, it is possible that a small number of decays due to $V^{\pm}$-particles may be interpreted as π-meson decays. In addition, a few tracks of type (*a*) may be classified as $\pi \rightarrow \mu$-decays, but probably not as $V^{\pm}$-decays since the transverse momenta of all but one of the $V^{\pm}$-particles are greater than 75 MeV/c, the minimum momentum for a visible recoil. The exceptional case (track 2) cannot be classified in category (*a*).

(ii) *Measurements on the Charged V-particle Tracks*

The data for the twenty-one charged V-tracks and the original one found by Rochester and Butler (1947) are given in table 1. The errors given in table 1 and throughout this paper are the probable errors of measurement. Only six of the events can be measured completely; the remainder have at least one track which is too short for measurement. It will be seen, however, that the primary momenta of these six events have large errors, and, therefore, they are not very valuable in a detailed analysis of the nature of the decay process.

Table 1. Measurements on 22 $V^{\pm}$-particles

(1) Sign of V		(2) Primary momentum (100 MeV/c)	(3) I/I_{min} of Primary	(4) Angle† ϕ (deg)	(5) Sign and momentum of secondary (100 MeV/c)	(6) Transverse momentum p_T (100 MeV/c)
1*	+	—	1	18·9	$+7{\cdot}7\pm1{\cdot}0$	$2{\cdot}5\pm0{\cdot}3$
2**	−	15 ± 5	1	24·5	$-0{\cdot}73\pm0{\cdot}20$	$0{\cdot}30\pm0{\cdot}10$
3	−	—	4–8	103	$-1{\cdot}85\pm0{\cdot}20$	$1{\cdot}80\pm0{\cdot}20$
4	?	—	1	43	—	—
5	?	—	1	11	>5	$>0{\cdot}95$
6	−	11 ± 4	1	12·5	$-5{\cdot}5\pm0{\cdot}8$	$1{\cdot}19\pm0{\cdot}17$
7	+	—	1	9·5	$+8{\cdot}3\pm1{\cdot}2$	$1{\cdot}37\pm0{\cdot}20$
8	?	—	4–6	68·5	—	—
9	−	$8{\cdot}0\pm2{\cdot}6$	1	15	$-5{\cdot}3\pm0{\cdot}8$	$1{\cdot}38\pm0{\cdot}20$
10	−	—	1	34	$-2{\cdot}9\pm0{\cdot}60$	$1{\cdot}62\pm0{\cdot}32$
11	+	$6{\cdot}0\pm0{\cdot}6$	1	22	—	—
12	−	15 ± 3	1	27·5	$-4{\cdot}4\pm1{\cdot}1$	$2{\cdot}04\pm0{\cdot}50$
13	+	50 ± 20	1	15	$+6{\cdot}2\pm1{\cdot}2$	$1{\cdot}62\pm0{\cdot}35$
14	+	—	1	20	—	—
15	+	$6{\cdot}0\pm1{\cdot}5$	1	9·5	$+5{\cdot}4\pm1{\cdot}8$	$0{\cdot}89\pm0{\cdot}30$
16	−	—	1	25	$-6{\cdot}5\pm1{\cdot}3$	$2{\cdot}50\pm0{\cdot}40$
17	−	—	1	20·5	$-6{\cdot}7\pm0{\cdot}7$	$2{\cdot}40\pm0{\cdot}20$
18	?	>40	1	7	—	—
19	−	>40	1	21·5	$-4{\cdot}0\pm1{\cdot}0$	$1{\cdot}47\pm0{\cdot}40$
20	+	—	1	34	$+5{\cdot}1\pm1{\cdot}0$	$2{\cdot}85\pm0{\cdot}57$
21	?	>40	1	5	—	—
22	+	$4{\cdot}1\pm0{\cdot}4$	1–3	60	—	—

* discovered by Rochester and Butler 1947.
** associated with V°-track.
† angle between direction of $V^{\pm}$-particle and direction of charged secondary particle.

Unfortunately, the masses of two of the heavily-ionizing V-particles (3 and 8) cannot be estimated directly, because the tracks are too short for their momenta to be measured. The estimated ionization of track 22, which is reproduced in Plate XXX,* is about twice minimum; using the measured momentum, the mass of the unstable particle is thus about 1 100 m_e. It is unlikely that the ionization exceeds three times the minimum value; therefore an upper limit to the mass is $(1\,580 \pm 160)\ m_e$, where the probable error is due to the uncertainty in the momentum measurement. The estimated ionization of track 11 (Plate XXXI) lies between the minimum value and an upper limit of twice the minimum ionization, which is consistent with an upper limit of $(1\,630 \pm 160)\ m_e$ for the mass of the unstable particle. Similarly the ionization of track 15 lies between minimum and twice minimum but, owing to the large error in the momentum of the primary, no reliable estimate of the upper limit to the mass can be made. The ionization estimates and momentum measurements for the remaining measurable primary tracks give considerably higher upper limits for the mass of the $V^{\pm}$-particles. Thus the preliminary measurements show that, if the $V^{\pm}$-particles are all of the same type, the smallest observed upper limit for their mass is 1 600 m_e. Therefore, the $V^{\pm}$-particle cannot be the charged counterpart of the V°_1-particle of mass 2 200 m_e, found by Armenteros *et al.* (1951 b). In the discussion of the decay process of $V^{\pm}$-particles in § 3, section (i), an upper limit for their mass is required; this is taken to be 1 900 m_e, which is about two probable errors above the value of 1 600 m_e.

It is impossible to determine the mass of any of the 14 charged secondary particles whose momenta have been measured, because their tracks are all approximately at minimum ionization. An ionization of twice the minimum value may be regarded as an upper limit for all of the secondaries; considering this value in conjunction with the momenta, it will be seen that the masses of the negatively-charged secondaries of tracks 2 and 3 were less than 500 m_e. If all the ionizations are close to the upper limit, then several of the positive secondaries could have been protons.

The secondary particle of event 1 penetrated 3·4 cm of lead and was certainly heavier than an electron. Seriff *et al.* (1950) observed a similar charged secondary particle; it penetrated their 2 cm lead plate without producing an interaction. Assuming that electrons and negative protons are not produced and that the charged secondary particles are all of the same type then, in terms of well-known particles, they are probably either μ- or π-mesons. Bridge and Annis (1951) have described an event in which a slow secondary particle from the decay of a charged V-particle was scattered through a large angle in an aluminium plate and subsequently behaved as a meson; they consider that this secondary was a π-meson.

Event 2 is of particular interest since it represents, either the chance occurrence of both a charged and a neutral decay on the same photograph, or the decay of a charged V-particle with the production of a secondary

* For plates see end of issue.

neutral *V*-particle. This photograph is described in detail in § 4; meanwhile the possibility that neutral *V*-particles may be produced by the decay of charged *V*-particles must be kept in mind.

(iii) *Non-decaying Particles with Mass about* 1 000 m_e

Since the decays of several heavily-ionizing charged *V*-particles have been observed in the cloud chamber, their mean range for decay cannot be very much shorter than the average chamber dimensions. Therefore, it is expected that some of these particles will cross the chamber without decaying. About 900 heavily-ionizing particles which did not decay have been examined; they were associated with nuclear interactions immediately above the chamber or in the lead plate. The great majority are protons, but some are mesons with estimated masses less than 350 m_e. Two particles, however, have estimated masses between those of the π-meson and the proton; the data are given in rows (1) and (2) of table 2. Butler *et al.* (1950) found two similar events, the data for which are given in rows (3) and (4) of table 2.

Table 2. Measurements on Four Slow Non-decaying Particles of Mass about 1 000 m_e

Sign	Measured momentum (MeV/c)	Estimated ionization (min.)	Mass range (m_e)
(1) —	160($\pm$10%)	6—10	900—1 500
(2) +	170($\pm$ 7%)	4— 6	800—1 000
(3) +	170($\pm$10%)	5— 7	900—1 140
(4) —	170($\pm$10%)	4— 6	800—1 020

Butler *et al.* (1950) discussed the problem of the detection of positive particles of mass about 1 000 m_e when they are associated with a very much larger number of protons. They showed that the error in the momentum measurements combined with the uncertainty of the ionization estimates makes it possible to separate these particles from protons only if they have momenta within the narrow band (100–200) MeV/c. This difficulty does not apply to the negative particles, providing it is assumed that there are no slow negative protons present. The measured momenta of the two negative particles (1) and (4) would have to be reduced by a factor of about four before they could be interpreted as π-mesons, this possibility is clearly very remote, because the errors in the momenta are only 10% of the measured values. Therefore, particles (1) and (4) are more massive than the π-meson; their average mass is approximately 1 000 m_e. Similarly the positive particles (2) and (3) cannot be π-mesons. If, however, their actual momenta are each about twice the measured values and the ionizations are close to the upper limits, they could be protons. Again the observed errors are 10% of the measured momenta so that this possibility is remote, but it cannot be excluded completely.

Some of these non-decaying 1 000 m_e particles are probably identical with the $V^{\pm}$- and κ-particles, but they may possibly be the same as the τ-particles, found in photographic emulsions. The latter particles can be identified unambiguously by their decay into three charged π-mesons (Fowler *et al.* (1951)).

§3. Interpretation of the Charged V-Tracks

The available data on charged V-tracks, given in table 1, can be used for a study of the dynamics of their decay process. For this purpose it is reasonable to assume that all the events, with the possible exception of event 2, can be explained in terms of a single process.

Fig. 1

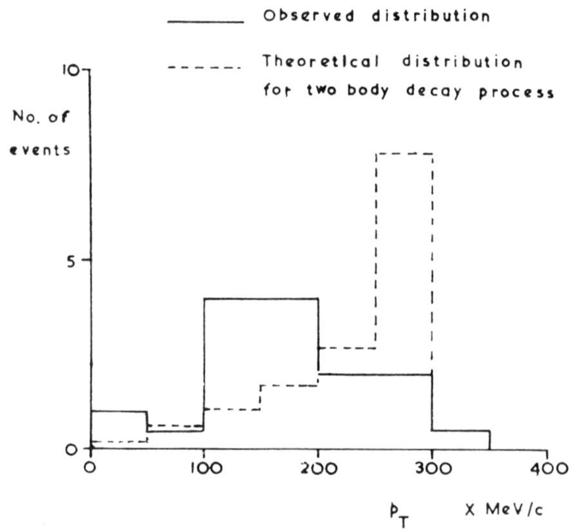

The distributions of the transverse momentum components (p^*=300 MeV/c for theoretical distribution)

(i) *The Number and Nature of the Secondary Particles*

The momenta of fourteen of the charged secondary particles are given in column (5) of table 1, and the momentum component of each secondary transverse to the direction of the V-particle is given in column (6). The observed distribution of the transverse momenta, shown in fig. 1, can be compared with the expected distribution on the assumption that all the $V^{\pm}$-particles are of the same kind and that they decay into only two secondary particles. Assuming that the secondary particles are emitted isotropically in the rest system, it has been shown by Armenteros *et al.* (1951 b) that the distribution of transverse momenta, p_T, has the form :

$$\mathscr{P}(p_T)\, dp_T = \frac{p_T}{p^*(p^{*2}-p_T{}^2)^{\frac{1}{2}}}\, dp_T, \qquad (1)$$

where p^* is the momentum of the secondary particles in the rest system and so is a constant for all the events. The value of p^* depends only on the mass of the unstable particle and the masses of the two secondary particles. These quantities, however, are unknown and so a value must be deduced indirectly from the relation $p_T = p^* \sin\theta^*$, where θ^* is the angle of emission of the two secondaries in the rest system with respect to the direction of the $V^\pm$-particle. Thus the maximum value of p_T is equal to p^*. Since the largest observed value of p_T is 285 MeV/c, p^* is assumed to have this value in the theoretical distribution (1). This is also shown in fig. 1, in the form of a histogram (dotted) normalized to the total number of events. The estimated errors in the values of p_T, given in table 1, are too small to reconcile the observations with the theoretical distribution, although the observed number of events in each range of p_T is small. Hence it is concluded that the assumption of a unique two-body decay scheme is probably false. It is necessary to assume, therefore, that either the $V^\pm$-particles have two or more modes of decay, each giving two secondary particles, or that there is a single decay process involving three or more secondary particles.

Assuming that at least three secondary particles are produced by the decay of $V^\pm$-particles, then the data given in table 1 can be used in a discussion of the nature of the secondary particles. It was shown in § 2, section (ii), that, if all the $V^\pm$-particles are of the same kind, the single charged secondary is probably either a μ- or a π-meson. It was also shown that 1 900 m_e is the upper limit to the mass of the $V^\pm$-particles. Therefore, neutrons, or $V^\circ{}_1$-particles of mass 2 200 m_e, cannot be produced.

Owing to the limited data available, the momentum distribution of the charged secondary in the rest system cannot be found. However, it is reasonable to assume, as previously, that the largest observed transverse momentum, p_T, is approximately the maximum possible value of the momentum of the charged meson in the rest system, namely $p^*_{\max}$. Assuming this value, the maximum energy, $E_{\max}$, of the charged μ- or π-meson in the rest system can be calculated. If a three-body process is considered, then the conditions that hold when the energy of the charged meson is $E_{\max}$ can be obtained. Assuming that M is the mass of the unstable particle, m_1, the mass of the charged meson, and m_2 and m_3, the masses of the neutral secondary particles, expressed in energy units, it can be shown that when the charged secondary is emitted with energy $E_{\max}$ the neutral secondaries move parallel to each other and with equal velocities in the opposite direction to the charged secondary. Using these conditions, it follows that the combined mass of the neutral particles is given by :

$$(m_2+m_3)^2 = M^2 + m_1{}^2 - 2ME_{\max}. \qquad (2)$$

A proof of this formula, which remains valid if either m_2 or m_3, or both, is zero, is given in the appendix. Values of (m_2+m_3) can be calculated, as a function of M, for various values of $p^*_{\max}$; they are shown graphically in fig. 2, where the charged secondary is assumed to be a μ-meson.

The observed upper limit of the p_T distribution is between 250 MeV/c and 300 MeV/c. The precise form of this distribution is unknown owing to the limited data available, and it is possible that the upper limit may be as high as 350 MeV/c. Using these values and the upper limit of 1 900 m_e for M, values of (m_2+m_3) can be obtained. They are summarized in table 3; those in column (2) are obtained if the charged secondary is a μ-meson while those in column (3) are obtained if it is a π-meson.

Fig. 2

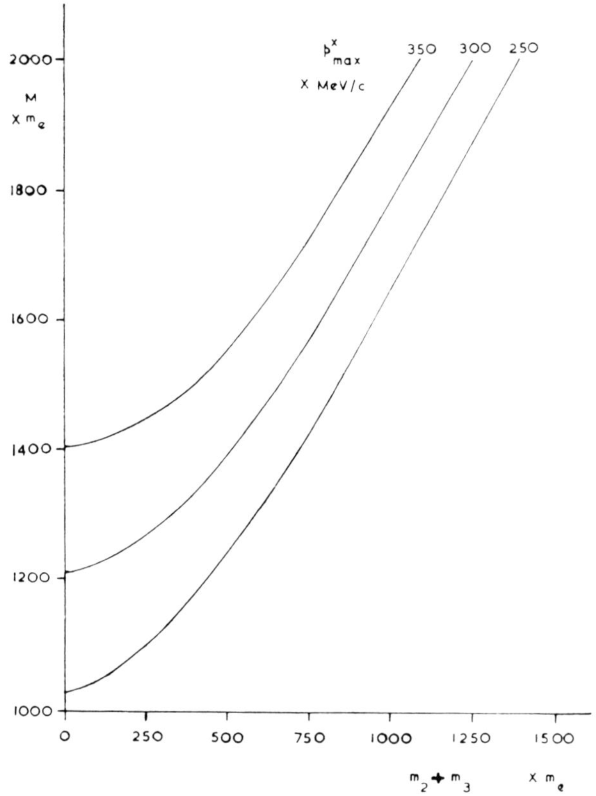

Relationships between M and (m_2+m_3) for $m_1=210\ m_e$

Table 3. The Values of (m_2+m_3), for $M=1\,900\ m_e$

(1) p^*_{max} (MeV/c)	(2) (m_2+m_3) for charged μ-meson (m_e)	(3) (m_2+m_3) for charged π-meson (m_e)
250	1 280	1 245
300	1 130	1 105
350	960	940

The calculated values of (m_2+m_3) show that a $V^\circ{}_2$-particle of mass about 800 m_e, described by Armenteros *et al.* (1951 b), may be produced by the decay of a $V^\pm$-particle, if the value of p^*_{max} lies in the range (250–350) MeV/c, but only if the value of M is close to the upper limit.

The above arguments, based on all the data at present available, show that no definite decay scheme for $V^\pm$-particles can be obtained. It is possible, however, to compare the properties of the $V^\pm$-particle with those of the $\kappa^\pm$-particles and to discuss possible decay schemes which may be used for the interpretation of both types of event.

(ii) *Possible Decay Schemes for $V^\pm$-particles*

Unstable charged particles, known as κ-particles, which decay into at least three secondary particles, only one of which is charged, have been found in photographic emulsions by O'Ceallaigh (1951), Menon (1951) and Trembley (1951). Bridge and Annis (1951) have observed the decay of several particles each into a single charged particle, one of which was heavier than the electron. These decays occurred when the particles were brought to rest in thin lead plates. These unstable particles are probably the same as the κ-particles. It is important to consider if the $V^\pm$-particles and these $\kappa^\pm$-particles are of the same type.

The masses of several κ-particles have been measured by O'Ceallaigh and Menon ; the average is about 1 200 m_e. In two cases, the charged secondary has been identified. The first came to the end of its range in the emulsion and produced an electron, a characteristic of the μ-meson ; the track of the second was very long and the particle was identified as a μ-meson from measurements of its grain density and scattering. Other κ-particles had charged secondaries with masses less than 400 m_e. The momenta of four of the secondary particles found at Bristol have been measured ; assuming they were μ-mesons, the values range from 35 MeV/c to 268 MeV/c. Since the κ-particles decayed at rest, it is impossible to explain the momentum distribution of the charged secondary particles by means of a unique two-body decay process. It is plausible to assume, however, that the κ-particles have a single decay process because their mass values are the same, within experimental error ; this decay process must have at least three secondary particles, the charged one being a μ-meson.

Several heavily-ionizing particles of mass about 1 000 m_e, which did not decay in the chamber, were described in § 2, section (iii). If these are the same as the $V^\pm$-particles, then the mass of the latter must be appreciably smaller than the upper limit of 1 900 m_e. Their mass could well be the same as the average mass of the κ-particles. In addition, the charged particles produced by the decay of $V^\pm$-particles could be μ-mesons. It is, therefore, plausible to assume, provisionally, that the κ- and $V^\pm$-particles are the same.

The highest observed momentum of the μ-meson in the rest system of a κ-particle is 268 MeV/c. This value is consistent with the range of values of p^*_{max} for the charged secondaries from the decay of $V^\pm$-particles, namely

(250–300) MeV/c. Using this result, values of (m_2+m_3) can be obtained from fig. 2 for values of M between 1 000 m_e and 1 400 m_e, which cover the range of the individual measurements of the masses of the κ-particles ; the values of (m_2+m_3) are given in table 4.

Table 4. Possible Values of (m_2+m_3) for κ- and $V^{\pm}$-particles

	$(m_1=210\ m_e)$				
M (m_e)	1 000	1 100	1 200	1 300	1 400
(m_2+m_3) if p^*_{max} is 250 MeV/c (m_e)	—	245	425	570	700
(m_2+m_3) if p^*_{max} is 300 MeV/c (m_e)	—	—	—	330	506

A dash in table 4 indicates that the particular set of conditions is not possible. A value of p^*_{max} of 350 MeV/c is inconsistent with the observed range of values of M. Thus, if the values of M are correct, a κ-particle cannot produce a $V^{\circ}{}_2$-particle. In § 2, section (ii), it was suggested that track 2, in which a V^--particle is associated with a V°-particle, may be unique and, since it cannot be interpreted with the κ-particles, if this association is real, it will be omitted from the analysis ; a possible interpretation is given in § 4. Should the mass of the κ-particles and the remaining $V^{\pm}$-particles lie between 1 000 m_e and 1 400 m_e then, if p^*_{max} is (250–300) MeV/c, the maximum mass of the neutral particle is about 500 m_e. The most likely combinations of neutral particles are therefore a π°-meson and a neutrino or alternatively two neutrinos.

If π°-mesons are produced they will decay into γ-rays which might be detected by their interaction in thin lead plates. Unfortunately very few of the cloud-chamber photographs of $V^{\pm}$-particles were taken with a suitable lead plate and so no direct evidence for the production of π°-mesons by $V^{\pm}$-particles has been obtained. O'Ceallaigh (1951) suggested a scheme for κ-particles involving a π°-meson in order to give them integral spin. This scheme is :

$$\kappa^{\pm}\rightarrow\mu^{\pm}+\pi^{\circ}+\text{neutrino}. \qquad (3)$$

Now the $V^{\pm}$-particles are nearly always observed in high-energy penetrating showers and, if they are not produced in pairs, they probably have integral spin and may possibly decay according to scheme (3). If p^*_{max} for the μ-meson is (250–300) MeV/c then, assuming scheme (3), the mass of the $V^{\pm}$-particles is about (1 100–1 260) m_e, which is consistent with the direct mass measurement on the tracks of κ-particles.

The only other plausible scheme for $V^{\pm}$-particles and κ-particles is :

$$V^{\pm} \text{ or } \kappa^{\pm}\rightarrow\mu^{\pm}+2 \text{ neutrinos}. \qquad (4)$$

Assuming this scheme and the same range for p^*_{max} as before, the mass of the unstable particles is within the range (1 025–1 225) m_e. If (m_2+m_3) $<300\ m_e$, the curves in fig. 2 show that the value of M is not strongly dependent on the precise value of (m_2+m_3). Thus the range of values of M for schemes (3) and (4) is about the same ; this is only the case if $p^*_{max} \gtrsim 250$ MeV/c.

§4. Description of Track 2 (Plate XXVIII)

Track 2, reproduced in Plate XXVIII, is so far unique because a V°-track is apparently associated with the charged decay. The spatial orientation of all the tracks on the photograph has been examined by reprojection. Two orthogonal projections are shown in fig. 3. The negative particle

Fig. 3

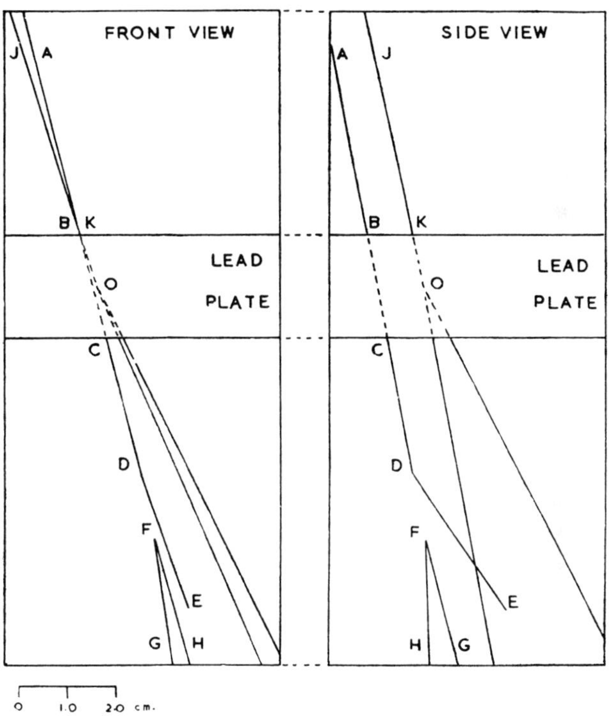

Projections of track 2

AB penetrated the lead plate with an angle of scattering $\leqslant 1^\circ$ and emerged as track CD. The particle decayed at D and the secondary particle DE was emitted at an angle of $(24{\cdot}5 \pm 2)^\circ$ to the direction CD. It can be shown from the ionization of this particle, which is between minimum and twice minimum, and the momentum of (73 ± 20) MeV/c, that the mass is less than $(200 \pm 50)\, m_e$. A V°-track was produced about 1·25 cm from D ; point D lies in the plane of the V°-track within experimental error. A second incident particle JK interacted at O in the lead and produced two charged secondary particles. The plane of the V°-track almost passes through the point O, in addition to point D. The line OF makes an angle of $(3{\cdot}5 \pm 2)^\circ$ with the plane of the V°-track GFH. It is, therefore, just possible that the V°-particle was produced at O and is only fortuitously related to the decay of the V^--particle. The first interpretation namely that the V^--particle produced the V°-particle, must, however, be given serious consideration.

The momenta of the primary and secondary tracks of the V^--particle 2 are given in columns (2) and (5) of table 1. No useful measurements can be made on the short secondary tracks of the V°-particle. FG makes an angle of 4° with the line of flight DF of the V°-particle. With this very limited data it is impossible to decide whether the fork was due to a ${V^\circ}_1$- or to a ${V^\circ}_2$-particle.

If it is assumed that the V^--particle of track 2 decayed into a meson and V°-particle, it cannot be interpreted by schemes (3) and (4). It is possible that track 2 and also the particle discovered by Bridge and Annis (§ 2, section (ii)) which decayed into a charged π-meson may be different from the κ-particles and the majority of the $V^\pm$-particles. A large fraction of the primary momentum of track 2, namely about (1–2) Bev/c, was given up to the neutral particles and, consequently, if the decay scheme involves only two secondaries, the neutral secondary was probably more massive than the charged one. A simple interpretation may be according to the scheme :

$$V^- \rightarrow V^\circ_{1\text{ or }2} + \text{meson}^-. \qquad . \quad . \quad . \quad . \quad . \quad . \quad (5)$$

If the event observed by Bridge and Annis is similar to track 2, the meson is a π-meson. Assuming that the neutral secondary is a ${V^\circ}_1$-particle of mass $2\,200\, m_e$, then the mass of the V^--particle (track 2) lies between $2\,500\, m_e$ and $2\,800\, m_e$ and the Q-value is about 60 Mev. For decay into a ${V^\circ}_2$-particle, the mass would be between $1\,100\, m_e$ and $1\,500\, m_e$.

If V°-particles were produced when the remaining twenty-one $V^\pm$-particles decayed in the chamber then, after correction for the relativistic dilatation factors and geometrical factors, the total path length of the V°-particles inside the chamber is greater than 50 cm. No V°-tracks were found, however, so that the lifetime of these V°-particles can hardly be less than 10^{-9} sec, which is longer than the lifetime of about 3×10^{-10} sec obtained from an examination of V°-tracks found in penetrating showers (Seriff *et al.* 1950). It is, therefore, unlikely that every $V^\pm$-particle produces a V°-particle.

§ 5. Conclusions

The decays of twenty-two $V^\pm$-particles have been examined. Each particle produces a charged secondary, which is probably a meson, and at least two neutral particles ; the combined mass of the latter particles is less than $1\,000\, m_e$. Additional information can be obtained from a comparison of the $V^\pm$-tracks with those of κ-particles. It is concluded, provisionally, that both particles are of the same type ; each produces a charged μ-meson and at least two neutral particles whose combined mass is less than 500 m_e. Two possible schemes are :

$$V^{\pm} \text{ or } \kappa^{\pm} \rightarrow \mu^{\pm} + \pi^{\circ} + \text{neutrino}, \qquad (6)$$

$$V^{\pm} \text{ or } \kappa^{\pm} \rightarrow \mu^{\pm} + 2 \text{ neutrinos}. \qquad (7)$$

Both schemes are consistent with a mass of about 1 200 m_e for the unstable particle.

One photograph (track 2, Plate XXVIII) shows the apparent association of a V°-track with a V^{-}-track. This may be fortuitous, but, if it is assumed that the V^{-}-particle did produce a V°-secondary, a possible explanation of the double-decay process is :

$$V^{-} \rightarrow V^{\circ}_{1 \text{ or } 2} + \pi^{-}. \qquad (8)$$

If a $V^{\circ}{}_1$-particle was produced, the mass of the V^{-}-particle is about 2 600 m_e ; if a $V^{\circ}{}_2$-particle was produced, then the mass is about 1 300 m_e.

Acknowledgments

It is with pleasure that we express our thanks to Professor P. M. S. Blackett for the excellent facilities he has given us and for the great interest he has taken in the analysis of the results. We are very much indebted to Dr. J. Rösch, the Director of the Observatoire du Pic-du-Midi, for his help during the experimental work and to Mr. A. H. Chapman who helped both with the work on the Pic-du-Midi and with the measurement of the tracks. Mr. M. G. Sowerby has been of great assistance during the measurement and analysis of the data. We also wish to thank Dr. J. Podolanski, Dr. G. D. Rochester, Prof. R. D. Sard, and Dr. J. G. Wilson for many helpful discussions. Financial assistance, which is very much appreciated, has been received from the Department of Scientific and Industrial Research, which has also given a grant to R. Armenteros. A. Cachon and C. M. York have received grants from the Centre National de la Recherche Scientifique and the United States Education Commission in the United Kingdom, respectively.

APPENDIX

When an unstable particle, of mass M, decays into three secondary particles, the relativistic energy equation, referring to conditions in the rest system of the unstable particle, may be written :

$$M = E_1 + E_2 + E_3 , \qquad (9)$$

where E_1, E_2 and E_3 are the total energies of the three secondary particles in the rest system and M is expressed in energy units. If one of the secondaries, of mass m_1, has its maximum energy, E_{max}, in the rest system, then it will be shown that E_{max} can be expressed simply in terms of M,

m_1 and (m_2+m_3), where m_2 and m_3 are the masses of the remaining secondary particles, and, for convenience, all the masses are expressed in energy units.

For the special conditions under consideration, eqn. (9) may be written :

$$M-E_{max}=E_2+E_3\,, \qquad \ldots \ldots \quad (10)$$

where the term (E_2+E_3) takes a minimum value. The two conditions that hold when this occurs will now be obtained.

The conservation of momentum in the decay process imposes the condition :

$$\overrightarrow{p^*_{max}}+\overrightarrow{p_2^*}+\overrightarrow{p_3^*}=0, \qquad \ldots \ldots \quad (11)$$

where $\overrightarrow{p^*_{max}}$ is the maximum momentum of the charged secondary in the rest system and $\overrightarrow{p_2^*}$ and $\overrightarrow{p_3^*}$ are the momenta of the remaining secondaries in the rest system. The vector eqn. (11) shows that the three vectors are coplanar. If the direction of p^*_{max} is along the x axis, then $\overrightarrow{p_2^*}$ may be resolved into two orthogonal components p^*_{2x} and p^*_{2y}. Similarly, the components of $\overrightarrow{p_3^*}$ are p^*_{3x} and p^*_{3y}. Now momentum is conserved in the y direction, so that

$$p^*_{2y}=-p^*_{3y}=p_t^*. \qquad \ldots \ldots \quad (12)$$

Substituting these momentum components in eqn. (10), we obtain :

$$M-E_{max}=(m_2^2+p^{*2}_{2x}+p_t^{*2})^{\frac{1}{2}}+(m_3^2+p^{*2}_{3x}+p_t^{*2})^{\frac{1}{2}}. \quad \ldots \quad (13)$$

The two right-hand terms of this equation take a minimum value if $p_t=0$; consequently, the first condition is that particles 2 and 3 move parallel to one another in the opposite direction to particle 1. Thus the vector eqn. (11) may be re-written as a scalar equation :

$$p^*_{max}=p_2^*+p_3^* \qquad \ldots \ldots \quad (14)$$

and eqn. (13) may be re-written in the form

$$M-E_{max}=(m_2^2+p_2^{*2})^{\frac{1}{2}}+(m_3^2+p_3^{*2})^{\frac{1}{2}}. \quad \ldots \quad (15)$$

The second condition that holds, if the right-hand side of eqn. (10) has a minimum value, can be obtained by eliminating p_3^* from eqn. (15) by means of eqn. (14) and differentiating with respect to the variable momentum, p_2^*. It will then be seen that the second condition may be written in the form :

$$p_2^*(m_2^2+p_2^{*2})^{-\frac{1}{2}}=p_3^*(m_3^2+p_3^{*2})^{-\frac{1}{2}}. \quad \ldots \ldots \quad (16)$$

Therefore particles 2 and 3 have equal velocities in the rest system of the unstable particle.

The unknown momenta p_2^* and p_3^* can now be eliminated from eqns. (14), (15) and (16) and the following value of (m_2+m_3) is obtained :

$$(m_2+m_3)^2=M^2+m_1^2-2ME_{max}. \quad \ldots \ldots \quad (17)$$

This equation, which remains valid if either m_2 or m_3, or both, is zero, has been used in a β-decay problem by Kofoed-Hansen (1951).

REFERENCES

ARMENTEROS, R., BARKER, K. H., BUTLER, C. C., CACHON, A., and CHAPMAN, A. H., 1951, *Nature, Lond.*, **167,** 501.

ARMENTEROS, R., BARKER, K. H., BUTLER, C. C., and CACHON, A., 1951, *Phil. Mag.*, **42,** 1113.

ASTBURY, P., CHIPPINDALE, P., NEWTH, J. A., PAGE, D. I., RYTZ, A., and SAHIAR, A. B., 1951, *Report given at Bristol Conference in December.*

BRIDGE, H. S., and ANNIS, M., 1951, *Phys. Rev.*, **82,** 445.

BUTLER, C. C., 1951, *Progress in Cosmic Ray Physics,* (North Holland Publishing Co.), 110.

BUTLER, C. C., ROSSER, W. G. V., and BARKER, K. H., 1950, *Proc. Phys. Soc.* A, **63,** 145.

FOWLER, P. H., MENON, M. G. K., POWELL, C. F., and ROCHAT, O., 1951, *Phil. Mag.*, **42,** 1040.

FRETTER, W. B., 1951, *Phys. Rev.*, **83,** 1053.

KOFOED-HANSEN, O., 1951, *Phil. Mag.*, **42,** 1411.

LEIGHTON, R. B., WANLASS, S. D., and ALFORD, W. L., 1951, *Phys. Rev.*, **83,** 843.

MENON, M. G. K., 1951, *Report given at Bristol Conference in December.*

O'CEALLAIGH, C., 1951, *Phil. Mag.*, **42,** 1032.

ROCHESTER, G. D., and BUTLER, C. C., 1947, *Nature, Lond.*, **160,** 855.

SERIFF, A. J., LEIGHTON, R. B., HSIAO, C., COWAN, E. W., and ANDERSON, C. D., 1950, *Phys. Rev.*, **78,** 290.

THOMPSON, R. W., COHN, H. O., and FLUM, R. S., 1951, *Phys. Rev.*, **83,** 175.

TREMBLEY, J., 1951, *Report given at Bristol Conference in December.*

DESCRIPTION OF THE PLATES

Plate XXVIII

(Track number 2)

This photograph is described in detail in § 4 and two projections are given in fig. 3. The unstable charged particle entered the chamber at (1) and decayed at level D into the charged particle (2) and probably a V°-particle which decayed at level F and produced tracks (3) and (4).

Plate XXIX

(Track number 6)

A high-energy negatively-charged particle (1), associated with a penetrating shower, decayed in the chamber producing one charged secondary particle, track (2). The measurements on the tracks are given in table 1. This photograph is a typical example of the decay of a high-energy charged V-particle.

Plate XXX

(Track number 22)

The data obtained from this photograph are discussed in § 2, section (ii). The unstable particle entered the chamber at (1) and produced a single charged secondary (2). The track of this secondary is too short for measurement. The momentum of the positively-charged primary is (400 ± 40) MEV/c and the estimated ionization is about twice the minimum value. The measurements are consistent with a value of 1100 m_e for the mass of the particle. The ionization may be compared with that of the neighbouring track (3) which was produced by a proton with a measured momentum of (400 ± 40) MeV/c. The momentum of the proton is the same as that of the V^+-particle but there is no doubt that it is more heavily ionizing. The estimated ionization for particle (3) is (3–5) minimum ; assuming it to be a proton, the expected ionization, allowing for the error in the momentum, is within the range (3·5–5·0) minimum. This comparison of the proton track (3) with that of the V^+-particle track (1) shows that an ionization of three times minimum is an upper limit for the ionization of the latter particle. This ionization is consistent with an upper limit of $(1580 \pm 160)\,m_e$ for the mass of the V^+-particle; the error in this mass arises from the error in the momentum measurement.

Track (4) is produced by a second proton, the measured momentum is (210 ± 20) MeV/c and the estimated ionization within the range (8–10) minimum ; the calculated ionization for a proton is (9–12) minimum, this range of values arises from the error in the momentum measurement.

Plate XXXI

(Track number 11)

The data obtained from this photograph are discussed in § 2, section (ii). The positively-charged particle entered the chamber at (1) and produced a charged secondary track (2) which is too short for measurement. The measured momentum of the primary particle is (600 ± 60) MeV/c and the ionization is between the minimum value and an upper limit of twice the minimum value. This range of values for the ionization is obtained by comparing the ionization of track (1) with the ionizations of the electrons and other fast particles in the photograph. If the upper value of twice the minimum is assumed for the ionization, then the upper limit for the mass of the unstable particle is $(1660 \pm 160)\,m_e$; the error arises from the uncertainty in the momentum. It will be noticed that the V^+-particle is associated with a high-energy penetrating shower.

R. ARMENTEROS *et al.* Phil. Mag. Ser. 7, Vol. 43, Pls. XXVIII & XXIX.

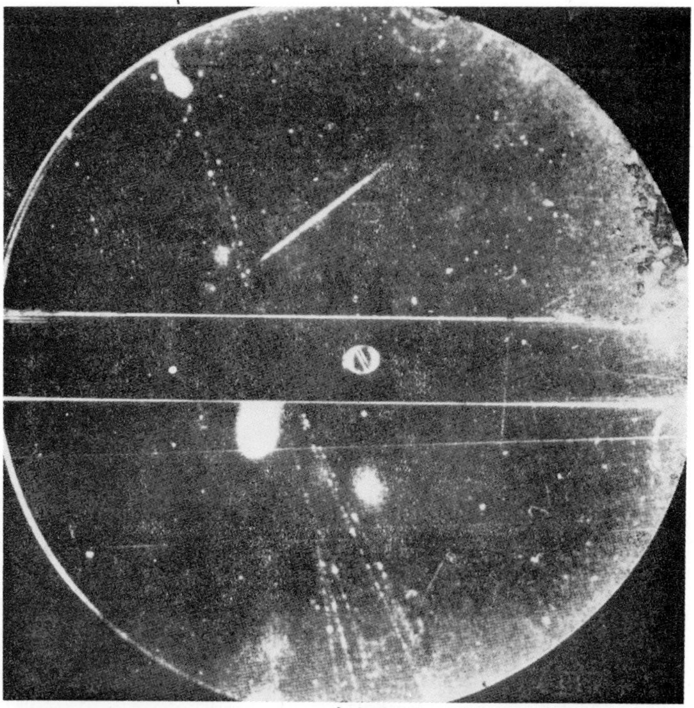

PLATE XXVIII

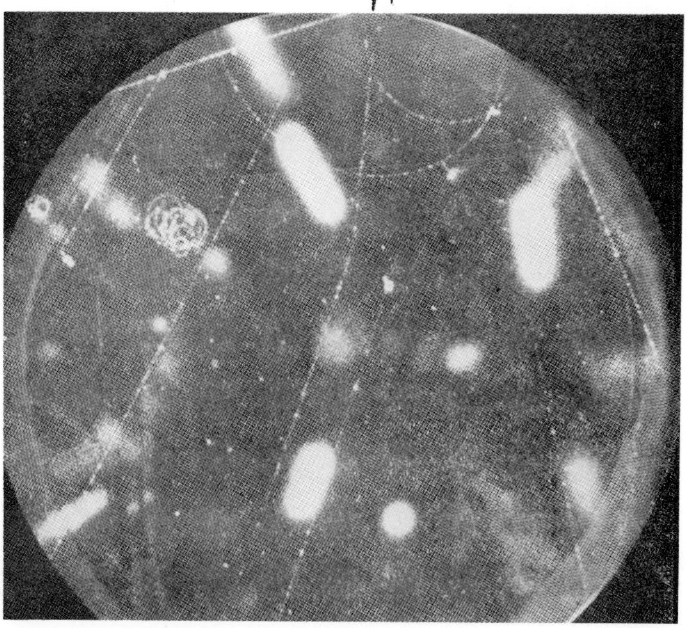

PLATE XXIX

R. ARMENTEROS *et al.* Phil. Mag. Ser. 7, Vol. 43, Pls. XXX & XXXI.

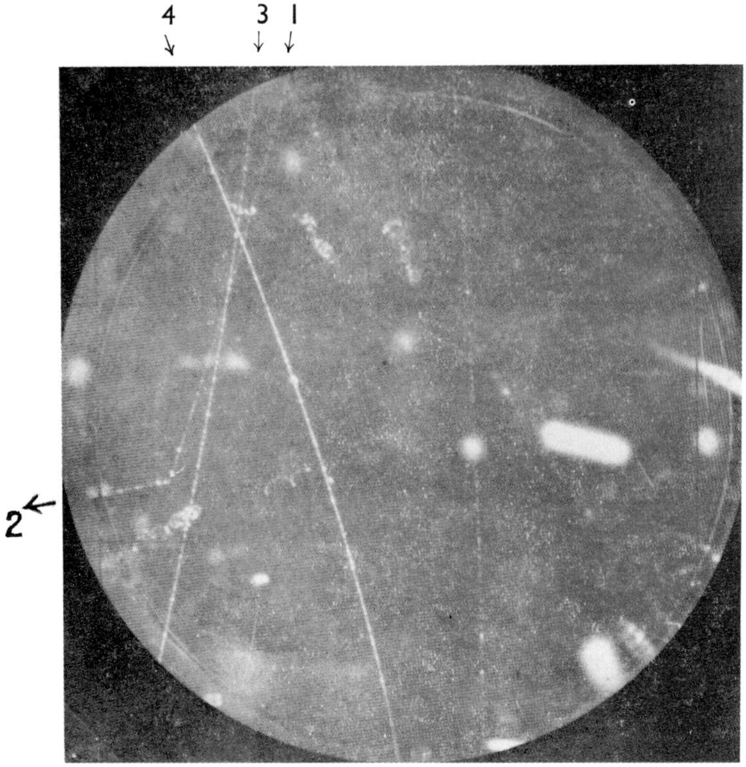

PLATE XXX

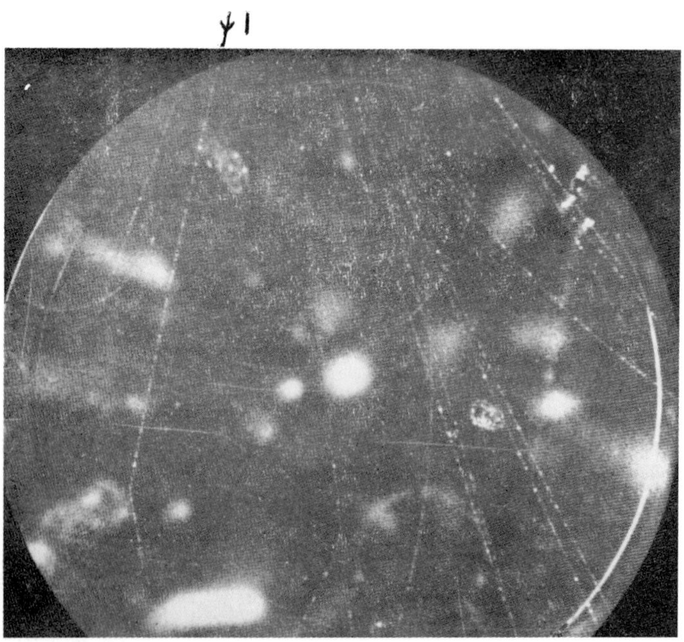

PLATE XXXI

Author Index

Subject Index

Name Index